AF353112

Systematic Approach to Agroforestry Policies and Practices in Asia

Systematic Approach to Agroforestry Policies and Practices in Asia

Editors

Mi Sun Park
Himlal Baral

MDPI • Basel • Beijing • Wuhan • Barcelona • Belgrade • Manchester • Tokyo • Cluj • Tianjin

Editors
Mi Sun Park
Graduate School of
International Agricultural
Technology
Seoul National University
Pyeongchang
Republic of Korea

Himlal Baral
Center for International
Forestry Research
CIFOR
Bogor Barat
Indonesia

Editorial Office
MDPI
St. Alban-Anlage 66
4052 Basel, Switzerland

This is a reprint of articles from the Special Issue published online in the open access journal *Forests* (ISSN 1999-4907) (available at: www.mdpi.com/journal/forests/special_issues/agroforestry_systems).

For citation purposes, cite each article independently as indicated on the article page online and as indicated below:

LastName, A.A.; LastName, B.B.; LastName, C.C. Article Title. *Journal Name* **Year**, *Volume Number*, Page Range.

ISBN 978-3-0365-5680-2 (Hbk)
ISBN 978-3-0365-5679-6 (PDF)

Cover image courtesy of Himlal Baral/CIFOR-ICRAF

Contents

About the Editors

Mi Sun Park

Associate Professor in the field of international forest policy from Graduate School of International Agricultural Technology, Seoul National University in the Republic of Korea. She studied forest science (bachelor degree) and environmental education (master degree) at the Seoul National University, Republic of Korea. She acheived a PhD in forest policy at the University of Göttingen, Germany. Her research interests include governance/regime discourse and communication, policy strategies and instruments, policy integration and international cooperation in forest policy. She works at the global academic society. As a Coordinating Lead Autor, she participated in writing the chapter on policy options of the thematic assessment, "Sustainable Use of Wild Species" from IPBES (Intergovernmental Science-Policy Platform on Biodiversity and Ecosystem Services) from 2018 to 2022. She works as a coordinator of Research Group 9.03.04 Traditional Forest Knowledge and a deputy coordinator of Working Party 9.01.07 Forests and the Media from IUFRO (International Union of Forest Research Organizations) since 2019. She also works as a board member of the International Journal of Forest Research (Taylor & Francis, ISSN 1341-6979) and as an editor of the Trees, Forests and People (Elsevier, ISSN 2666-7193).

Himlal Baral

Dr Himlal Baral is a Forest and Landscape Restoration (FLR) Scientist with CIFOR-ICRAF's Climate Change, Energy and Low-Carbon Development team. Based in Bali, Indonesia, Dr Baral coordinates CIFOR's work on FLR, planted forests and bioenergy in Asia and the Pacific, specifically, (i) promoting FLR for multiple ecosystem goods and services (EGS), (ii) quantifying and valuing EGS under different landscape management scenarios; (iii) investigating and developing locally appropriate landscape restoration models; and (iv) contributing to land-resource planning for food, biomaterials, and energy security. Prior to joining CIFOR, he worked in the forestry and natural resources management field for over 25 years in both the private and public sectors in Asia and the Pacific. Dr Baral has achieved a PhD in Land and Environment and two Master degrees: in Forest Science and in Social Science.

Preface to "Systematic Approach to Agroforestry Policies and Practices in Asia"

Agroforestry is an intensive land management system involving the integration of tree management into crop and animal farming. It provides diverse ecosystem services by bridging agriculture, forestry, and husbandry to offer environmental, economic, and social benefits. In order to improve the benefits of agroforestry to meet development and climate goals, a systematic approach is necessary for understanding agroforestry practices, designing agroforestry policies and associated outcomes. Multiple methodologies, including systematic reviews and landscape restoration approaches, can be applied to analyze agroforestry policies and ecosystem services derived from agroforestry practices. Therefore, this Special Issue focuses on systematic approaches to agroforestry policies, strategies, and practices. It includes case studies from several Asian countries to explore the economic, social, and environmental dimensions.

Mi Sun Park and Himlal Baral

Editors

forests

Editorial

Systematic Approach to Agroforestry Policies and Practices in Asia

Mi Sun Park [1,2,*], Himlal Baral [3,4] and Seongmin Shin [3]

[1] Graduate School of International Agricultural Technology, Seoul National University, 1447 Pyeongchang-daero, Daehwa, Pyeongchang 25354, Gangwon, Republic of Korea
[2] Institutes of Green Bio Science and Technology, Seoul National University, 1447 Pyeongchang-daero, Daehwa, Pyeongchang 25354, Gangwon, Republic of Korea
[3] Center for International Forestry Research (CIFOR), Jalan CIFOR, Bogor 16115, Situ Gede, Indonesia; h.baral@cgiar.org (H.B.); seongmin@snu.ac.kr (S.S.)
[4] School of Ecosystem and Forest Sciences, University of Melbourne, Parkville, VIC 3010, Australia
* Correspondence: mpark@snu.ac.kr; Tel.: +82-33-339-5858

Abstract: This paper introduces the Special Issue "Systematic Approach to Agroforestry Policies and Practices in Asia". This Special Issue contains eleven papers on agroforestry at national, regional, and global levels. These papers discuss research trends; dominant services and functions of agroforestry; multiple case studies from Asian countries including Nepal, Lao PDR, Indonesia, Vietnam, Bangladesh, and Timor-Leste; and the benefits of agroforestry including income generation and carbon sequestration. They also interpret the goals, challenges, and social and cultural norms in agroforestry policies in national and local contexts. The research results can support policy design for the systematization and stabilization of agroforestry. This Special Issue provides us with scientific evidence and practical lessons on agroforestry policies and practices in Asia. It contributes to expanding the knowledge base for agroforestry and towards establishing and implementing agroforestry policies and practices in the region.

Keywords: agroforestry; policy; Asia; landscape restoration

Citation: Park, M.S.; Baral, H.; Shin, S. Systematic Approach to Agroforestry Policies and Practices in Asia. *Forests* **2022**, *13*, 635. https://doi.org/10.3390/f13050635

Received: 6 April 2022
Accepted: 11 April 2022
Published: 19 April 2022

Publisher's Note: MDPI stays neutral with regard to jurisdictional claims in published maps and institutional affiliations.

1. Introduction

Agroforestry, a traditional land-use practice found throughout the world, focuses on maximizing the benefits of biological interaction by intentionally linking trees, crops, and animals under agroecological systems [1]. According to a World Agroforestry Centre (ICRAF) working paper [2], agroforestry covers around one billion hectares or 43% of agricultural lands globally, and involves more than 900 million people. In Asia, agroforestry is also prevalent and has played a critical role in local livelihoods since ancient times [3]. If defined as being more than 10% tree cover on agricultural land [2], then agroforestry covers 77.8% of all agricultural land in Southeast Asia, 50.5% in East Asia, 27.0% in South Asia, and 23.6% in Northern and Central Asia.

Asian peoples have long reflected and implemented inventive wisdom and strategies in diverse agroforestry systems for their basic fuel, food, medicine, and cash income needs [4]. However, since the early 20th century, growing populations and food requirements have led to agroforestry land in Asia being converted into intensive agriculture and monoculture tree plantations, causing environmental and social challenges, such as a loss of food and homes for forest-dependent species and peoples [5]. Scientists have begun to acknowledge high levels of complexity and environmental heterogeneity in the interactions of trees, crops, and animals [6] and the various ecosystem services they provide. Sustainable agroforestry landscapes came into the spotlight in the late 20th century with growing global concern over environmental issues, natural resource depletion, and climate change [4]. Since agroforestry began to gain more attention in Asia, agroforestry research has been increasing steadily in the region (Figure 1) [7].

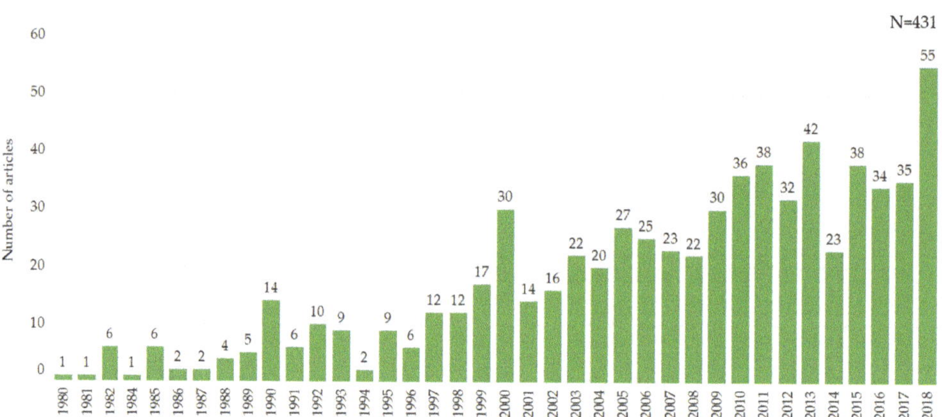

Figure 1. Numbers of academic publications on agroforestry and ecosystem services in Asia, based on data from Shin et al. (2020).

Following an increase in awareness of the benefits it provides, agroforestry has become an important topic in global and national agendas for sustainable development during international policy meetings. It is now recognized as a climate-smart agricultural system by the Intergovernmental Panel on Climate Change (IPCC) and the United Nations Framework Convention on Climate Change (UNFCCC); as a Nationally Appropriate Mitigation Action (NAMA) and National Adaptation Programme of Action (NAPA) for the agriculture sector; as an important strategy to combat desertification by the United Nations Convention to Combat Desertification (UNCCD); as being beneficial to biodiversity by the Convention on Biological Diversity (CBD); and as being able to enhance ecosystem services by the United Nations Forest Forum (UNFF) [8].

The World Agroforestry Centre (ICRAF) has been introducing agroforestry agendas and leading international discussions on agroforestry development since 2003. The ICRAF-led World Congress on Agroforestry (WCA) shares agroforestry ideas and knowledge and provides opportunities to strengthen cooperation between various stakeholders (Garrity, 2012). The first World Congress on Agroforestry (WCA 1), held in 2004 in Orlando, FL, USA, involved nearly 600 stakeholders from 82 countries. These stakeholders announced the Orlando Declaration, which focuses on increasing household earnings from agroforestry; promoting gender equality and women's participation; improving health and welfare; and promoting environmental sustainability. The second World Congress on Agroforestry (WCA 2) was held in 2009 in Nairobi, Kenya; the third in Delhi, India, in February 2014, and the fourth in Montpellier, France in May 2019. This world congress series has led to notable outcomes for strengthening links between science, environment, society, and policy in diverse aspects of agroforestry and subjects such as climate change, biodiversity, soil protection, food security, livelihoods, and Sustainable Development Goals (SDGs) [5].

In Asia, agroforestry has gained attention in the fields of forestry and agriculture. In most countries, agroforestry is now included in agriculture or forestry policies [9]. In recent years, some Asian countries have introduced specialized national agroforestry policies. In 2014, for example, India became the first country in the world to adopt a national agroforestry policy [10], followed by Nepal in 2019 [11]. At the regional level, Asian countries have cooperated in facilitating agroforestry. At the 37th ASEAN Ministers of Agriculture and Forestry Meeting in 2015, ministers endorsed the Vision and Strategic Plan for ASEAN Cooperation in Food, Agriculture and Forestry 2016–2025. This strategic plan identified seven 'strategic thrusts' and related action programs. Strategic Thrust 4 focuses on increasing resilience to climate change, natural disasters, and other shocks, and includes action programs for expanding resilient agroforestry systems where ecologically and economically appropriate [12]. Recognizing the contributions of agroforestry to achieving food security, sustainable forest management and SDGs, participants in the 20th ASEAN Senior Officials of Forestry meeting agreed to develop agroforestry guidelines through the ASEAN Guidelines for Agroforestry Development (ASEAN GAD), which were adopted at the 39th ASEAN Ministers of Agriculture and Forestry Meeting in 2017. The ASEAN GAD includes 75 guidelines under 14 principles for the institutional, economic, environmental, socio-cultural, technical design and communication, and scaling-up areas of agroforestry [13].

As mentioned above, national and regional agroforestry strategies have been established in Asia. In this context, systematic approaches and methodologies can help to better understand Asian agroforestry systems, policies and practices. This Special Issue, entitled "Systematic Approach to Agroforestry Policies and Practices in Asia", explores environmental, economic, and social dimensions of agroforestry in Asia through multiple research methods. It can contribute to maximizing benefits from agroforestry, to designing agroforestry policies, and ultimately to achieving SDGs.

2. Overviews of Papers in the Special Issue

Papers included in the Special Issue cover an extensive variety of topics: farmers' willingness to adopt agroforestry systems; ecosystem services; social and cultural norms; agroforestry management policies; and bamboo and wild-simulated ginseng; as well as different research methodologies (systematic reviews, surveys, topic modeling, policy analysis and case studies) (Table 1). These papers provide an overall picture of agroforestry as well as case studies of specific agroforestry systems in Asia.

Providing a comprehensive overview of agroforestry systems in Asia, the article by Shin et al. [7] explores research trends on agroforestry practices and ecosystem services in the Asia-Pacific region. It presents a systematic review of 431 articles published between 1970 and 2018 included in the specialized international academic database, SCOPUS. Research results show: (1) countries where silvoarable practices and regulating and supporting ecosystem services are most prominent (China, Indonesia, and Australia); (2) research trends by decade; and (3) the intersection of agroforestry practices and ecosystem services in academic research targeting the Asia-Pacific region.

Dhyani et al. [14] review published papers to investigate evidence of the climate change adaptation and mitigation potential of agroforestry systems in South Asia. The paper looks at agroforestry systems being developed in South Asian countries as enabling conditions for commitments to international agreements, and a consensus on establishing agroforestry policies to tackle climate change through the South Asian Association for Regional Cooperation's Resolution on Agroforestry. It also discusses constraints impeding agroforestry system climate goals, including data uniformity, stakeholder rights, financial support, governance, and water scarcity. The paper provides strong recommendations on incentives and sustainable local livelihoods, as well as carbon neutrality scenarios.

As the first of two papers documenting research in Nepal, Dhakal and Rai [15] investigate factors that influence farmers' decisions to adopt agroforestry practices in the country. The Multinomial Logistic Regression (MNL) model was used to analyze 288 households' agroforestry and conventional agriculture choices. Results show head of household gender, off-farm income, home-to-forest distance, farmer group membership, livestock herd size, extension services, and awareness of the environmental benefits of agroforestry being factors that affect the likelihood of farmers adopting agroforestry. It also shows farmland area, household size, and irrigation being major constraints to agroforestry adoption.

Lee et al. [16] discuss the potential of the bamboo industry to be a key component in agroforestry development by comparing an existing model of supplying semi-processed bamboo splits with a new model involving locally produced bamboo handcrafts. A value chain assessment of the two models uses statistical data and interview responses collected in eastern Lao PDR. Results show the new model of selling handcrafts being able to generate 234–244 times more revenue for farmers than the old bamboo split model. The new handcraft production model could be a credible income generator and ensure sustainable use of bamboo, though it would necessitate specific skills training.

In a study on specific agroforestry species, Shin et al. [17] explore trends in global research on wild-simulated ginseng (WSG), which has been widely used in agroforestry systems across the world for food and medicine. Results show key subjects in global WSG research being growth conditions, effects of WSG, and its components; key countries being the Republic of Korea, China, and the USA; key words being ginsenosides, antioxidant activity, and metabolites; and key topics being medicinal effects, metabolite analysis, genetic diversity, cultivation conditions, and bioactive compounds. They also show research topics changing towards precise identification and characterization of the bioactive metabolites of WSG. This trend suggests increasing academic interest in the value-added utilization of WSG and the potential of non-timber forest products for human well-being.

Nurrochmat et al. [18] examine changes in agroforestry management policies in Indonesia by evaluating local development goals, barriers, and risks to agroforestry implementation. They show the key elements of local development goals being increased stability, public order, and legal and political awareness; increased environmental quality; and increased social and community protection. Results also show challenges to agroforestry implementation and management being lack of financial resources; inappropriate communication with stakeholders; low levels of human capital; and weak coordination. The research stresses the importance of understanding policies and stakeholder interaction and awareness for ensuring successful agroforestry implementation and management.

In a case study in Vietnam, Nguyen et al. [19] apply a knowledge-based approach to investigate how the social and cultural norms of different ethnic groups influence agroforestry adoption in Vietnam. It shows farmers having different ideas about the benefits of agroforestry, and ethnic groups' different social and cultural values affecting farmers' willingness to adopt agroforestry practices. The results indicate the importance of understanding social and cultural preferences before developing agricultural policies and interventions.

Siarudin et al. [20] examine the carbon stock potential of six different agroforestry systems on degraded land in Indonesia using survey, interview, and Rapid Carbon Stock Appraisal (RaCSA) approaches. The study shows all six agroforestry systems having significant capacity for carbon capture, but mixed tree lots having the highest carbon stock. Farmer interview and survey outcomes indicate agroforestry helping soil erosion control and landscape restoration on degraded lands.

Rahman et al. [21] assess tree cover and the direct and mediated effects of tree diversity on carbon storage through stand structure in homegardens in southwestern Bangladesh. Results indicate tree diversity increasing stand basal area and improving total carbon storage in homegardens, and privately managed homegardens represent a potential nature-based solution for biodiversity conservation and climate change mitigation in Bangladesh.

Khadka et al. [22] studied how the adoption of agroforestry would influence the farmers' livelihood through interviews with a structured questionnaire in Nepal. Agroforestry systems are limited with a lack of technical and functional knowledge about agroforestry uses compared to community forestry although farmers in the region adopted agri-silviculture, agri-silvi-pastoral, and horti-agri-silvicultural practices. The authors suggest proper extension services and market development to fill the gaps which hinder agroforestry adoption.

Paudel et al. [23] examined that agroforestry presents opportunities and challenges as a viable way to balance ecological and socio-economic functions in Timor-Leste, where farmers have traditionally implemented agroforestry systems. However, challenges for agroforestry in Timor-Leste still remain, such as lack of knowledge, institutional capacity, and funding. This research suggests sustainable forest management and promotes capacity building and other initiatives.

Table 1. Highlights of research papers presented in the Special Issue.

Category	Key Topic	Target Area	Methodology	Contributing Manuscript
Article	Research trends on agroforestry and ecosystem services.	Asia-Pacific region	Systematic review	[7]
	Factors that influence farmers' decisions to adopt agroforestry.	Nepal	Interviews with a structured questionnaire	[15]
	Bamboo-based income generation.	Lao PDR	Value chain analysis (survey questionnaire, field observations and stakeholder interviews)	[16]
	Global trends in research on wild-simulated ginseng.	Global region	Systematic review Keyword analysis Topic modeling	[17]
	Transformation of agro-forest management policy.	Indonesia	Interpretive Structural Modeling (ISM) Survey questionnaire	[18]
	Social and cultural norms in agroforestry.	Vietnam	Interviews Group discussion	[19]
	The carbon sequestration potential of agroforestry systems in degraded landscapes.	Indonesia	Interview Rapid Carbon Stock Appraisal (RaCSA)	[20]
	Assessing tree cover and the direct and mediated effects of tree diversity on carbon storage through stand structure in homegardens in southwestern Bangladesh.	Bangladesh	Carbon calculation	[21]
	Contribution of agroforestry to farmers' livelihood.	Nepal	Interviews with a structured questionnaire	[22]
Review	The climate change adaptation and mitigation potential of agroforestry systems	South Asia	Review	[14]
Opinion	Agroforestry opportunities and challenges	Timor-Leste	Review essay	[23]

3. Expanding the Knowledge Base for Agroforestry in Asia

There are several collections of agroforestry research at the regional level. A Special Issue of the journal *Agroforestry Systems* published in 2012, entitled "Agroforestry Landscapes", presented nine articles focusing on fruit trees in West African parklands in the African dimension [24]. A Special Issue of the same journal published in 2018, entitled "Advances in European agroforestry", contained 23 papers based on AGFORWARD project results in the European dimension [25]. This Special Issue focuses specifically on Asian research on agroforestry. It provides diverse approaches and new knowledge on agroforestry policies and practices in Asia. By reviewing previous studies, it explains agroforestry research trends focusing primarily on dominant services and functions of agroforestry and specific crops. Above all, this Special Issue offers multiple case studies from Asian countries including Nepal, Lao PDR, Indonesia, Vietnam, Bangladesh, and Timor-Leste. It examines notable research areas in the social, economic, and environmental benefits of agroforestry, including income generation and carbon sequestration. In this Special Issue, goals, challenges, and social and cultural norms in agroforestry practices and policies are interpreted in national and local contexts. The research can support policy design for the systematization and stabilization of agroforestry.

In conclusion, this Special Issue provides us with scientific evidence and practical lessons on agroforestry policies and practices in Asia. In doing so, it contributes to expanding the knowledge base for agroforestry in the region. The expectation is that the results of this issue can be applied to establishing and implementing agroforestry policies and practices in Asia.

Author Contributions: Conceptualization, M.S.P. and H.B.; writing—original draft preparation, M.S.P., H.B. and S.S.; writing—review and editing, M.S.P., H.B. and S.S. All authors have read and agreed to the published version of the manuscript.

Funding: This research received no external funding.

Acknowledgments: This Special Issue was supported by the Creative-Pioneering Researchers Program at Seoul National University (SNU) and the CGIAR Research Program on Forests, Trees and Agroforestry.

Conflicts of Interest: The authors declare no conflict of interest.

References

1. Association for Temperate Agroforestry (AFTA). *The Status, Opportunities and Needs for Agroforestry in the United States*; AFTA: Columbia, MO, USA, 1997.
2. Zomer, R.J.; Trabuco, A.; Coe, R.; Place, F.; Noordwijk, M.; Xu, J. *Trees on Farms: An Update and Reanalysis of Agroforestry's Global Extent and Socio-Ecological Characteristics*; Working Paper 179; World Agroforestry Centre (ICRAF) Southeast Asia Regional Program: Bogor, Indonesia, 2014.
3. Cannell, M.G.R. Agroforestry—A decade of development. *ICRAF* **1987**, *24*, 393.
4. Kumar, B.M.; Singh, A.K.; Dhyani, S.K. South Asian Agroforestry: Traditions, Transformations, and Prospects. In *Agroforestry—The Future of Global Land Use*; Springer: Dordrecht, The Netherlands, 2012; pp. 359–389.
5. Garrity, D. Agroforestry and the future of global land use. In *Agroforestry—The Future of Global Land Use*; Springer: Dordrecht, The Netherlands, 2012; pp. 21–27.
6. Palma, J.; Graves, A.R.; Bunce, R.G.H.; Burgess, P.J.; de Filippi, R.; Keesman, K.J.; van Keulen, H.; Liagre, F.; Mayus, M.; Moreno, G.; et al. Modeling environmental benefits of silvoarable agroforestry in Europe. *Agric. Ecosyst. Environ.* **2007**, *119*, 320–334. [CrossRef]
7. Shin, S.; Soe, K.T.; Lee, H.; Kim, T.H.; Lee, S.; Park, M.S. A systematic map of agroforestry research focusing on ecosystem services in the Asia-Pacific region. *Forests* **2020**, *11*, 368. [CrossRef]
8. FAO. *Advancing Agroforestry on the Policy Agenda: A Guide for Decision-Makers*; Agroforestry Working Paper No. 1; Buttoud, G., Ajayi, O., Detlefsen, G., Place, F., Torquebiau, E., Eds.; Food and Agriculture Organization of the United Nations (FAO): Rome, Italy, 2013; p. 37.
9. De Zoysa, M. *A Review of Forest Policy Trends in Sri Lanka*; Policy Trend Report 2001; Institute for Global Environmental Strategies (IGES): Kanagawa, Japan, 2001.
10. Chavan, S.B.; Keerthika, A.; Dhyani, S.K.; Handa, A.K.; Newaj, R.; Rajarajan, K. National agroforestry policy in India: A low hanging fruit. *Curr. Sci.* **2015**, *108*, 1826–1834.

11. Atreya, K.; Subedi, B.P.; Ghimire, P.L.; Khanal, S.C.; Charmakar, S.; Adhikari, R. Agroforestry for mountain development: Prospects, challenges and ways forward in Nepal. *Arch. Agric. Environ. Sci.* **2021**, *6*, 87–99. [CrossRef]
12. ASEAN. The Vision and Strategic Plan for ASEAN Cooperation in Food, Agriculture and Forestry 2016–2025. 2015. Available online: https://asean-crn.org/vision-and-strategic-plan-for-asean-cooperation-in-food-agriculture-and-forestry-2016-2025/ (accessed on 13 April 2022).
13. Catacutan, D.; Finlayson, R.; Gassner, A.; Perdana, A.; Lusiana, B.; Leimona, B.; Simelton, E.; Öborn, I.; Galudra, G.; Roshetko, J.M.; et al. *ASEAN Guidelines for Agroforestry Development*; ASEAN Secretariat: Jakarta, Indonesia, 2018; Available online: https://asean-crn.org/wp-content/uploads/2019/09/2018-ASEANGuideline-agroforestry.pdf (accessed on 13 April 2022).
14. Dhyani, S.; Murthy, I.K.; Kadaverugu, R.; Dasgupta, R.; Kumar, M.; Adesh Gadpayle, K. Agroforestry to Achieve Global Climate Adaptation and Mitigation Targets: Are South Asian Countries Sufficiently Prepared? *Forests* **2021**, *12*, 303. [CrossRef]
15. Dhakal, A.; Rai, R.K. Who adopts agroforestry in a subsistence economy—Lessons from the Terai of Nepal. *Forests* **2020**, *11*, 565. [CrossRef]
16. Lee, B.; Rhee, H.; Kim, S.; Lee, J.-W.; Koo, S.; Lee, S.-J.; Alounsavath, P.; Kim, Y.-S. Assessing sustainable bamboo-based income generation using a value chain approach: Case study of Nongboua village in Lao PDR. *Forests* **2021**, *12*, 153. [CrossRef]
17. Shin, S.; Park, M.S.; Lee, H.; Lee, S.; Lee, H.; Kim, T.H.; Kim, H.J. Global trends in research on wild-simulated ginseng: Quo vadis? *Forests* **2021**, *12*, 664. [CrossRef]
18. Nurrochmat, D.R.; Pribadi, R.; Siregar, H.; Justianto, A.; Park, M.S. Transformation of agro-forest management policy under the dynamic circumstances of a two-decade regional autonomy in Indonesia. *Forests* **2021**, *12*, 419. [CrossRef]
19. Nguyen, M.P.; Pagella, T.; Catacutan, D.C.; Nguyen, T.Q.; Sinclair, F. Adoption of agroforestry in Northwest Viet Nam: What roles do social and cultural norms play? *Forests* **2021**, *12*, 493. [CrossRef]
20. Siarudin, M.; Rahman, S.A.; Artati, Y.; Indrajaya, Y.; Narulita, S.; Ardha, M.J.; Larjavaara, M. Carbon sequestration potential of agroforestry systems in degraded landscapes in West Java, Indonesia. *Forests* **2021**, *12*, 714. [CrossRef]
21. Rahman, M.M.; Kundu, G.K.; Kabir, M.E.; Ahmed, H.; Xu, M. Assessing Tree Coverage and the Direct and Mediation Effect of Tree Diversity on Carbon Storage through Stand Structure in Homegardens of Southwestern Bangladesh. *Forests* **2021**, *12*, 1661. [CrossRef]
22. Khadka, D.; Aryal, A.; Bhatta, K.P.; Dhakal, B.P.; Baral, H. Agroforestry systems and their contribution to supplying forest products to communities in the Chure range, Central Nepal. *Forests* **2021**, *12*, 358. [CrossRef]
23. Paudel, S.; Baral, H.; Rojario, A.; Bhatta, K.P.; Artati, Y. Agroforestry: Opportunities and challenges in Timor-Leste. *Forests* **2022**, *13*, 41. [CrossRef]
24. Ræbild, A. Improved management of fruit trees in West African parklands. *Agrofor. Syst.* **2012**, *85*, 425–430. [CrossRef]
25. Burgess, P.J.; Rosati, A. Advances in European agroforestry: Results from the AGFORWARD project. *Agrofor. Syst.* **2018**, *92*, 801–810. [CrossRef]

Article

A Systematic Map of Agroforestry Research Focusing on Ecosystem Services in the Asia-Pacific Region

Seongmin Shin [1], Khaing Thandar Soe [2], Haeun Lee [1], Tae Hoon Kim [1], Seongeun Lee [1] and Mi Sun Park [1,3,*]

[1] Graduate School of International Agricultural Technology, Seoul National University, 1447 Pyeongchang-daero, Daehwa, Pyeongchang, Gangwon 25354, Korea; seongmin@snu.ac.kr (S.S.); haeunl630@snu.ac.kr (H.L.); switch-thkim@snu.ac.kr (T.H.K.); lse1229@snu.ac.kr (S.L.)

[2] Research Institute of Agriculture and Life Sciences, Seoul National University, 1 Gwanakro, Gwanakgu, Seoul 08826, Korea; khaing9588@snu.ac.kr

[3] Center for International Agricultural Development, Institutes of Green Bio Science and Technology, Seoul National University, 1447 Pyeongchang-daero, Daehwa, Pyeongchang, Gangwon 25354, Korea

* Correspondence: mpark@snu.ac.kr; Tel.: +82-33-339-5858

Received: 20 February 2020; Accepted: 21 March 2020; Published: 26 March 2020

Abstract: Agroforestry is an intensive land management system that integrates trees into land already used for crop and animal farming. This provides a diverse range of ecosystem services by bridging the gaps between agriculture, forestry, and animal husbandry. It is an important approach to improve the environmental, economic, and social benefits of complex social–ecological systems in the Asia-Pacific region. This paper aims to examine the research trends in agroforestry and the current state of knowledge, as well as the research gaps in the ecosystem services of agroforestry in this region. A systematic mapping methodology was applied, where analysis units were academic articles related to agroforestry practices in the Asia-Pacific region. The articles published between 1970 and 2018 were collected through the international specialized academic database, SCOPUS. They were coded according to the types of agroforestry practices and ecosystem services. The research result indicates silvorable systems, especially plantation crop combinations, tree management, habitats for species, biological controls, and maintenance of genetic diversity and gene-pools, are the most prominent in the agroforestry research from the Asia-Pacific region. Approximately 60% of all research articles include case studies from India, China, Indonesia, and Australia. Research on agroforestry has changed following the international discourse on climate change and biodiversity. Therefore, this systematic map improves our understanding of the nature, volume, and characteristics of the research on ecosystem services with regard to agroforestry in the Asia-Pacific region. It provides scholars with a springboard for further meta-analysis or research on agroforestry and ecosystem services.

Keywords: agroforestry; ecosystem services; Asia-Pacific region; systematic map

---

1. Introduction

Agroforestry is defined as agriculture that incorporates trees [1]. However, this definition is oversimplified. Agroforestry is in fact much more complex. Geographic information system (GIS) data show that 43% of all agricultural land, globally, is used for agroforestry, which is more than 1 billion hectares [2]. Agroforestry systems in the Asia Pacific region are abundant under various agro-ecological environments, especially in Indonesia, Malaysia, India, Sri Lanka, and Bangladesh, as the practices have played important roles there since ancient times [3]. As agroforestry practices in the region have evolved over a long period of time, they utilize many novel and historic strategies to fulfill the basic needs of the smallholder farmers for food, fodder, medical products, fuelwood, and as a cash income [4]. In particular, multifunctional home gardens managed in a traditional way in the Asia-Pacific region are important for enhancing food security [5], diversity, and cultural and ecological functions, by providing natural fertilizer from trees, keeping soil resources, and favoring habitats for improved agrobiodiversity environments, which can mitigate climate change [6,7]. The experiences from these traditional systems highlight the importance of agroforestry in the twenty-first century to tackle land management problems, such as food insecurity, deforestation and forest degradation, biodiversity loss, and climate change.

However, socio-economic and technological factors have led to changes in traditional land use systems over time, in line with the paradigm shifts from traditional cultivation and monocropping to sustainable land use [4,8]. The paradigm shifts resulting from market economies, and specifically, exotic commercial crops have brought about the decline of traditional land-use systems [7,9]. For instance, the historic predominant land-use systems, such as home gardens, have been converted for land-use intensification in Asia [9–11]. As the population pressure and food requirements have driven mono specific production systems [12], agricultural and forest policies also began to favor fast-growing and commercial species, which had adverse impacts on agroforestry in Asia. The industrial agriculture systems have resulted in new social and environmental challenges, by altering the biotic patterns and interactions [13,14].

For this reason, the traditional methods have recently been revisited for their ecosystem services, including contribution to economic, cultural, and social values [11]. Several countries have enacted agroforestry-related policies. The national agricultural policy of India supports farmers adopting agroforestry practices to increase their incomes by providing them with credit and technologies. The national forestry policy of Sri Lanka stresses the important role of agroforestry in offering timber and nontimber forest products (NTFPs), and bio-energy conserving environments at the same time [15]. However, in most countries, public policies have lacked incentives and support to drive uptake of agroforestry systems and do not consider ecosystem services derived from agroforestry as sustainable production methods worth promoting to conserve agrobiodiversity [9,16–19]. The forest policies of Pakistan, for example, mention farm forestry and its importance but offer almost no practical implementations [20]. In this context, stabilizing agroforestry in the Asia-Pacific region should be warranted, owing to the global challenges related to food security [5], biodiversity preservation, climate change, and the great number of stakeholders that rely on agroforestry for their subsistence [11].

International organizations and governments have been investing in agroforestry research and projects in lower-middle-income countries since the 1960s and 1970s [5,21,22], exploring the interactions between agroforestry and ecosystem oriented services in the world, as well as specifically in the Asia-Pacific region [18]. For instance, Chang et al. [19] analyzed bird biodiversity in a coffee agroforestry system in India, Yang et al. [23] described the impact of riparian buffers on biomass in China, and Tiwari et al. [24] calculated the ecosystem services of maize from hillside agroforestry systems in Nepal, but each of these studies focused on only one type of agroforestry. There have been more general studies, such as Bohra et al. [25] who reviewed the socio-economic impacts of agroforestry, Basu [26] who conducted interviews to figure out the impacts of agroforestry on climate change mitigation, and Goswami et al. [27] who explained the estimated biomass and carbon sequestration in diverse agroforestry systems. These studies however, are limited in scope and there is a lack of coherent evidence of their effectiveness. This is partly because it is a complex process to synthesize the regional research and it is difficult to interpret the interrelations of agroforestry and ecosystem services. There are number of reasons for this. First, ecosystem services are interrelated, and multifaceted trade-offs vary depending on the service and spatial scale [28]. Second, the results reporting the benefits [17,29] and disservices [30,31] are inconsistent. Third, agroforestry practices in different disciplines resulted in a lack of synthesis of the evidence.

One of the solutions for synthesizing the information, when there is disparate evidence, is a systematic map, providing a robust method to identify and present research evidence extracted from peer-reviewed papers and grey literature [22,32–34]. The rigorous methodology describes the characteristics and trends of research on a broad scale, and, in part, fills the knowledge gaps by bringing together repeatable and quantitative evidence [33,34].

Previous works to synthesize the knowledge of agroforestry and ecosystem services have been conducted in some regions, including Europe [35], high-income countries [22], and low- and middle-income countries [36], but no study has systematically identified and described the nature, volume, and characteristics of the research in the field of agroforestry and ecosystem services in the Asia-Pacific region. To make up for the limits of the systematic maps, several previous papers have reported meta-analysis investigations [17,31,37]. This study provides comprehensive and systematic evidence about agroforestry and its impacts on ecosystem services by mapping the results of previously published investigations from around the Asia-Pacific agroforestry region. The systematic map aims to catalog the available knowledge and show the research characteristics and trends from the existing findings in the field, providing the first synthesis of ecosystem services of the Asia-Pacific agroforestry.

The following research questions were addressed:

1. What is the quantitative distribution of evidence-based research regarding agroforestry and ecosystem services in the Asia-Pacific region?
2. What types of agroforestry practices and ecosystems have been studied and how do they change over time?
3. How are research approaches between agroforestry systems and ecosystem services interlinked?

2. Theoretical Background

2.1. Agroforestry

Agroforestry is generally said to be a 'new term for an old practice', since the name was not recognized in the literature until the mid-1970s, but it has been in practice for a much longer period of time [3]. Agroforestry is a collection of land-use practices, systems, and technologies that integrates woody perennials into crop- and animal-based agricultural practices [38]. The main requirements for agroforestry are that at least two plants or animal species are included in the land-use system and one of these should be a woody perennial. Moreover, there are economic and ecological interactions between two or more production systems, such as tree–crop, tree–livestock, or tree–fish. Compared with intensified agriculture systems, the cycle of agroforestry systems may take one or more years and be more complex structurally, economically, ecologically, and functionally [39].

Agroforestry types consist of agrisilviculture, silvopasture, agrosilvipasture, forest farming, and urban agroforestry, amongst others, including tree integration with fisheries or beekeeping. Agrisilviculture is defined as integrating trees with cropping, also called silvorable, while silvopasture is an integrated system of trees and livestock, and agrosilvipasture is a tree-integrated system with livestock and crops together. Forest farming is an operation that raises livestock or produces crops and NTFPs within forests [38]. Agroforestry systems are classified using different criteria such as the different configuration and structure of the system's components. They are also identified based on the temporal sequences to bring in different species to the systems and the function of trees in agroforestry practices, such as providing shade, breaking wind, or conserving soil. As some agroforestry types are more suitable for certain environmental conditions than other systems, ecological appropriateness and the level or scale of inputs for the systems management for commercial uses can also differ [39]. Table 1 describes and identifies common agroforestry practices. These practices are mutually exclusive.

Table 1. Major agroforestry practices [38,40,41].

Agroforestry Practices	Category	Definition	Components
Silvorable system/ agrosilviculture/ agrisilviculture	Improved or rotational fallow	Woody species planted and left to grow during the 'fallow phase'	Fast-growing preferably leguminous with agricultural crops
	Taungya	A combined stand of woody and agricultural species during the early stages of plantation establishment	Usually tree species for plantation with agricultural crops
	Forest farming	Operations grow food, herbal, botanical, or decorative crops under the protection of a managed forest canopy; also called multistory cropping	Trees and nontimber products
	Alley-cropping	Woody species in hedges; agricultural species in alleys in between hedges; microtonal or strip arrangement	Fast-growing and leguminous, which coppice vigorously with agricultural crops
	Multistory agroforestry	Multispecies, multilayer dense plant associations with no organized planting arrangements	Different woody components of varying forms and growth habits, shade-tolerant ones sometimes present
	Multipurpose trees	Trees scattered haphazardly or according to some systematic patterns on bunds terraces or plot/field boundaries	Multipurpose trees and other fruit trees with agricultural crops
	Plantation crop combinations and tree management	(1) Integrated a multistory (mixed and dense) mixture of plantation crops; (2) mixtures of plantation crops in alternate or another regular arrangement; (3) shade trees for plantation crops, shade trees scattered; (4) intercropping with agricultural crops	Plantation crops like coffee, cacao, coconut, fruit trees; fuelwood/fodder species, shade-tolerant species
	Home garden	Intimate, multistory combinations of various trees and crops around homesteads	Fruit trees, predominate; also, other woody species, vines, shade-tolerant crops
	Shelterbelts and windbreak/ hedgerows/ live hedges	Trees around farmland/plot	Combination of tall-growing spreading types with agricultural crops of the locality
	Fuelwood production	Interplanting firewood species on or around agricultural lands	Firewood species
Silvopasture/ silvopastoral systems/ forest grazing	Trees on rangeland or pastures	Trees scattered irregularly or arranged according to some systematic pattern	Multipurpose usually of fodder value
	Protein banks	Production of protein-rich tree fodder on-farm/rangelands for cut-and-carry fodder production	Leguminous fodder trees
	Plantation crops with pasture and animals	Livestock under woody perennials	Plantation crops, for example, cattle under coconuts
Agrosilvopastoral systems/ agrosilvipasture	Apiculture with trees/ entomoforestry	Tree fruits, leaves, flowers being used for insects	Trees and shrubs preferred by insects such as bees
	Aquaforestry/ silvofishery	Trees lining fishponds, tree leaves being used as 'forage' for fish	Trees and shrubs preferred by fish
	Multipurpose woody hedgerows	Woody hedges for browse, mulch, green manure, soil conservation, etc.	Fast-growing and coppicing fodder shrubs
	Home gardens involving animals	Intimate, multistory combination of various trees and crops, and animals, around homesteads	Fruit trees predominate; also, other woody species
Riparian buffer strips		Areas along rivers and streams planted with trees, shrubs, and grasses. Serving as sponges to filter farm runoff, and the roots stabilize stream banks to prevent erosion	Adjacent to perennial, intermittent, and ephemeral streams, lakes, and estuarine marine shorelines

2.2. Ecosystem Services

The term 'ecosystem services' was initially employed by Ehrlich and Mooney [42] to bring attention to the human activities causing land degradation and consequently the diminishing functions and services delivered by ecosystems [42]. With the rising awareness of the importance of ecosystem services and the inseparable relationship between ecosystem services and human activities, research on ecosystem services has been carried out from a multitude of analytical angles [32,43]. Multiple ways to classify ecosystem services exist to assess and reflect the ecosystem [44]. According to the theoretical structure of the Millennium Ecosystem Assessment (MEA) [45], which was created by many experts, ecosystem services contain direct services affecting human provisioning, regulating, and cultural services. Indirect services, such as the supporting services, underpin other services including photosynthesis, nutrient cycling, and primary production [45]. When properly managed, forests and trees provide diverse provisioning services, such as food, timber, raw material, and medical products [46,47]. Other investigations have assessed services forests provide at the local level, including pollination [48,49], biodiversity [9,16–19,50,51], pest control [52,53], moderation of nutrient run-off [54], and soil nutrient enhancement [55]. National and global scale studies explain how forest ecosystem services work and contribute to watershed protection, climate regulation, carbon storage [56], and the challenges in integrating the ecosystem services in decision making and implementation [57]. Ecosystem services are classified by using a spatial and functional scale (See Table 2) based on previously defined references [45,57–59]. This helps in defining local, national, and global ecosystem services that forests and agroforestry systems provide. The classification is considered relevant and important in the context of this systematic review.

Ecosystem services act as a transformative lens, revealing the agricultural systems and driving forces that should be considered when trying to understand the relationship between nature and human development [60–62]. Agroforestry has been recognized as an environmentally friendly and cost-effective land-use system for landscape restoration by reconciling production and environmental conservation/enhancement at the landscape level [17,29]. Given that the integrated agroforestry systems can be more efficient at capturing agricultural resources, including water and solar radiation, agroforestry systems in certain conditions can produce more than monoculture systems [18,63]. The combination of production and conservation/enhancement introduced a revitalization of agroforestry.

Each type of agroforestry practice contributes to ecosystem services in a different way. The detailed impacts of agroforestry practices on ecosystem services are listed below by the types of ecosystem services (Table 2). First, well managed agroforestry provides provisioning services; food [5,17,29,45,64], fiber [17], freshwater [18], raw materials [17], fuel wood [29], NTFPs [17,29], medicinal resources [17], genetic resources, and ornamental resources [17]. Second, agroforestry also delivers regulating ecosystem services; erosion control, climate change moderation [29], nutrient retention [17,29], carbon storage and sequestration [17], and pest control [18]. Moreover, agroforestry systems play a critical role in biodiversity enhancement [9,16–19,50,51] and climate regulation [18]. More specifically, soil can be more fertile in agroforestry systems where leguminous trees collect the nitrogen in their leaves and roots and provide it to the crops [18]. Trees planted and interspersed with crops can incorporate their leaves into the soil, which increases yields [18]. In silvopastoral systems, animals that inhabit a forest can make the soil rich by providing manure [41]. In this way, the lands can be more productive and give farmers more stable yields, which improves food security [5,45]. Cultural services are a corresponding outcome from agroforestry and include: social relations [3], cultural heritage values [18], ecotourism recreation [16], spiritual values, experience and knowledge systems, and educational values [3]. For example, communities can share the byproducts of agroforestry, such as cultural and household goods and utilities [64]. However, it has also been reported that agroforestry has neutral and negative impacts on ecosystem services [30,31].

Table 2. Ecosystem service types and components [45,57–59].

Ecosystem Services	Category	Definition	Components
Provisioning services (products obtained from ecosystems)	Food	Presence of edible plants and animals	Seafood, gamete, crops, wild foods, and species
	Fiber	Presence of species or abiotic components with potential use for timber or textile	Timber, cotton, hemp, silk
	Freshwater/ water/ drinking water/ irrigation water	Presence of water reservoirs	Water
	Raw materials/ fuel wood/ biofuels/ bioenergy/energy/ hydroelectricity/ biomass/charcoal/ firewood/NTFPs	Presence of species or abiotic components with potential use as a fuel or raw material	Lumber, skins, fuel wood, organic matter, fodder, and fertilizer
	Biochemicals/ pharmaceuticals/ medicinal resources	Presence of species or abiotic components with potentially useful chemicals and/or medicinal uses	Pharmaceuticals, chemical models, and test and assay organisms
	Genetic resources	Presence of species with (potentially) useful genetic material	Crop improvement genes, and health care
	Ornamental resources	Presence of species or abiotic	Fashion, handicrafts, jewelry, pets, worship, decoration, and souvenirs
Regulating and supporting (benefits obtained from the regulation of ecosystem processes and underpinning services that enable other services to function)	Air quality maintenance	The capacity of ecosystems to extract aerosols and chemicals from the atmosphere	The ability of the atmosphere to cleanse itself
	Carbon sequestration and storage	Regulation of the global climate by storing and sequestering greenhouse gases. Removing carbon dioxide from the atmosphere and locking it away in their tissues	Net source of carbon sequestration
	Water regulation/ water flows/ water purification and waste treatment/ waste-water treatment	Role in water infiltration and gradual release of water, and in biotic and abiotic processes of removal or breakdown of organic matter, xenic nutrients, and compounds	Chemical condition of freshwaters and saltwater
	Regulation of human diseases	Control of pest populations through trophic relations	Disease control
	Pollination	Contribute to abundance and effectiveness of pollinators	Pollination and seed dispersal
	Moderation of extreme events/ storm protection/ erosion control/ climate regulation	Influence on local and global climate through land-cover and biologically mediated processes, and the role of forests in dampening extreme events	Storm and flood protection, micro and regional climate regulation
	Soil formation/ soil composition/ maintenance of soil fertility/ nutrient cycling	Role of natural processes in soil formation, regeneration, and composition	Buffering and attenuation of mass flows

Table 2. Cont.

Ecosystem Services	Category	Definition	Components
	Primary production	The assimilation or accumulation of energy and nutrients by organisms	Products supporting microorganisms, algae, plants, and animals
	Habitats for species/ biological control/ maintenance of genetic diversity/ gene-pool	Provision of breeding, feeding, or resting habitat for transient species/maintenance of a given ecological balance and evolutionary processes	Maintenance of nursery populations and habitats
Cultural services (nonmaterial benefits obtained from ecosystems that enrich lives)	Spiritual and religious values/ inspiration/spiritual experience	Landscape features or species with spiritual and religious value/landscape features or species with inspirational value to human	Use of nature for religious, heritage, or natural values
	Knowledge systems/ educational values	Features with special educational and scientific value/interest	Use of natural systems for school excursions, and scientific discovery
	Social relations	Influence on the types of social relations that are established in particular cultures. e.g., difference between fishing, nomadic herding, or agricultural societies, in many respects of their social relations	
	Cultural heritage values/ cultural diversity	Culturally important landscape features or species	Use of nature as a motif in books, film, painting, folklore, national symbols, architecture, advertising, etc.
	Ecotourism/ tourism/ recreation/ aesthetic values	Aesthetic quality of the landscape, based on e.g., structural diversity, "greenness", tranquility	Ecotourism, outdoor sports, and recreation

3. Materials and Methods

Evidence-based methods, including systematic reviews, were developed to support policy decision making by providing scientific information [65]. The methods are now employed in a diverse array of fields, such as conservation and environmental management [66,67]. The methods have advantages such as transparency, robustness, independence, and comprehensiveness and help to develop a comprehensive picture and new knowledge, by analyzing previous rigorous studies [34,68]. As one of the evidence-based methods, systematic review, which focuses on providing science-based knowledge through existing evidence, was applied to this study on agroforestry and ecosystem services in the Asia-Pacific region. The PRISMA (preferred reporting items for systematic reviews and meta-analyses) flow diagram was adapted from Moher et al. [69] by following four steps: identification, screening, eligibility, and inclusion. The details are explained in the following sections and Figure 1.

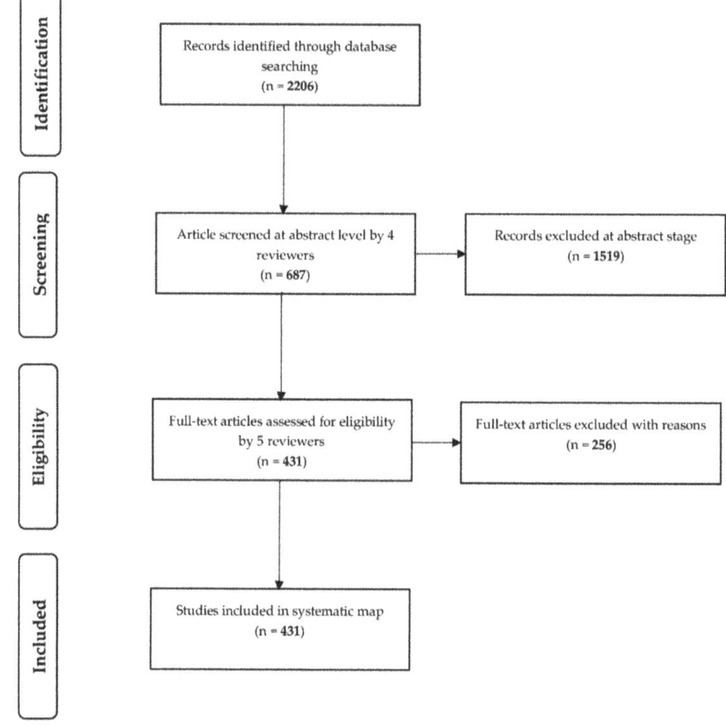

Figure 1. Systematic mapping process of the study, illustrating articles from the initial search to screening for synthesis (identification, screening, eligibility, and inclusion). Articles were found through database search at the identification stage. Then the articles captured were screened based on the categories of agroforestry practices, ecosystem services, and the Asia-Pacific region (through titles, keywords, abstracts, and full text articles) at the screening and eligibility stages. Finally, the articles satisfied with the eligibility criteria were included for the study.

3.1. Data Collection (Identification)

The literature search was conducted in English, in February 2019, in a bibliographic database, SCOPUS, which is one of the largest citation databases comprising peer-reviewed articles in life, social, physical, and health science fields, but it does not include grey literature. Search fields included article title, abstract, and keywords. The search strings were a combination of three major topics: agroforestry practices, ecosystem services, and the Asia-Pacific countries and regions (Table A1 of Appendix A). We included single ecosystem services (e.g., irrigation water, hydroelectricity, and nutrient cycling) in the search words to get a wide-range of results using methods mentioned above [57–59,64]. Despite efforts to cover a broad range of terms representing the Asia-Pacific agroforestry practices and their ecosystem services, it is possible that some articles were not captured. All of the extracted articles were stored and shared in a Mendeley database and duplicates were removed.

3.2. Article Screening, Study Eligibility Criteria and Inclusion

A total of 2206 articles resulted from the literature search were manually screened in April 2019 through two stages, according to the procedure of the systematic review [65]. The criteria were adjusted for intercoder reliability, and the screening criteria were different at each stage. At the first screening stage, three researchers examined the titles, keywords, and abstracts of the literature to check correspondence with the criteria; covering at least one agroforestry practice and assessing the impacts of agroforestry on the ecosystem services. The intercoder agreement was tested using Fleiss' Kappa statistical measure. The eligible articles were included in the study and the others were recorded and excluded. Uncertain articles regarding eligibility were kept for a full text assessment. At the second stage, all included articles were reviewed by five coders who were trained, and the intercoder agreement was measured by the percent agreement method using approximately 10% of included articles. Through the second stage of full text review, data was coded, and noneligible articles were excluded.

3.3. Data Coding Strategy

To categorize the context of the agroforestry practices and ecosystem services in the selected articles, each study was classified according to the study country and types of agroforestry practices and ecosystem services. To categorize the agroforestry practices, a coding category system was developed from previous categorizations (Table 3). The categories of the agroforestry practices included traditional and modern systems modified by the authors. For the ecosystem services, each individual study was coded according to the type of ecosystem service, including provisional, regulating and supporting, and cultural services, according to the Millennium Ecosystem Assessment and previous literature [45,57] (Table 3).

Table 3. Coding category system [38,40,41,45,57–59].

Category		Sub-Category
Published date		Year/Month/Day
The study target country		Name of country
Agroforestry	Silvorable	Improved or rotational fallow (IR), taungya (TA), forest farming (FF), alley-cropping (AC), multistory agroforestry (MA), multipurpose trees (MT), plantation crop combinations and tree management (PC), home garden (HG), shelterbelts and windbreak/hedgerows/live hedges (SW), and fuelwood production (FP)
	Silvopasture	Trees on rangeland or pastures (TR), protein banks (PB), and plantation crops with pasture and animals (PP)
	Agrosilvopasture	Apiculture with trees (entomoforestry) (AF), aquaforestry/silvofishery (AS), home gardens involving animals (HA), multipurpose woody hedgerows (HW), and riparian buffer strips (RB)
Ecosystem services	Provisioning	Food (FO), fiber (FI), fresh water/ water/ drinking water/ irrigation water (FW), raw materials/ fuel wood/ biofuels/ bioenergy/ energy/ hydroelectricity/ biomass/ charcoal/ firewood/ NTFPs (RM), biochemicals/ pharmaceuticals/ medicinal resources (BC), genetic resources (GR), ornamental resources (OR), and other (PO)
	Regulating/ supporting	Air quality maintenance (AQ), carbon sequestration and storage (CS), water regulation/ water flows/ water purification and waste treatment/ waste-water treatment (CR), regulation of human diseases (RH), pollination (PL), moderation of extreme events/ storm protection/ erosion control/ climate regulation (EE), soil formation/ soil composition/ maintenance of soil fertility/ nutrient cycling (SF), primary production (PR), habitats for species/biological control/maintenance of genetic diversity/ gene-pool (HS), and other (RS)
	Cultural	Spiritual and religious values/ inspiration/ spiritual experience (SR), knowledge systems/ educational values (KS), social relations (SO), cultural heritage values/ cultural diversity (CH), ecotourism/ tourism/ recreation/ aesthetic values (ET), and other (CO)

There was no quality appraisal of the individual articles in the review, because of the large scope and size encompassed by the literature involved. Analysis on the distribution of the evidence base with linkages between the agroforestry practices and ecosystem services was carried out.

Lastly, a systematic map was created indicating evidence distribution and linkages between agroforestry practices and ecosystem services. Rows represent individual agroforestry practices, and columns are divided into categories of ecosystem services. If the article covered more than one practice or service, then multiple practices or services from one article were mapped.

4. Results

4.1. Number of Articles

This research follows the systematic mapping process, the PRISMA (preferred reporting items for systematic reviews and meta-analyses) flow diagram adapted from Moher et al. [69] (Figure 1). It illustrated articles from the initial search to screening for synthesis (identification, screening, eligibility, and inclusion). The articles identified through database searching were recorded at the 'identification' stage. Then, the articles captured were screened based on our included agroforestry practices, ecosystem services, and Asia-Pacific region (through title, keywords, abstract, and full text article) during 'screening and eligibility' stages. After full text articles were assessed, the articles with the eligibility criteria were included for the study. The 2206 articles from the SCOPUS database were extracted with the search strings for the agroforestry practices, ecosystem services, and Asia-Pacific countries (Table A1 of Appendix A) [69]. However, a large portion of the selected article were excluded due to irrelevance after checking the titles, keywords, and abstracts (Figure 1). Five reviewers screened the full-text of 687 articles and reduced the number of articles to 431 by excluding the inappropriate studies, such as articles in which the abstracts were written in English, but the main text was in another language. Ultimately, our final map comprised 431 articles. There was a 92.2% intercoder agreement for the sampled articles selected in the full-screening testing.

There were a few agroforestry-related articles that were published before 1990 and the publication numbers began to rise after this point (Figure 2). Among the selected articles, the articles on the silvorable type of agroforestry practices and the regulating and supporting service type of ecosystem services were dominant.

Figure 3 displays the geographical distribution of the included literature in the systematic review. There were many studies focused on South Asia, Oceania, and North-east Asia, and fewer studies on Western Asia, when compared. At the country level, there were four countries that were dominant, accounting for approximately 60% of all research articles included in the map: India ($n = 122$), China ($n = 70$), Indonesia ($n = 54$), and Australia ($n = 35$).

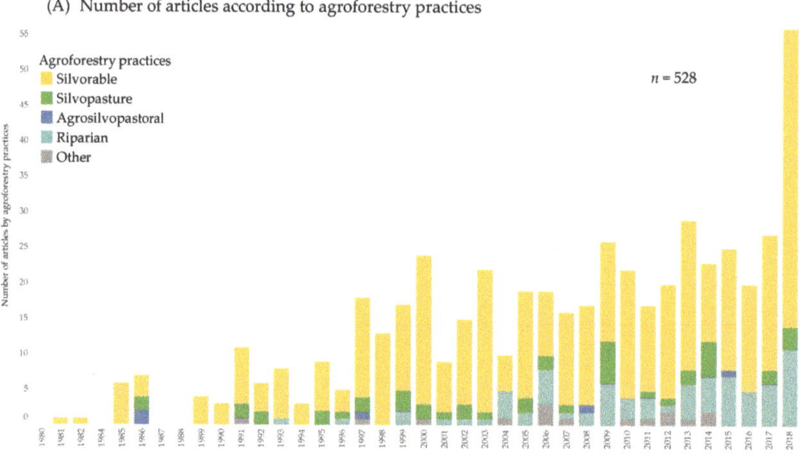

Figure 2. *Cont.*

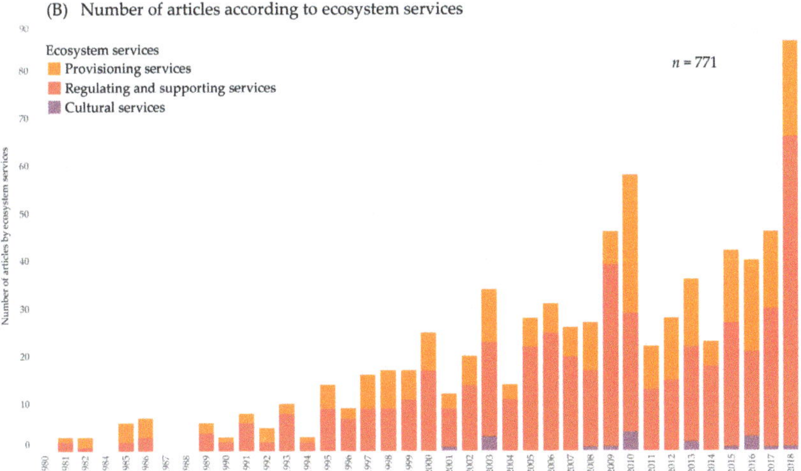

Figure 2. Distribution of the published articles on agroforestry practices and ecosystem services. (**A**) shows the number of publications related to agroforestry practices by year and (**B**) shows the number of articles related to ecosystem services by year.

Figure 3. Geographic distribution of the selected articles on agroforestry practices and ecosystem services. Note: The darker the color of the green, the higher the number of the articles focusing on agroforestry practices and ecosystem services in the Asia-Pacific region.

4.2. Characteristics of Agroforestry Practices and Ecosystem Services

We classified the practices and services of the articles into multiple subcategories, as agroforestry practices and ecosystem services can be complex. The results show that most of the investigations focused on agroforestry practices categorized as silvorable systems/ agrosilviculture, as constituted about 73% of all the articles. Articles about riparian buffers and the silvopasture practices were the second largest portion of all the studies, making up 12%, and agrosilvopastoral practices were 1.5 % of the investigations. The most studied practices among the individual agroforestry practices were as follows: plantation crop combinations and tree management (PC, n = 246), riparian buffer strips (RB, n = 146), home garden (HG, n = 116), shelterbelts and windbreak/hedgerows/live hedges (SW, n = 104), and alley-cropping (AC, n = 82) (Figure 4).

Overall, most of the articles examined the impacts of agroforestry on regulating and supporting services (64.5%), with the majority focusing on soil formation/soil composition/maintenance of soil fertility/nutrient cycling (SF, n = 349), and habitats for species/biological control/maintenance of genetic diversity/gene-pool (HS, n = 226). Less than 3% of the articles disaggregated the cultural services (n = 36).

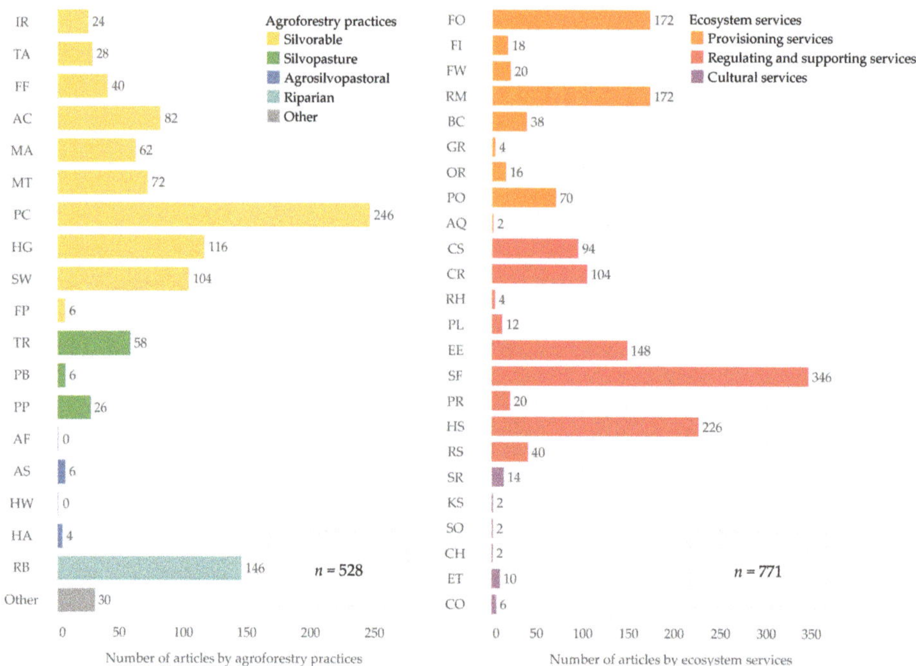

Figure 4. The number of articles by agroforestry practices and ecosystem services.

4.3. Research Trends for Agroforestry and Ecosystem Services by Decade

The results showed the research trends for agroforestry and ecosystem services by decade (Figure 5). In the 1980s, the studies on home gardens (HG, 21 %) and taungya (TA, 16 %) for silvorable practices were prominent. Shelterbelts and windbreak/hedgerows/live hedges (SW), alley-cropping (AC), and home garden (HG) as silvorable practices were mostly studied in the 1990s. After the 2000s, the publications on riparian buffers (RB) increased, and the plantation crop combinations and tree management (PC) were dominant.

The results of analysis of the ecosystem services in the 1980s showed that it was the provisioning services that were mainly addressed in the agroforestry research, followed by the regulating and supporting services. Soil formation/soil composition/maintenance of soil fertility/nutrient cycling (SF, 28 %), food (FO, 25%), and raw materials/ fuel wood/ biofuels/ bioenergy/ energy/ hydroelectricity/ biomass/ charcoal/ firewood/ NTFPs (RM, 25%) were mainly addressed. In the 1990s and 2000s, the focus was on soil formation/ soil composition/ maintenance of soil fertility/ nutrient cycling (SF) and the moderation of extreme events/ storm protection/ erosion control/ climate regulation (EE). During 2010–2018, the majority of the studies focused on habitats for species/ biological control/ maintenance of genetic diversity/ gene-pool (HS) and soil formation/ soil composition/ maintenance of soil fertility /nutrient cycling (SF), and a few studies focused on cultural services, such as spiritual and religious values/ inspiration/ spiritual experience (SR), and ecotourism/ tourism/ recreation/ aesthetic values (ET).

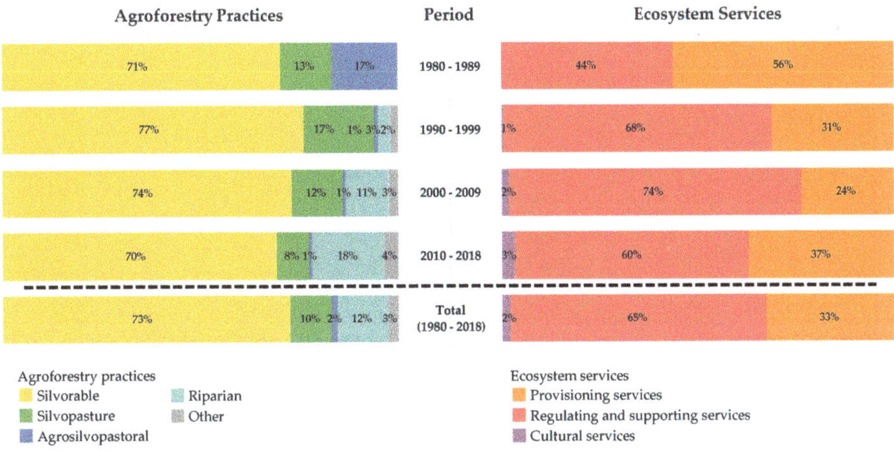

Figure 5. Ratio of the articles on agroforestry practices and ecosystem services by decade.

4.4. *The intersection of Agroforestry Practices and Ecosystem Services*

The linkage between the agroforestry and ecosystem services in the selected articles (Figure 6) was analyzed. The number of articles that included specific types of agroforestry practices and simultaneously specific types of ecosystem services, was calculated. Plantation crop combinations and tree management (PC) and their links to soil formation/soil composition/maintenance of soil fertility/nutrient cycling (SF) and habitats for species/biological control/maintenance of genetic diversity/gene-pool (HS) were most commonly studied in these investigations. Most of the links are between silvorable practices and their relationship to regulating and supporting services. Other pathways were less studied, especially in relation to cultural services. For example, there is no article in the linkage between agrosilvopasture practices and social services. One of the most studied agroforestry practices, home garden (HG), was highly linked to food (FO) and raw materials/ fuelwood/ biofuels/ bioenergy/ energy/ hydroelectricity/ biomass/ charcoal/ firewood/ NTFPs (RM), which are providing services. Unlike home garden (HG), riparian buffer (RB) has a strong connection to regulating and supporting services, including soil formation/ soil composition/ maintenance of soil fertility/ nutrient cycling (SF) and habitats for species/ biological control/ maintenance of genetic diversity/ gene-pool (HS).

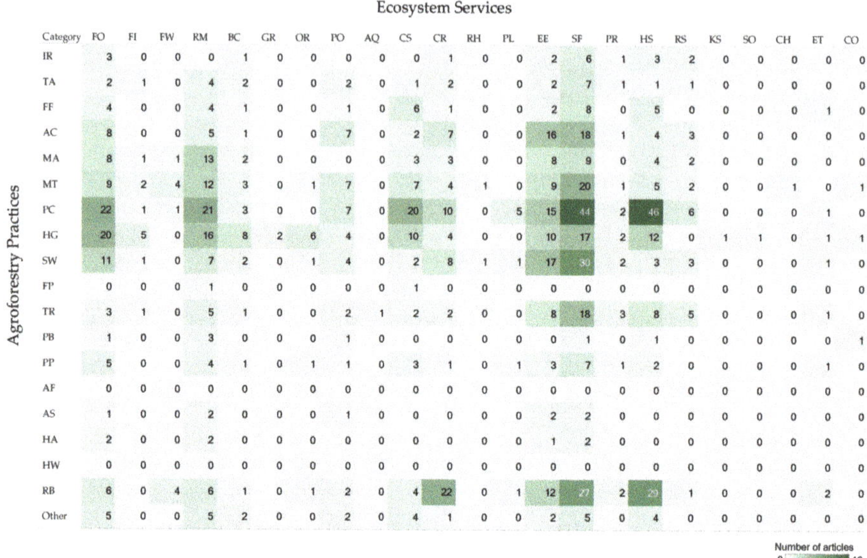

Category	FO	FI	FW	RM	BC	GR	OR	PO	AQ	CS	CR	RH	PL	EE	SF	PR	HS	RS	KS	SO	CH	ET	CO
IR	3	0	0	0	1	0	0	0	0	0	1	0	0	2	6	1	3	2	0	0	0	0	0
TA	2	1	0	4	2	0	0	2	0	1	2	0	0	2	7	1	1	1	0	0	0	0	0
FF	4	0	0	4	1	0	0	1	0	6	1	0	0	2	8	0	5	0	0	0	0	1	0
AC	8	0	0	5	1	0	0	7	0	2	7	0	0	16	18	1	4	3	0	0	0	0	0
MA	8	1	1	13	2	0	0	0	0	3	3	0	0	8	9	0	4	2	0	0	0	0	0
MT	9	2	4	12	3	0	1	7	0	7	4	1	0	9	20	1	5	2	0	0	1	0	1
PC	22	1	1	21	3	0	0	7	0	20	10	0	5	15	44	2	46	6	0	0	0	1	0
HG	20	5	0	16	8	2	6	4	0	10	4	0	0	10	17	2	12	0	1	1	0	1	1
SW	11	1	0	7	2	0	1	4	0	2	8	1	1	17	30	2	3	3	0	0	0	1	0
FP	0	0	0	1	0	0	0	0	0	1	0	0	0	0	0	0	0	0	0	0	0	0	0
TR	3	1	0	5	1	0	0	2	1	2	2	0	0	8	18	3	8	5	0	0	0	1	0
PB	1	0	0	3	0	0	0	1	0	0	0	0	0	0	1	0	1	0	0	0	0	0	1
PP	5	0	0	4	1	0	1	1	0	3	1	0	1	3	7	1	2	0	0	0	0	1	0
AF	0	0	0	0	0	0	0	0	0	0	0	0	0	0	0	0	0	0	0	0	0	0	0
AS	1	0	0	2	0	0	0	1	0	0	0	0	0	2	2	0	0	0	0	0	0	0	0
HA	2	0	0	2	0	0	0	0	0	0	0	0	0	1	2	0	0	0	0	0	0	0	0
HW	0	0	0	0	0	0	0	0	0	0	0	0	0	0	0	0	0	0	0	0	0	0	0
RB	6	0	4	6	1	0	1	2	0	4	22	0	1	12	27	2	29	1	0	0	0	2	0
Other	5	0	0	5	2	0	0	2	0	4	1	0	0	2	5	0	4	0	0	0	0	0	0

Figure 6. The linkage between agroforestry practices and ecosystem services. Note: The darker the color of the cells, the higher the frequency of articles.

5. Discussion

This systematic map was designed to synthesize the knowledge on agroforestry and ecosystem services in the Asia-Pacific region. The results highlight the gaps and limitations of previous agroforestry research works, concentrating on specific practices and ecosystem services and geographic areas. In particular, there were several gaps where there were only a few or no evidence that existed between all types of the agroforestry practices and their impacts on social services. In the following sections, we will explore these gaps and limitations.

5.1. Geographical Gaps

The previous research was skewed towards several countries. India was the country with the most research, and this probably reflects a research boom in line with the agroforestry project, including the All India Coordinated Research Project on Agroforestry (ACRPA) in 1983 and policy implementations promoting agroforestry practices [4,15]. As India has a long history of agroforestry since the Mesolithic period (8000–2000 BC), agroforestry systems, especially home gardens (HG), have been predominant there over time. Likewise, in this study, home garden (HG) related papers were centered on India, with a high focus on food (FO), raw materials/ fuel wood/ biofuels/ bioenergy/ energy/ hydroelectricity/ biomass/ charcoal/ firewood/ NTFPs (RM), habitats for species/ biological control/ maintenance of genetic diversity/ gene-pool (HS), and soil formation/ soil composition/ maintenance of soil fertility/ nutrient cycling (SF). Case studies in China were intensively conducted, with the focus on the riparian buffer (RB), including design of RB and site-specific nutrient management (SSNM) [70]. Western Asian countries however, were rarely studied, and only 0–3 articles were found per country. This asymmetry creates an evidence gap across this geographical area. The gap indicates the limitations of the contextual diversity, as well as the limitations of the applications of research insights to decisions or practices [71]. In this regard, it is highly recommended that the scope of agroforestry research be broadened, to build the geographical evidence base.

5.2. Practice and Services Gap

The research efforts were vastly weighted towards measuring the impacts of silvorable practices on regulating and supporting services, with an emphasis on soil formation/ soil composition/ maintenance of soil fertility/ nutrient cycling (SF) and habitats for species/ biological control/ maintenance of genetic diversity/ gene-pool (HS), obtained from plantation crop combinations and tree management (PC) (Figure 4). In addition, the hit maps (Figures 6–9) reveal an extreme emphasis on several practices and services. We can easily find areas with less evidence from studies about agrosilvopastoral practices and cultural services (Figure 6). We set forth two possible reasons for the concentrations in the works. First, that plantation crop combinations and tree management (PC)-related crops, such as rubber and coffee, have been widely cultivated in the Asia-Pacific regions. This is because the top six producers of natural rubber in the world are all in Asia and most Southeast Asian countries cultivate coffee, which likely leads to active studies on the topic. In Europe, silvopastoral systems are well-explored rather than silvorable practices, which are dominant in the Asia-Pacific region, as these systems have traditionally been formed in European landscapes [35]. Second, there were fewer articles measuring social services than articles measuring the regulating and supporting services. The articles about agroforestry practices and cultural services were published later than the articles about other ecosystem services. It was assumed that the term ecosystem services was rarely interpreted in the field of social science. Social services like aesthetic and cultural values should be measured based on a deep understanding of social and cultural contexts. For comprehensive designs of future agroforestry practices, the lack of evidence regarding social services should be addressed and supplemented.

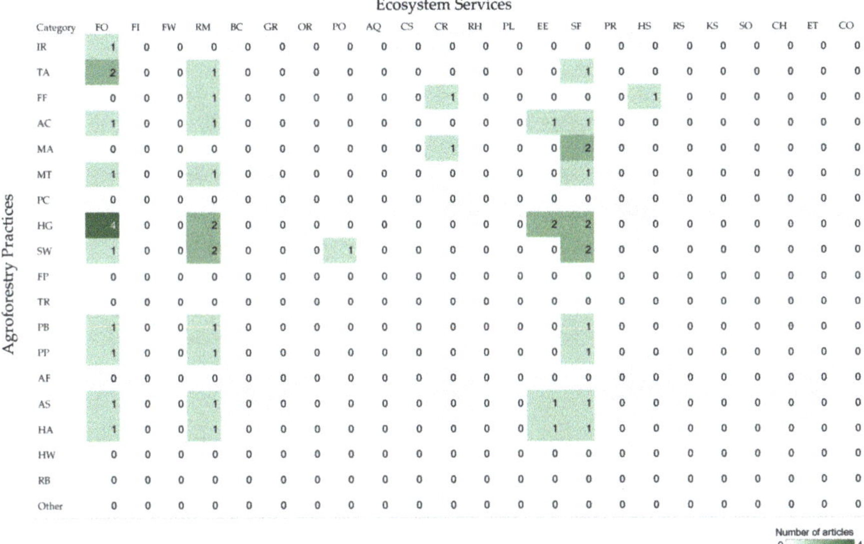

Category	FO	FI	FW	RM	BC	GR	OR	PO	AQ	CS	CR	RH	PL	EE	SF	PR	HS	RS	KS	SO	CH	ET	CO
IR	1	0	0	0	0	0	0	0	0	0	0	0	0	0	0	0	0	0	0	0	0	0	0
TA	2	0	0	1	0	0	0	0	0	0	0	0	0	0	1	0	0	0	0	0	0	0	0
FF	0	0	0	1	0	0	0	0	0	0	1	0	0	0	0	0	1	0	0	0	0	0	0
AC	1	0	0	1	0	0	0	0	0	0	0	0	0	1	1	0	0	0	0	0	0	0	0
MA	0	0	0	0	0	0	0	0	0	0	1	0	0	0	2	0	0	0	0	0	0	0	0
MT	1	0	0	1	0	0	0	0	0	0	0	0	0	0	1	0	0	0	0	0	0	0	0
PC	0	0	0	0	0	0	0	0	0	0	0	0	0	0	0	0	0	0	0	0	0	0	0
HG	4	0	0	2	0	0	0	0	0	0	0	0	0	2	2	0	0	0	0	0	0	0	0
SW	1	0	0	2	0	0	0	1	0	0	0	0	0	0	2	0	0	0	0	0	0	0	0
FP	0	0	0	0	0	0	0	0	0	0	0	0	0	0	0	0	0	0	0	0	0	0	0
TR	0	0	0	0	0	0	0	0	0	0	0	0	0	0	0	0	0	0	0	0	0	0	0
PB	1	0	0	1	0	0	0	0	0	0	0	0	0	0	1	0	0	0	0	0	0	0	0
PP	1	0	0	1	0	0	0	0	0	0	0	0	0	0	1	0	0	0	0	0	0	0	0
AF	0	0	0	0	0	0	0	0	0	0	0	0	0	0	0	0	0	0	0	0	0	0	0
AS	1	0	0	1	0	0	0	0	0	0	0	0	0	1	1	0	0	0	0	0	0	0	0
HA	1	0	0	1	0	0	0	0	0	0	0	0	0	1	1	0	0	0	0	0	0	0	0
HW	0	0	0	0	0	0	0	0	0	0	0	0	0	0	0	0	0	0	0	0	0	0	0
RB	0	0	0	0	0	0	0	0	0	0	0	0	0	0	0	0	0	0	0	0	0	0	0
Other	0	0	0	0	0	0	0	0	0	0	0	0	0	0	0	0	0	0	0	0	0	0	0

Figure 7. The linkage between agroforestry practices and ecosystem services in the 1980s. Note: The darker the color of the cells, the higher the frequency of articles.

Ecosystem Services

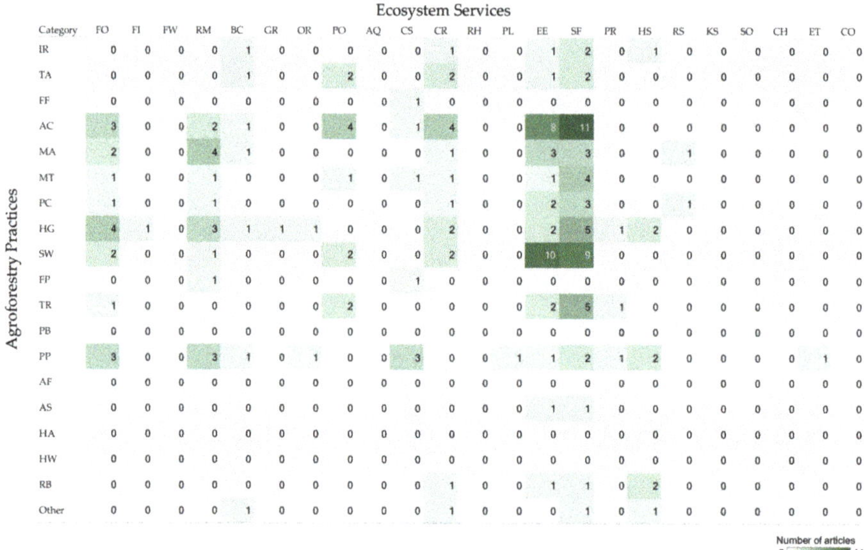

Category	FO	FI	FW	RM	BC	GR	OR	PO	AQ	CS	CR	RH	PL	EE	SF	PR	HS	RS	KS	SO	CH	ET	CO
IR	0	0	0	0	1	0	0	0	0	0	1	0	0	1	2	0	1	0	0	0	0	0	0
TA	0	0	0	0	1	0	0	2	0	0	2	0	0	1	2	0	0	0	0	0	0	0	0
FF	0	0	0	0	0	0	0	0	0	1	0	0	0	0	0	0	0	0	0	0	0	0	0
AC	3	0	0	2	1	0	0	4	0	1	4	0	0	8	11	0	0	0	0	0	0	0	0
MA	2	0	0	4	1	0	0	0	0	0	1	0	0	3	3	0	0	1	0	0	0	0	0
MT	1	0	0	1	0	0	0	1	0	1	1	0	0	1	4	0	0	0	0	0	0	0	0
PC	1	0	0	1	0	0	0	0	0	0	1	0	0	2	3	0	0	1	0	0	0	0	0
HG	4	1	0	3	1	1	1	0	0	0	2	0	0	2	5	1	2	0	0	0	0	0	0
SW	2	0	0	1	0	0	0	2	0	0	2	0	0	10	9	0	0	0	0	0	0	0	0
FP	0	0	0	1	0	0	0	0	0	1	0	0	0	0	0	0	0	0	0	0	0	0	0
TR	1	0	0	0	0	0	0	2	0	0	0	0	0	2	5	1	0	0	0	0	0	0	0
PB	0	0	0	0	0	0	0	0	0	0	0	0	0	0	0	0	0	0	0	0	0	0	0
PP	3	0	0	3	1	0	1	0	0	3	0	0	1	1	2	1	2	0	0	0	0	1	0
AF	0	0	0	0	0	0	0	0	0	0	0	0	0	0	0	0	0	0	0	0	0	0	0
AS	0	0	0	0	0	0	0	0	0	0	0	0	0	1	1	0	0	0	0	0	0	0	0
HA	0	0	0	0	0	0	0	0	0	0	0	0	0	0	0	0	0	0	0	0	0	0	0
HW	0	0	0	0	0	0	0	0	0	0	0	0	0	0	0	0	0	0	0	0	0	0	0
RB	0	0	0	0	0	0	0	0	0	0	1	0	0	1	1	0	2	0	0	0	0	0	0
Other	0	0	0	0	1	0	0	0	0	0	1	0	0	0	1	0	1	0	0	0	0	0	0

Number of articles 0 – 11

Figure 8. The linkage between agroforestry practices and ecosystem services in the 1990s. Note: The darker the color of the cells, the higher the frequency of articles.

Ecosystem Services

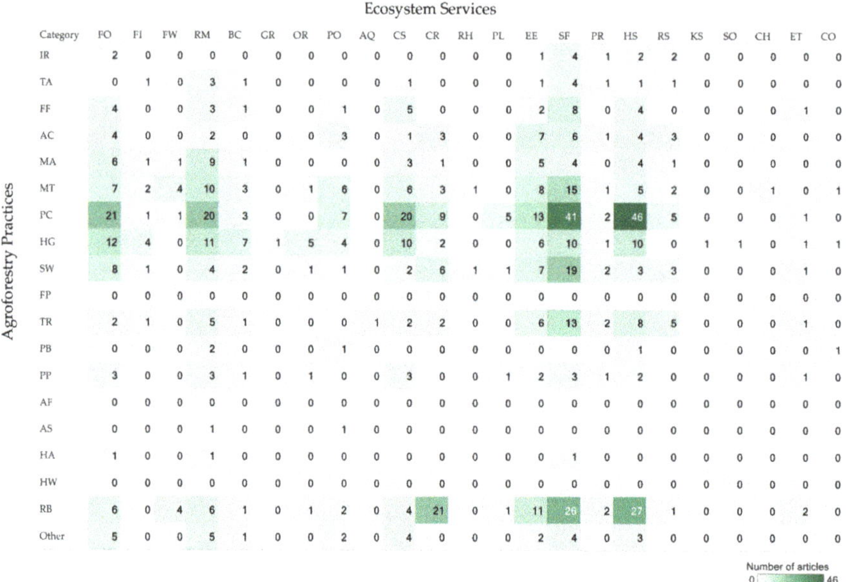

Category	FO	FI	FW	RM	BC	GR	OR	PO	AQ	CS	CR	RH	PL	EE	SF	PR	HS	RS	KS	SO	CH	ET	CO
IR	2	0	0	0	0	0	0	0	0	0	0	0	0	1	4	1	2	2	0	0	0	0	0
TA	0	1	0	3	1	0	0	0	0	1	0	0	0	1	4	1	1	1	0	0	0	0	0
FF	4	0	0	3	1	0	0	1	0	5	0	0	0	2	8	0	4	0	0	0	0	1	0
AC	4	0	0	2	0	0	0	3	0	1	3	0	0	7	6	1	4	3	0	0	0	0	0
MA	6	1	1	9	1	0	0	0	0	3	1	0	0	5	4	0	4	1	0	0	0	0	0
MT	7	2	4	10	3	0	1	6	0	6	3	1	0	8	15	1	5	2	0	0	1	0	1
PC	21	1	1	20	3	0	0	7	0	20	9	0	5	13	41	2	46	5	0	0	0	1	0
HG	12	4	0	11	7	1	5	4	0	10	2	0	0	6	10	1	10	0	1	1	0	1	1
SW	8	1	0	4	2	0	1	1	0	2	6	1	1	7	19	2	3	3	0	0	0	1	0
FP	0	0	0	0	0	0	0	0	0	0	0	0	0	0	0	0	0	0	0	0	0	0	0
TR	2	1	0	5	1	0	0	0	1	2	2	0	0	6	13	2	8	5	0	0	0	1	0
PB	0	0	0	2	0	0	0	1	0	0	0	0	0	0	0	0	1	0	0	0	0	0	1
PP	3	0	0	3	1	0	1	0	0	3	0	0	1	2	3	1	2	0	0	0	0	1	0
AF	0	0	0	0	0	0	0	0	0	0	0	0	0	0	0	0	0	0	0	0	0	0	0
AS	0	0	0	1	0	0	0	1	0	0	0	0	0	0	0	0	0	0	0	0	0	0	0
HA	1	0	0	1	0	0	0	0	0	0	0	0	0	0	1	0	0	0	0	0	0	0	0
HW	0	0	0	0	0	0	0	0	0	0	0	0	0	0	0	0	0	0	0	0	0	0	0
RB	6	0	4	6	1	0	1	2	0	4	21	0	1	11	26	2	27	1	0	0	0	2	0
Other	5	0	0	5	1	0	0	2	0	4	0	0	0	2	4	0	3	0	0	0	0	0	0

Number of articles 0 – 46

Figure 9. The linkage between agroforestry practices and ecosystem services after 2000. Note: The darker the color of the cells, the higher the frequency of articles.

5.3. Changing Trends

Figure 5 confirms the changes in the research trends over time. Although the plantation crop combinations and tree management (PC) have been constantly dealt with as a key topic, the focus has changed with the decades. For more inclusive considerations, we applied a conceptual framework, international forest discourses, to help understand research trend changes [72] (Table 4). First, according to Arts et al. (2010), in the 1970s, the Asian countries prioritized economic growth over poverty, and in the 1980s, they valued forests for timber to reduce poverty and create revenues, which are related to the provisioning services, including raw materials/ fuel wood/ biofuels/ bioenergy/ energy/ charcoal/ firewood. In the same way, the systematic map shows that in the 1980s, home gardens (HG) and services including food (FO) and raw materials/ fuel wood/ biofuels/ bioenergy/ energy/ hydro-electricity/ biomass/ charcoal/ firewood/ NTFPs (RM) were the primary topics to study (Figure 7). This indicates that researchers were mainly interested in provisioning the services of agroforestry in the Asia-Pacific region. In the 1990s, there was a shift in valuing ecosystem services. At a global level, policy discourses on climate change emerged in the 1990s and have been prevalent since signing the United Nations Framework Convention on Climate Change (UNFCCC) in 1992. These results also demonstrate the same trend of science discourse on agroforestry in the Asia-Pacific Region. The articles on the moderation of extreme events/ storm protection/ erosion control/ climate regulation (EE), which is interrelated with the discourse on climate change, increased since the 1990s (Figure 8). The discourse on biodiversity became animated later than climate change in the world and in Asia. In more detail, anthropologists started mentioning biodiversity and its benefits and traditional forestry systems like agroforestry came into the spotlight [72]. Globally, since the United Nations Conference on Environment and Development (UNCED) in 1992, forest biodiversity became of higher value and was considered social justice [73]. A multilateral treaty, Convention on Biological Diversity (CBD), has been signed by many countries, including numerous Asian countries. The policy discourse on biodiversity influenced the research on agroforestry in the Asia-Pacific region. Investigations into agroforestry and biodiversity, including habitats for species/ biological control/ maintenance of genetic diversity/ gene-pool (HS), have been numerous since 2000, but prior to this, there were only a few studies that focused on this. Based on these results, it was concluded that science discourses on agroforestry in Asia have been changed following the international discourses on forest policy. Policy and science discourses are inter-related in the field of agroforestry.

Table 4. The change of forest policy discourse and science discourse on agroforestry practices and ecosystem services in the Asia-Pacific region [72].

Decade	Policy Discourse	Science Discourse	
		Agroforestry Practice	Ecosystem Services
1980s	Poverty and economic growth	Home garden (HG)	Food (FO), Raw materials (RM)
1990s	Climate change	Alley-cropping (AC), Shelterbelts and windbreak (SW)	Erosion control/ Climate regulation (EE)
2000s	Biodiversity	Riparian buffer strips (RB)	Habitats for species (HS)

5.4. Limitations of the Map

Though the systematic map was designed to be as robust as possible, by attempting to capture the relevant articles, it was not perfectly inclusive, because of the limited resources and time. We acknowledge the possibility of missed or biased evidence for several reasons, despite our effort to utilize diverse search and screening strategies. First, documents only written in English were covered in the systematic map owing to finite language barriers; however, it is likely to exist in a large volume of articles in other languages, for instance in Chinese, Indian, Malaysian, and so forth. Second, publication searches were conducted in only one database, SCOPUS, and we did not carry out a citation screening because of the time limitation and for better efficiency. Third, our search string may not include all relevant literature, because the preceding systematic reviews [22,35,37,71] that we referred to were typically about European agroforestry and other regions, and Asian agroforestry may have unique characters. Keywords (Tables 1 and 2) are also generated from major precedent studies [38,40,41,45,57–59], also significantly reflecting the characteristics of European agroforestry systems and ecosystem services.

6. Conclusions

The number of agroforestry-related articles in the Asia-Pacific region has been fast-growing but is still small compared to all the literature in the world. India and China are hotspots of the research field with supporting policies and institutes [4,15,70] (Figure 3), whereas Western Asian countries have received little attention. Therefore, more targeted and comprehensive research is required to reduce the geographical gaps. Among the agroforestry practices, silvorable systems, especially plantation crop combinations and tree management (PC) and habitats for species/ biological control/ maintenance of genetic diversity/ gene-pool (HS) were the most popular. However, agrosilvopastoral and silvopastoral systems in the Asia-Pacific region have not received as much attention (Figure 4), in contrast with Europe where silvopastoral practices are predominant [35]. The linkage map expresses the occurrence of the evidence base but that does not mean that high frequency equals high or positive impacts, nor quality.

This map will contribute to designing policies, research, practical implementations, and save resources and time for decision making, by providing systematic evidence and frameworks. In particular, the heatmap offers insights to integrate ecosystem services around agroforestry systems into decision making, which is challenging [74,75]. We propose a wider range of additional studies for decision making on agroforestry works. Specifically, multiscale and upscale methods and approaches are essential to assess ecosystem services beyond biophysical approaches, in order to cover a broader range of ecosystem services including social services.

In conclusion, the systematic map identifies and describes the nature, volume, and characteristics of the research in the field of agroforestry and ecosystem services in the Asia-Pacific region. It pictures an existing evidence base on agroforestry and ecosystem service in the Asia-Pacific region. This comprehensive map could also be useful as a resource to enhance the knowledge of agroforestry–ecosystem service linkages. Furthermore, this map points out the gaps where further studies and investments should be focused.

Author Contributions: Conceptualization, M.S.P.; data coding, S.S., K.T.S., H.L., T.H.K., S.L.; writing—original draft preparation, S.S., M.S.P.; writing—review and editing, S.S., K.T.S., H.L., T.H.K., S.L., M.S.P.; supervision, M.S.P. All authors have read and agreed to the published version of the manuscript.

Funding: This work was supported by the Creative-Pioneering Researchers Program through Seoul National University (SNU).

Conflicts of Interest: The authors declare no conflict of interest.

Appendix A

Table A1. List of search terms in English language.

Search String	
	Search String
Agroforestry Practices	("agro forest*" OR "agro-forest*" OR "agrisilvicultur*" OR "agrosilvicultur*" OR "silvorable" OR "forest farm*" OR "farm forest*" OR "forest grazing" OR "riparian buffer*" OR "riparian-buffer*" OR "improved fallow*" OR "rotational tree fallow" OR "multipurpose tree*" OR "silvopastur*" OR "agrosilvipastur*" OR "agrosilvopast*" OR "agrosilvopast* system*" OR "hedgerow*" OR hedge-row* OR shelter belt* OR shelter-belt* OR (tree* AND (farmland* OR plot*)) OR "hedgerow intercrop*" OR "hedge-row intercrop*" OR "wood* hedge row*" OR "wood* hedge-row*" OR windbreak* OR wind-break* OR "live hedge*" OR "alley-crop*" OR "alley crop*" OR "meadow orchard*" OR orchard* OR "home garden*" "homegarden*" OR ("home garden*" NEAR/3 animal*) OR ("homegarden*" NEAR/3 animal*) OR (multi-stor* farm* NEAR/5 (tree* AND crop*)) OR (multistor* farm* NEAR/5 (tree* AND crop*)) OR entomo-forest* OR entomoforest* OR aquasilvofisher* OR aqua-silvo-fisher* OR aqua-silvofisher* OR taungya OR "taungya farming" OR "shifting cultivat*" OR "taungya cultivation" OR (woody* AND agriculture* crop*) OR multilayer tree garden* OR plantation crop combination* OR shade tree* OR intercrop* OR (integrated NEAR/2 (farm* OR system*)) OR (tree* AND pasture*) OR (tree* AND rangeland*) OR (fodder* AND (farm* OR rangeland*)) OR (apiculture AND tree*) OR "aquaforest*" OR "aqua-forest*" OR interplant* OR (interplant* NEAR/5 firewood*) OR ("soil conservation" AND tree*) OR (tree* AND (bund* OR terrace* OR raiser*)))
	AND
Ecosystem Services	("ecosystem" OR "service*" OR "ecosystem service*" OR "provision*" OR "provision* service*" OR "food*" OR "fiber*" OR "fresh water" OR "water" OR "drink* water*" OR "irrigate* water*" OR "hydro-electricity*" OR "hydroelectricity*" OR "raw material*" OR "fuel*" OR "wood*" OR "fuelwood*" OR "fuel-wood*" OR "charcoal" OR "firewood*" OR "non-timber forest product*" OR "NTFP*" OR "non-wood forest product*" OR "NWFP*" OR "biofuel*" OR "bio-fuel*" OR "bioenerg*" OR "bio-energ*" OR "biomass" OR "bio-mass" OR "energy" OR "biochemical*" OR "bio-chemical*" OR "pharmaceutical*" OR "medicin* resource*" OR "gene* resource*" OR "ornament* resource*" OR "regulat*" OR "regulat* service*" OR "air quality" OR "air quality maintain*" OR "climate regulat*" OR "carbon" OR "carbon sequest*" OR "carbon stor*" OR "water regulat*" OR "water flow*" OR "erosion control" OR "maintain* NEAR/2 soil ferti*" OR "water purify*" OR "waste treat*" OR "waste-water treat*" OR "regulat* NEAR/2 human disease*" OR "biology* control*" OR "pollinat*" OR "moderat* NEAR/2 extreme event*" OR "storm protect*" OR "support*" OR "support* service*" OR "habitat*" OR "soil form*" OR "nutrient* cycl*" OR "primary product*" OR "habitat* NEAR/2 specie*" OR "maintain* NEAR/2 genetic diversity" OR "gene-pool*" OR "gene pool*" OR "cultur*" OR "cultur* service*" OR "cultur* diversity" OR "spirit* value*" OR "religi* value*" OR "knowledge* system*" OR "education* value*" OR "inspiration* experience*" OR "spirit* experience*" OR "aesthetic value*" OR "social relation*" OR "sense of place" OR "culture* heritage value*" OR "recreat*" OR "ecotouris*" OR "touris*")
	AND
Geographical terms	(Oceania OR Australia OR "Commonwealth of Australia" OR Fiji OR "Republic of Fiji" OR "Kiribati" OR "Republic of Kiribati" OR "Marshall Islands" OR "Republic of the Marshall Islands" OR Micronesia OR "Federated States of Micronesia" OR Nauru OR "Republic of Nauru" OR "Pleasant Island" OR "New Zealand" OR Palau OR "Republic of Palau" OR "Papua New Guinea" OR "Independent State of Papua New Guinea" OR PNG OR Samoa OR "Independent State of Samoa" OR "Western Samoa" OR "Solomon Islands" OR Tonga OR "Kingdom of Tonga" OR "Friendly Islands" OR Tuvalu OR "Ellice Islands" OR Vanuatu OR "Republic of Vanuatu" OR "Pacific island territories of Cook Islands" OR "Cook Islands" OR "New Caledonia" OR "American Samoa" OR Tokelau OR "Union Islands" OR "French Polynesia" OR Niue OR Guam OR "Territory of Guam" OR "Commonwealth of the Northern Mariana Islands" OR CNMI OR "Northern Mariana Islands" OR "Pitcairn Islands" OR "Pitcairn, Henderson, Ducie and Oeno Islands" OR "Wallis and Futuna" OR "Territory of the Wallis and Futuna Islands" OR "Oceanic and sub-Antarctic islands in the Pacific region" OR "Pacific and Indian Ocean regions" OR "South-East Asia" OR "South East Asia" OR "Southeast Asia" OR "Brunei Darussalam" OR "Nation of Brunei, the Abode of Peace" OR Brunei OR Cambodia OR "Kingdom of Cambodia" OR Kampuchea OR Indonesia OR "Republic of Indonesia" OR "Lao People's Democratic Republic" OR Laos OR "Muang Lao" OR Malaysia OR Myanmar OR "Republic of the Union of Myanmar" OR Burma OR Philippines OR "Republic of the Philippines" OR Singapore OR "Republic of Singapore" OR Thailand OR "Kingdom of Thailand" OR Siam OR "Timor-Leste" OR "Timor Leste" OR "Democratic Republic of Timor-Leste" OR "Democratic Republic of Timor Leste" OR "East Timor" OR Vietnam OR "Socialist Republic of Vietnam" OR SRV OR "North-East Asia" OR "North East Asia" OR "Northeast Asia" OR China OR "People's Republic of China" OR PRC OR "Democratic People's Republic of Korea" OR "North Korea" OR DPRK OR Korea OR Japan OR Mongolia OR "Republic of Korea" OR "South Korea" OR ROK OR "South Asia" OR "South-Asia" OR Afghanistan OR "Islamic Republic of Afghanistan" OR Bangladesh OR "People's Republic of Bangladesh" OR Bhutan OR "Kingdom of Bhutan" OR India OR "Republic of India" OR Iran OR "Islamic Republic of Iran" OR Persia OR Maldives OR "Republic of Maldives" OR Nepal OR "Federal Democratic Republic of Nepal" OR Pakistan OR "Islamic Republic of Pakistan" OR "Sri Lanka" OR "Democratic Socialist Republic of Sri Lanka" OR "Western Asia" OR "Western-Asia" OR Bahrain OR "Kingdom of Bahrain" OR Kuwait OR "State of Kuwait" OR Oman OR "Sultanate of Oman" OR Qatar OR "State of Qatar" OR Saudi Arabia OR "Kingdom of Saudi Arabia" OR KSA OR "United Arab Emirates" OR Emirates OR UAE OR Yemen OR "Arabian peninsula" OR "Republic of Yemen" OR Iraq OR "Republic of Iraq" OR Jordan OR "The Hashemite Kingdom of Jordan" OR Lebanon OR "Lebanese Republic" OR "State of Palestine" OR Palestine OR Mashreq OR Mashrek OR Syria OR "Syrian Arab Republic")

References

1. World Agroforestry. What is Agroforestry? Available online: http://www.worldagroforestry.org/about/agroforestry (accessed on 10 November 2019).
2. Zomer, R.; Trabucco, A.; Coe, R.; Place, F.; van Noordwijk, M.; Xu, J. *Trees on Farms: An Update and Reanalysis of Agroforestry's Global Extent and Socio-ecological Characteristics*; Working Paper 179; World Agroforestry Center: Bogor, Indonesia, 2014; pp. 1–33. [CrossRef]
3. Cannell, M.G.R. Agroforestry—A Decade of Development. *ICRAF* **1987**, *24*, 393.
4. Kumar, B.M.; Singh, A.K.; Dhyani, S.K. *South Asian Agroforestry: Traditions, Transformations, and Prospects. Agroforestry—The Future of Global Land Use*; Springer: Dordrecht, The Netherlands, 2012; pp. 359–389.
5. Waldron, A.; Garrity, D.; Malhi, Y.; Girardin, C.; Miller, D.C.; Seddon, N. Agroforestry Can Enhance Food Security While Meeting Other Sustainable Development Goals. *Trop. Conserv. Sci.* **2017**, *10*, 194008291772066. [CrossRef]
6. Das, T.; Das, A.K. Inventorying plant biodiversity in homegardens: A case study in Barak Valley, Assam, North East India. *Curr. Sci.* **2005**, *89*, 155–163.
7. Kumar, B.M. Land use in Kerala: Changing scenarios and shifting paradigms. *J. Trop. Agric.* **2005**, *42*, 1–12.
8. Kumar, B.M.; Takeuchi, K. Agroforestry in the Western Ghats of peninsular India and the satoyama landscapes of Japan: A comparison of two sustainable land use systems. *Sustain. Sci.* **2009**, *4*, 215–232. [CrossRef]
9. Guillerme, S.; Kumar, B.M.; Menon, A.; Hinnewinkel, C.; Maire, E.; Santhoshkumar, A.V. Impacts of public policies and farmer preferences on agroforestry practices in Kerala, India. *Environ. Manag.* **2011**, *48*, 351–364. [CrossRef]
10. Kumar, B.M. Carbon sequestration potential of tropical homegardens. *J. Trop. Agric.* **2006**, *1*, 185–204.
11. Kumar, B.M.; Nair, P.K.R. The enigma of tropical homegardens. *Agrofor. Syst.* **2004**, *61–62*, 135–152.
12. Capistrano, A.D.; Marten, G.G. Agriculture in Southeast Asia. In *Traditional Agriculture in Southeast Asia: A Human Ecology Perspective*; Marten, G.G., Ed.; Westview Press: Boulder, CO, USA, 1986; pp. 6–19.
13. Matson, P.A.; Parton, W.J.; Power, A.G.; Swift, M.J. Agricultural intensification and ecosystem properties. *Science* **1997**, *277*, 504–509. [CrossRef]
14. Lobao, L.; Meyer, K. The Great Agricultural Transition: Crisis, Change, and Social Consequences of Twentieth Century US Farming. *Annu. Rev. Sociol.* **2001**, *27*, 103–124. [CrossRef]
15. De Zoysa, M. *A Review of Forest Policy Trends in Sri Lanka. Policy Trend Report 2001*; Institute for Global Environmental Strategies (IGES): Kanagawa, Japan, 2001.
16. Garrett, H.G.; Buck, L. Agroforestry practice and policy in the United States of America. *For. Ecol. Manag.* **1997**, *91*, 5–15. [CrossRef]
17. De Oliveira, R.E.; Carvalhaes, M.A. Agroforestry as a tool for restoration in atlantic forest: Can we find multi-purpose species? *Oecologia Aust.* **2016**, *20*, 425–435. [CrossRef]
18. Jose, S. Agroforestry for ecosystem services and environmental benefits: An overview. *Agrofor. Syst.* **2009**, *76*, 1–10. [CrossRef]
19. Chang, C.H.; Karanth, K.K.; Robbins, P. Birds and beans: Comparing avian richness and endemism in arabica and robusta agroforests in India's Western Ghats. *Sci. Rep.* **2018**, *8*, 1–9.
20. Akbar, G.; Baig, M.B.; Asif, M. Social aspects in launching successful agroforestry projects in developing countries. *Sci. Vis.* **2000**, *5*, 52–58.
21. Garrity, D.P.; Akinnifesi, F.K.; Ajayi, O.C.; Weldesemayat, S.G.; Mowo, J.G.; Kalinganire, A.; Larwanou, M.; Bayala, J. Evergreen Agriculture: A robust approach to sustainable food security in Africa. *Food Secur.* **2010**, *2*, 197–214. [CrossRef]
22. Brown, S.E.; Miller, D.C.; Ordonez, P.J.; Baylis, K. Evidence for the impacts of agroforestry on agricultural productivity, ecosystem services, and human well-being in high-income countries: A systematic map protocol. *Environ. Evid.* **2018**, *7*, 24. [CrossRef]
23. Yang, S.; Bai, J.; Zhao, C.; Lou, H.; Zhang, C.; Guan, Y.; Zhang, Y.; Wang, Z.; Yu, X. The assessment of the changes of biomass and riparian buffer width in the terminal reservoir under the impact of the South-to-North Water Diversion Project in China. *Ecol. Indic.* **2018**, *85*, 932–943. [CrossRef]
24. Tiwari, T.P.; Brook, R.M.; Wagstaff, P.; Sinclair, F.L. Effects of light environment on maize in hillside agroforestry systems of Nepal. *Food Secur.* **2012**, *4*, 103–114. [CrossRef]

25. Bohra, B.; Sharma, N.; Saxena, S.; Sabhlok, V.; Ramakrishna, Y.B. Socio-economic impact of Biofuel Agroforestry Systems on Smallholder and Large-holder Farmers in Karnataka, India. *Agrofor. Syst.* **2018**, *92*, 759–774. [CrossRef]

26. Basu, J.P. Agroforestry, climate change mitigation and livelihood security in India. *N. Zeal. J. For. Sci.* **2014**, *44*, S11. [CrossRef]

27. Goswami, S.; Verma, K.S.; Kaushal, R. Biomass and carbon sequestration in different agroforestry systems of a Western Himalayan watershed. *Biol. Agric. Hortic.* **2014**, *30*, 88–96. [CrossRef]

28. Mupangwa, W.; Twomlow, S.; Walker, S.; Hove, L. Effect of minimum tillage and mulching on maize (*Zea mays* L.) yield and water content of clayey and sandy soils. *Phys. Chem. Earth* **2007**, *32*, 1127–1134. [CrossRef]

29. Hillbrand, A.; Borelli, S.; Conigliaro, M.; Olivier, E. *Agroforestry for Landscape Restoration. Exploring the Potential of Agroforestry to Enhance the Sustainability and Resilience of Degraded Landscapes*; FAO: Rome, Italy, 2017.

30. Jose, S.; Gillespie, A.R.; Pallardy, S.G. Interspecific interactions in temperate agroforestry. *Agrofor. Syst.* **2004**, *61*, 237–255.

31. Rivest, D.; Paquette, A.; Moreno, G.; Messier, C. A meta-analysis reveals mostly neutral influence of scattered trees on pasture yield along with some contrasted effects depending on functional groups and rainfall conditions. *Agric. Ecosyst. Environ.* **2013**, *165*, 74–79. [CrossRef]

32. Foli, S.; Reed, J.; Clendenning, J.; Petrokofsky, G.; Padoch, C.; Sunderland, T. To what extent does the presence of forests and trees contribute to food production in humid and dry forest landscapes? A systematic review protocol. *Environ. Evid.* **2014**, *3*, 15. [CrossRef]

33. Thorn, J.P.R.; Friedman, R.; Benz, D.; Willis, K.J.; Petrokofsky, G. What evidence exists for the effectiveness of on-farm conservation land management strategies for preserving ecosystem services in developing countries? A systematic map. *Environ. Evid.* **2016**, *5*, 13. [CrossRef]

34. Haddaway, N.R.; Bernes, C.; Jonsson, B.G.; Hedlund, K. The benefits of systematic mapping to evidence-based environmental management. *Ambio* **2016**, *45*, 613–620. [CrossRef]

35. Torralba, M.; Fagerholm, N.; Burgess, P.J.; Moreno, G.; Plieninger, T. Do European agroforestry systems enhance biodiversity and ecosystem services? A meta-analysis. *Agric. Ecosyst. Environ.* **2016**, *230*, 150–161. [CrossRef]

36. Miller, D.C.; Ordonez, P.J.; Baylis, K.; Rana, P. Protocol for an evidence and gap map The impacts of agroforestry on agricultural productivity, ecosystem services, and human well-being in low-and middle-income countries: An evidence and gap map. *Campbell Syst. Rev.* **2017**, *13*, 1–27. [CrossRef]

37. Santos, P.Z.F.; Crouzeilles, R.; Sansevero, J.B.B. Can agroforestry systems enhance biodiversity and ecosystem service provision in agricultural landscapes? A meta-analysis for the Brazilian Atlantic Forest. *For. Ecol. Manag.* **2019**, *433*, 140–145. [CrossRef]

38. Atangana, A.; Khasa, D.; Chang, S.; Egrande, A. *Tropical Agroforestry*; Springer Netherlands: Dordrecht, Netherland, 2014; ISBN 978-94-007-7723-1.

39. Nair, P.K.R. *An Introduction to Agroforestry*; Kluwer Academic Publishers: Dordrecht, The Netherlands, 1993; ISBN 0792321340.

40. Nair, P.K.R. Agroforestry systems in major ecological zones of the tropics and subtropics. In Proceedings of the In International Workshop on the Applications of Meteorology to Agroforestry Systems Planning and Management, Nairobi, Kenya, 9–13 February 1987; ICRAF: Nairobi, Kenya, 1989.

41. USDA Agroforestry. *USDA Reports to America, Fiscal Years 2011–2012—Comprehensive Version*; U.S. Department of Agriculture: Washington, DC, USA, 2013.

42. Ehrlich, P.R.; Mooney, H.A. Extinction, substitution, and ecosystem services. *Bioscience* **1983**, *33*, 248–254. [CrossRef]

43. Daily, G.C. *Nature's Services: Societal Dependence on Natural Ecosystems*; Island Press: Washington, DC, USA, 1997.

44. Boyd, J.; Banzhaf, S. What are ecosystem services? The need for standardized environmental accounting units. *Ecol. Econ.* **2007**, *63*, 616–626. [CrossRef]

45. Millennium Ecosystem Assessment. *Millenium Ecosystem Assessment: Ecosystems and Human Well-Being*, 5th ed.; Island Press: Washington, DC, USA, 2005; ISBN 1597260401.

46. Kalaba, F.K.; Quinn, C.H.; Dougill, A.J. The role of forest provisioning ecosystem services in coping with household stresses and shocks in Miombo woodlands, Zambia. *Ecosyst. Serv.* **2013**, *5*, 143–148. [CrossRef]

47. Maass, J.M.; Balvanera, P.; Castillo, A.; Daily, G.C.; Mooney, H.A.; Ehrlich, P.; Quesada, M.; Miranda, A.; Jaramillo, V.J.; García-Oliva, F.; et al. Ecosystem services of tropical dry forests: Insights from long-term ecological and social research on the Pacific Coast of Mexico. *Ecol. Soc. A J. Integr. Sci. Resil. Sustain.* **2005**, *10*, 1–23. [CrossRef]

48. Balvanera, P.; Kremen, C.; Martínez-Ramos, M. Applying community structure analysis to ecosystem function: Examples from pollination and carbon storage. *Ecol. Appl.* **2005**, *15*, 360–375. [CrossRef]

49. Ricketts, T.H. Tropical forest fragments enhance pollinator activity in nearby coffee crops. *Conserv. Biol.* **2004**, *18*, 1262–1271. [CrossRef]

50. Maes, J.; Paracchini, M.L.; Zulian, G.; Dunbar, M.B.; Alkemade, R. Synergies and trade-offs between ecosystem service supply, biodiversity, and habitat conservation status in Europe. *Biol. Conserv.* **2012**, *155*, 1–12. [CrossRef]

51. Polasky, S.; Nelson, E.; Pennington, D.; Johnson, K.A. The impact of land-use change on ecosystem services, biodiversity and returns to landowners: A case study in the state of Minnesota. *Environ. Resour. Econ.* **2011**, *48*, 219–242. [CrossRef]

52. Karp, D.S.; Mendenhall, C.D.; Sandí, R.F.; Chaumont, N.; Ehrlich, P.R.; Hadly, E.A.; Daily, G.C. Forest bolsters bird abundance, pest control and coffee yield. *Ecol. Lett.* **2013**, *16*, 1339–1347. [CrossRef]

53. Klein, A.M.; Steffan-Dewenter, I.; Tscharntke, T. Rain forest promotes trophic interactions and diversity of trap-nesting Hymenoptera in adjacent agroforestry. *J. Anim. Ecol.* **2006**, *75*, 315–323. [CrossRef] [PubMed]

54. Logan, T.J. Agricultural best management practices for water pollution control: Current issues. *Agric. Ecosyst. Environ.* **1993**, *46*, 223–231. [CrossRef]

55. Power, A.G. Ecosystem services and agriculture: Tradeoffs and synergies. *Philos. Trans. R. Soc. B Biol. Sci.* **2010**, *365*, 2959–2971. [CrossRef] [PubMed]

56. Daily, G.C.; Matson, P.A. Ecosystem services: From theory to implementation. *Proc. Natl. Acad. Sci. USA* **2008**, *105*, 9455–9456. [CrossRef]

57. De Groot, R.S.; Alkemade, R.; Braat, L.; Hein, L.; Willemen, L. Challenges in integrating the concept of ecosystem services and values in landscape planning, management and decision making. *Ecol. Complex.* **2010**, *7*, 260–272. [CrossRef]

58. TEEB Ecosystem Services. Available online: http://www.teebweb.org/resources/ecosystem-services/#.XC8cEKfVmH0.gmail (accessed on 12 November 2019).

59. FAO. *Valuing Forest Ecosystem Services A Training Manual for Planners and Project Developers*; FAO: Rome, Italy, 2019.

60. Swinton, S.M.; Lupi, F.; Robertson, G.P.; Landis, D.A. Ecosystem services from agriculture: Looking beyond the usual suspects. *Am. J. Agric. Econ.* **2012**, *88*, 1160–1166. [CrossRef]

61. O'Farrell, P.J.; Anderson, P.M.L. Sustainable multifunctional landscapes: A review to implementation. *Curr. Opin. Environ. Sustain.* **2010**, *2*, 59–65. [CrossRef]

62. Tscharntke, T.; Klein, A.M.; Kruess, A.; Steffan-Dewenter, I.; Thies, C. Landscape perspectives on agricultural intensification and biodiversity—Ecosystem service management. *Ecol. Lett.* **2005**, *8*, 857–874. [CrossRef]

63. Cannell, M.G.R.; Van Noordwijk, M.; Ong, C.K. The central agroforestry hypothesis: The trees must acquire resources that the crop would not otherwise acquire. *Agrofor. Syst.* **1996**, *34*, 27–31. [CrossRef]

64. McAdam, J.H.; Burgess, P.J.; Graves, A.R.; Rigueiro-Rodríguez, A.; Mosquera-Losada, M.R. Classifications and Functions of Agroforestry Systems in Europe. In *Agroforestry in Europe: Current Status and Future Prospects*; Rigueiro-Rodróguez, A., McAdam, J., Mosquera-Losada, M.R., Eds.; Springer Netherlands: Dordrecht, The Netherlands, 2009; pp. 21–41. ISBN 978-1-4020-8272-6.

65. Pullin, A.S.; Stewart, G.B. Guidelines for systematic review in conservation and environmental management. *Conserv. Biol.* **2006**, *20*, 1647–1656. [CrossRef]

66. Pullin, A.S.; Knight, T.M. Effectiveness in conservation practice: Pointers from medicine and public health. *Conserv. Biol.* **2001**, *15*, 50–54. [CrossRef]

67. Fazey, I.; Salisbury, J.G.; Lindenmayer, D.B.; Maindonald, J.; Douglas, R. Can methods applied in medicine be used to summarize and disseminate conservation research? *Environ. Conserv.* **2004**, *31*, 190–198. [CrossRef]

68. Stewart, L.A.; Tierney, J.F. To IPD or not to IPD? *Eval. Health Prof.* **2002**, *25*, 76–97. [CrossRef] [PubMed]

69. Moher, D.; Liberati, A.; Tetzlaff, J.; Altman, D.G. Preferred reporting items for systematic reviews and meta-analyses: The PRISMA statement. *Ann. Intern. Med.* **2009**, *154*, 264–269. [CrossRef]

70. Reidsma, P.; Feng, S.; van Loon, M.; Luo, X.; Kang, C.; Lubbers, M.; Kanellopoulos, A.; Wolf, J.; Van Ittersum, M.K.; Qu, F. Integrated assessment of agricultural land use policies on nutrient pollution and sustainable development in Taihu Basin, China. *Environ. Sci. Policy* **2012**, *18*, 66–76. [CrossRef]

71. Cheng, S.H.; MacLeod, K.; Ahlroth, S.; Onder, S.; Perge, E.; Shyamsundar, P.; Rana, P.; Garside, R.; Kristjanson, P.; McKinnon, M.C.; et al. A systematic map of evidence on the contribution of forests to poverty alleviation. *Environ. Evid.* **2019**, *8*, 3. [CrossRef]

72. Arts, B.; Appelstrand, M.; Kleinschmit, D.; Pülzl, H.; Vissen-Hamakers, I.; Eba'a Atyi, R.; Enters, T.; Mcginley, K.; Yasmi, Y. Discources, actors and instruments in international forest governance. *Embrac. Complex. Meet. Chall. Int. For. Gov.* **2010**, *28*, 57–73.

73. Zhouri, A. Global–Local Amazon Politics: Conflicting Paradigms in the Rainforest Campaign. *Theory Cult. Soc.* **2004**, *21*, 69–89. [CrossRef]

74. Maes, J.; Egoh, B.; Willemen, L.; Liquete, C.; Vihervaara, P.; Schägner, J.P.; Grizzetti, B.; Drakou, E.G.; La Notte, A.; Zulian, G.; et al. Mapping ecosystem services for policy support and decision making in the European Union. *Ecosyst. Serv.* **2012**, *1*, 31–39. [CrossRef]

75. Nieto-Romero, M.; Oteros-Rozas, E.; González, J.A.; Martín-López, B. Exploring the knowledge landscape of ecosystem services assessments in Mediterranean agroecosystems: Insights for future research. *Environ. Sci. Policy* **2014**, *37*, 121–133. [CrossRef]

Review

Agroforestry to Achieve Global Climate Adaptation and Mitigation Targets: Are South Asian Countries Sufficiently Prepared?

Shalini Dhyani [1,*], Indu K Murthy [2], Rakesh Kadaverugu [3], Rajarshi Dasgupta [4], Manoj Kumar [5] and Kritika Adesh Gadpayle [2]

[1] Water Technology and Management Division, CSIR-National Environmental Engineering Research Institute (NEERI), Nehru Marg, Maharashtra, Nagpur 440020, Maharashtra, India

[2] Adaptation and Risk Analysis Division, Center for Study of Science, Technology and Policy, Nagashetty Halli, R.M.V. 2nd Stage, Bengaluru 560094, Karnataka, India; indukmurthy@cstep.in (I.K.M.); kritika@cstep.in (K.A.G.)

[3] Cleaner Technology and Modeling Division, CSIR-National Environmental Engineering Research Institute (NEERI), Nehru Marg, Nagpur 440020, Maharashtra, India; rakesh927@gmail.com

[4] Natural Resources and Ecosystem Services Division, Institute for Global Environmental Strategies (IGES), 2108-11 Kamiyamaguchi, Hayama, Kanagawa 240-0115, Japan; dasgupta@iges.or.jp

[5] GIS Center, Forest Research Institute, Dehradun 248006, Uttarakhand, India; manojfri@gmail.com

* Correspondence: s_dhyani@neeri.res.in or shalinidhyanineeri@gmail.com

Citation: Dhyani, S.; Murthy, I.K; Kadaverugu, R.; Dasgupta, R.; Kumar, M.; Adesh Gadpayle, K. Agroforestry to Achieve Global Climate Adaptation and Mitigation Targets: Are South Asian Countries Sufficiently Prepared? *Forests* **2021**, *12*, 303. https://doi.org/10.3390/f12030303

Academic Editors: Mi Sun Park and Himlal Baral

Received: 6 January 2021
Accepted: 26 February 2021
Published: 6 March 2021

Publisher's Note: MDPI stays neutral with regard to jurisdictional claims in published maps and institutional affiliations.

Abstract: Traditional agroforestry systems across South Asia have historically supported millions of smallholding farmers. Since, 2007 agroforestry has received attention in global climate discussions for its carbon sink potential. Agroforestry plays a defining role in offsetting greenhouse gases, providing sustainable livelihoods, localizing Sustainable Development Goals and achieving biodiversity targets. The review explores evidence of agroforestry systems for human well-being along with its climate adaptation and mitigation potential for South Asia. In particular, we explore key enabling and constraining conditions for mainstreaming agroforestry systems to use them to fulfill global climate mitigation targets. Nationally determined contributions submitted by South Asian countries to the United Nations Framework Convention on Climate Change acknowledge agroforestry systems. In 2016, South Asian Association for Regional Cooperation's Resolution on Agroforestry brought consensus on developing national agroforestry policies by all regional countries and became a strong enabling condition to ensure effectiveness of using agroforestry for climate targets. Lack of uniform methodologies for creation of databases to monitor tree and soil carbon stocks was found to be a key limitation for the purpose. Water scarcity, lack of interactive governance, rights of farmers and ownership issues along with insufficient financial support to rural farmers for agroforestry were other constraining conditions that should be appropriately addressed by the regional countries to develop their preparedness for achieving national climate ambitions. Our review indicates the need to shift from planning to the implementation phase following strong examples shared from India and Nepal, including carbon neutrality scenarios, incentives and sustainable local livelihood to enhance preparedness.

Keywords: agroforestry; South Asia; climate change; mitigation; adaptation; policy; REDD+; national determined contributions; climate neutrality

1. Introduction

Climate change is a reality and it is well established that the planet is facing climate emergency [1]. Emissions from the agriculture sector alone emits 6 billion metric tons of greenhouse gases (GHG) into the environment per annum [2]. Climate change impacts in certain regions have been more damaging and devastating because of the enhanced exposure to climatic hazards, already prevailing vulnerabilities and lower adaptive capacity [3,4]. Climate change mitigation, food security, conservation of biodiversity, restoration of ecosystems and localizing the sustainable development goals (SDGs) are the fundamental global challenges of present times [5]. With increasing natural disasters and climate variability there is growing urgency for recognizing and supporting efforts for climate adaptation and mitigation [6]. Of these, adaptation efforts to improve land and water management related practices have been identified as central to boosting capacity for overall resilience to climate vulnerability [7].

The South Asia region includes the countries of Bangladesh, Bhutan, India, the Maldives, Nepal, Pakistan, and Sri Lanka. S. Asia has huge range of human, cultural, and ecosystem diversity [8]. S. Asia's rapid population growth, widespread poverty, large dependence on natural resources and inadequate adaptive capacity has made the region highly vulnerable to climate change. The region is home to more than one fifth of the world's population, and is one of the most climate disaster-prone areas on earth [9–11]. Agriculture and pasture land in the region accounts for one third of the total land cover [2]. Fulfilling the food requirements of a fast-growing population without affecting land use is a primary challenge due to sustenance agriculture, and this has resulted in widespread food shortages [12,13]. Agriculture expansion and intensification are drivers of deforestation and biodiversity loss in the region. Due to low per capita land available for agriculture, production of food with a marginal ecological footprint becomes essential [12]. There are growing expectations on multifunctional land use systems, to fulfill mounting regional land and food demands while addressing emerging climate hazards, as they support sustenance of productive landscapes, habitats, social, economic, and also regulatory aspirations [14].

Adaptation is an urgent requirement under the present climate change scenario, particularly in developing and underdeveloped countries, which are anticipated to be severely impacted by climate extremes [15]. The contribution made by agriculture to achieve the SDGs will require climate adaptation followed by cropland advances that are affordable and profitable to the poor [16]. The Intergovernmental Panel on Climate Change (IPCC) in its first, second, and third assessment reports (1990, 1996 and 2001) have acknowledged the South Asian region for its capacity to incorporate adaptation and mitigation approaches that can also facilitate pro-poor development through carbon-offset arrangements such as farmer managed natural regeneration, agroforestry, and adaptive agriculture practices [17]. While synergies in adaptation and mitigation approaches need to be addressed, they should not be limited to income diversification from tree or forest-based products. Adaptation and mitigation approaches should ideally include approaches for improving soil health and biodiversity, and reducing fire risks, through restoration of natural ecosystems [18]. Intended Nationally Determined Contributions (INDCs) have emerged as the principal tool for benchmarking and reporting under the Paris Agreement. Likewise, removing atmospheric carbon and storing it in terrestrial vegetation is a feasible adaptation and mitigation option that contributes to the NDCs. Researchers have identified agroforestry among critical landscapes as an approach that can fulfill NDC commitments, particularly in developing countries [19,20].

Trees outside forests (TOFs) substantively contribute to livelihood improvement, while also enhancing biomass and carbon stocks. In the last few decades, policy makers have recognized the significance of TOFs, and included them in the national forest inventories [21]. Indigenous and traditional resource management by agroforestry is proven to benefit livelihood benefits in terms of provisioning, regulating, and supporting ecosystem services [22]. Trees on arable land have the potential to support carbon sinks under Nature-based Solutions (NbS) contributing to climate change adaptation and mitigation through carbon sequestration [23–26].

Understanding the regional agroforestry status, creating opportunities for further promotion to fulfill climate promises, and ensuring successful acceptance of agroforestry practices are all crucial and pertinent, in light of climate change [27]. For this paper, we performed an initial bibliometric analysis to understand the existing published literature on regional agroforestry practices and their importance in addressing global climate adaptation and mitigation targets. Based on the limitations of the analysis, we then conducted a detailed review of available literature on Scopus, Web of Science and Google Scholar to obtain a detailed overview of the potential for agroforestry systems (AFS) in supporting country-specific mitigation targets as well as supporting NDCs as proposed by countries in S. Asia. Additionally, this paper discusses the need for integrating AFS into MRV (Monitoring, Reporting and Verification) while providing a critical understanding of key gap areas, existing policies and concerns that need specific attention to be scaled up by adoption and promotion of agroforestry in the region. The review paper critically tries to address the following questions:

1. What is the substantial evidence that AFS and its practices deliver diverse ecosystem services, thereby ensuring human well-being in S. Asia?
2. What are the important climate discussions including agroforestry for climate adaptation and mitigation?
3. What are the key capabilities, and constraints when looking to include agroforestry into climate adaptation and mitigation?

2. Traditional Agroforestry Systems in South Asia

Agroforestry systems are dynamic, sustainable food production, and natural resource management systems with high prevalence and acceptance in developing countries in the tropics of South-East Asia, South Asia, and Central, and South America. These systems occupy more than 50% of the land coverage [28–30]. Despite global recognition and the presence of AFS, it is still a challenge to find reliable and accurate information on the extent for S. Asia. A list of land areas that are under agroforestry in different countries of the world including S. Asia was prepared by The International Assessment of Agricultural Knowledge, Science and Technology for Development (IAASTD) [31]. Nair et al. [32] estimated global agroforestry cover to be 1023 million hectares followed by Zomer et al. [33]. Zomer [29] projected global agroforestry cover to be 1020 million hectares [22], thereby agreeing with Nair et al. [32] (Table 1).

Table 1. Overview of global agroforestry cover.

% of Tree Cover Present in the Agricultural Lands	Global Agricultural Land with Trees (in km^2)	% of All Agricultural Land with Trees
>10	10,120,000	46
>20	5,960,000	27
>50	1,670,000	7.5

Source: [29,33].

South Asia is recognized for its AFS and its long history of acceptance and adoption of traditional practices across diverse agro-ecological conditions and agro-climatic zones. The diverse AFS in the region showcase the accumulated knowledge related to climate adaptation and mitigation approaches developed by millions of smallholding farmers and marginalized communities over centuries [34]. Approximately 60% of the research on AFS in the Asia-Pacific region has been carried out in India, China, Indonesia, and Australia, with a clear focus on silvi-pastoral systems. Shin et al. [35] provided details on the extensive research on AFS in India from 1970–2018. Nair et al. [36] provided a detailed overview on traditional AFS in S. Asia, along with other regions of the world.

Home gardens are the dominant AFS across S. Asian countries. Traditional AFS in S. Asia are trusted for their diverse benefits from the small land holdings (Table 2). In India, Nepal, Bhutan, Bangladesh, the Maldives and Sri Lanka, growing fuelwood, fodder and fruit trees on cropland bunds by local people is a common practice to fulfill energy and food demands, and are these practices that constitute important livelihood options for the region's rural poor [37,38]. However, in Pakistan, local farmers are hesitant to plant trees on cropland bunds to avoid competition between trees and crops. Hence, their fuelwood and fodder needs are mostly met from natural forests or wasteland vegetation.

Table 2. Traditional agroforestry systems accepted/adopted in South Asia.

Type of AFS	Agro-Ecological Adaptation
Agri-silvicultural systems	
Shifting cultivation, Chena, Taungya, Bewat, dhya, dippa, erka, jhum, kumara, peenda, pothur, podu, rep syrti, zabo	In tropical forest areas in North-East India, Sri Lanka
Plantation-based cropping system	Mainly humid tropical countries (India, Bangladesh, Maldives, Sri Lanka)
Scattered trees on farms, parklands	All regions, especially semiarid, and arid regions
Shelterbelts and windbreaks	In wind-prone areas, especially coastal, arid, and alpine regions of India, Bangladesh, Maldives, Sri Lanka
Boundary Planting and live hedges	In all countries of the region
Woodlots for soil conservation	In hilly areas, along sea coast and ravine lands of the region
Industrial plantations with crops	Intensively cropped area having plantation on bunds
Silvi-pastoral systems	
Silvi-pastures	Sub tropics and tropics with bio-edaphic sub- climaxes
Horti- pastoral	In hilly and non-hilly orchards for soil conservation
Tree on rangelands	In all countries of the region
Plantation crops with pastures	Mostly humid and sub-humid regions with less grazing pressure on plantation lands
Seasonal forestry Grazing	Semi- arid and mountainous ecosystems
Agro-silvi-pastoral systems	
Home gardens	In all countries of the region especially Sri Lanka, India, Maldives, Bangladesh
Others	
Aqua forestry	Low lands
Apiculture with trees	In all countries of the region

Source: [39,40].

The magnitude of agroforestry in the region at present is highly underestimated, because of technical constraints to recognize low-density tree cover common the small landholdings of local farmers [20]. Agroforestry cover reported from different parts of Asia shows that there are fewer areas with trees in S. Asia region, compared to other regions in Asia (Table 3).

Table 3. Extent of agroforestry systems in different parts of Asia.

Regions of the World	Agricultural Area with Trees (in Million km²)	% Of All Agricultural Area with Trees
South-East Asia	1.34	82
Northern and Central Asia	0.65	27
East Asia	0.41	23
South Asia	0.38	21
Western Asia and North Africa	0.1	9
Total (Global)	10.12	46

Source: [29,33].

The Central Agroforestry Research Institute (CAFRI) based in Jhansi, India estimated agroforests to span 13.75 million hectares in the country [41]. In the biennial State of Forest Report (ISFR) of India for 2019, AFS are located under trees outside forests (TOF) category, spanning an area of 293,840 km^2, or about 8.94% of the geographical area of the country. More than 65% of the country's timber and more than 50% of the fuelwood requirements are supported by AFS. Oli et al. [42] reported higher tree species richness in agroforests of Nepal compared to natural forests. Chakraborty et al. [43] stressed the value of agroforests in Bangladesh. Agroforests in Bangladesh support household fuelwood needs and thus, help in reducing household expenses and dependence on wood from natural forests. The National Research Centre for Agroforestry projected the livelihood potential of 943 million person-days/annum from 25.4 million ha agroforests in India [44]. The Agroforests with species such as teak (*Tectona grandis* L.f.) or Silver Oak (*Grevillea robusta* A. Cunn. ex R.Br.) are an investment option for the region providing significant economic, and ecological returns, for ensuring long and short term diverse ecological and social benefits for local communities [39]. Fast growing high biomass yielding species like Poplar (*Populus* spp.) and Eucalyptus (*Eucalyptus* spp.) have gained larger acceptance and recognition in industrial plantations of Pakistan and India. Fast growing trees (*Eucalyptus* spp., *Populus* spp., *Tectona grandis*, *Casuarina equisetifolia* L. etc) are preferred in industrial agroforestry plantations and shelterbelts because of their economic and ecological values and fast growth rates [45]. Agroforestry trees that have market value are preferred by farmers in the region, as they have less susceptibility to fail as annual crops. *Moringa oleifera* trees are preferred in India because of the medicinal properties and market value of its all plant parts. Similarly, many traditional fodder trees like *Grewia optiva* J. R. Drumm. ex Burret, *Carpinus viminea* Wall. ex Lindl. etc., that can be harvested multiple times a year [22,46].

Noticeable examples of AFS include multifunctional landscapes such as home gardens that secure food and support conservation of lesser known underutilized biodiversity in Sri Lanka, Maldives, Bangladesh and India [47]. These tree-based land management practices (spice gardens in Kerala, India, and in Sri Lanka) have proven their potential in providing livelihood opportunities for rural industrialization. Integrated agri-silvi-horti production systems that favor resource conservation and support conservation of traditional agro-biodiversity also ensures climate adaptation and mitigation in the region [34].

2.1. Agroforestry Systems and Human Well-Being

Ecosystem services from natural ecosystems (or semi-natural) largely support and contributes various benefits for human well-being (environmental, material as well as psychological benefits) [48–50]. Agroforests on croplands or pasture lands as an important traditional land management practice and thus provide diverse socio-economic and ecological benefits including NbS for climate change adaptation [35,51]. Agroforestry delivers diverse provisioning, regulating and supporting ecosystem services, and climate adaptation is an important one to address global climate change [5]. Historically, AFS across S. Asian countries have been designed to capitalize and harness diverse benefits for human well-being [52]. The presence of multifunctional landscapes, ensures the conservation of lesser known wild species, encourages traditional agrobiodiversity and also improves pollinator benefits [53]. These well-managed multifunctional sustainable AFS provide considerable livelihood benefits as well as safeguarding diverse ecological functions [42]. It is important to mention here that decisions by farmers for adoption of a land use is not dependent on a benefit cost ratio, but essentially rests on how much net income will be earned. Hence, horticulture-based agroforestry is preferred by farmers in Bangladesh over cropland and homestead agroforestry [54].

AFS have the potential to serve in the restoration and rehabilitation of degraded ecosystems, and could help to reinstate ecosystem services [55]. Food security, land tenure security, enhanced farm-based incomes, management of terrestrial and soil biodiversity, carbon sinks, hydrological functions, wildlife corridors, reduced soil erosion, biodiversity conservation, microclimate improvement, increased nutrient retention via root capture and cycling, etc. are some of the diverse benefits of AFS reported from the region [20,38,56–58]. Supporting agroforestry interventions to ensure food security in Nepal includes high biomass of fodder, meat, and production by Non Timber Forest Produces (NTFPs) [59]. Areas under agroforestry are reported to result in reduced soil erosion and improved nitrogen fixation in Bhutan [60]. In Bangladesh, there was comparatively less nutrient depletion from soil erosion in AFS than in jhum/slash and burn agriculture [61]. There is considerable evidence that AFS support sustainable production, providing subsidiary household provisions with diversified products, conservation of natural resources, aquifer recharge, etc. [35,62]. According to Muschler [63] agroforests support "sustainable intensification" within a land use archetype that that are based more on ecology than on chemistry and climate science. Article 2 of the Paris Agreement proposed to strengthen global efforts to reduce climate impacts with reference to sustainable development and poverty alleviation. Hence, it is vital to recognize and acknowledge the role of agroforestry and to mainstream it at country level to address global climate targets. Leveraging the mitigation potential of land use sectors is crucial, in meeting emission reduction targets [64]. By endorsing the benefits of diverse AFS practiced across S. Asia, less fertile marginal croplands with low productivity can be included for income diversification. This can be achieved by restoring soil health, improving irrigation efficiency, creating carbon sinks [52,65–67], thereby also strengthening adaptive rainfed dryland agriculture [68].

2.2. Bibliometric Analysis Agroforestry Systems in S. Asia Region

Bibliometric analysis was carried out to take stock of existing information on AFS in S. Asia. A total of 52 published works were retrieved from the Web of Science (WoS) database according to the keywords "Agroforestry" and "South Asia". The retrieved literature spans the period 1991 to 2019, covering 30 journal articles, 7 review papers, 5 proceedings. The metadata of the retrieved literature contains information about the author names, journal, title, abstract, author defined keywords, machine learning generated keywords (known as keyword-plus), local and global citations, referred articles, year of publication, etc. Analysis of the metadata associated with articles provide useful insights about the research structure and themes. In this study we used the *bibliometrix* library of R programming language for the analysis (https://www.bibliometrix.org, accessed on 25 June 2019). The annual scientific production pertaining to the study followed an average growth rate of 6.21%. Most relevant sources (and their h-index) in terms of journals from where maximum papers originated, are Agroforestry Systems (5), New Forests (3), and Society and Natural Resources (2). A word tree-map of keywords is a simple method to visualize the overall spread of the research field. The word tree-map for author keywords is shown in Figure 1, in which the area of the rectangle labelled with the keyword is proportional to the frequency of its occurrence in retrieved literature. Frequency analysis of author keywords indicate that author keywords-conservation, agroforestry, biodiversity, and management have appeared most frequently. Conservation and biodiversity, agricultural management, biomass, carbon sequestration, and climate change topics are also associated with the overall theme of agroforestry in South Asia. Topics related to socio-cultural aspects such as livelihoods of local people and shifting cultivation also appeared in the literature.

The temporal evolution of the research topics can be understood by plotting the most frequent author keywords or keywords-plus with respect to the year of appearance. The trend of author keywords is shown in Figure 2 containing the keywords that have appeared at least twice in any year (considered between 2004–2019). Results indicate that there has been a shift in topics from the physical aspects related to agroforestry such as soil and water conservation, land productivity, and forestry to land use changes, forest disturbances, socio-economic development from 2004–2011. Studies in the last decade were related to shifting cultivation, livelihood of people, rubber plantations, oil palm farming, along with carbon sequestration. The trend in keywords do not reflect aspects related to climate change adaption and mitigation strategies, and the research momentum has not yet gained traction as expected.

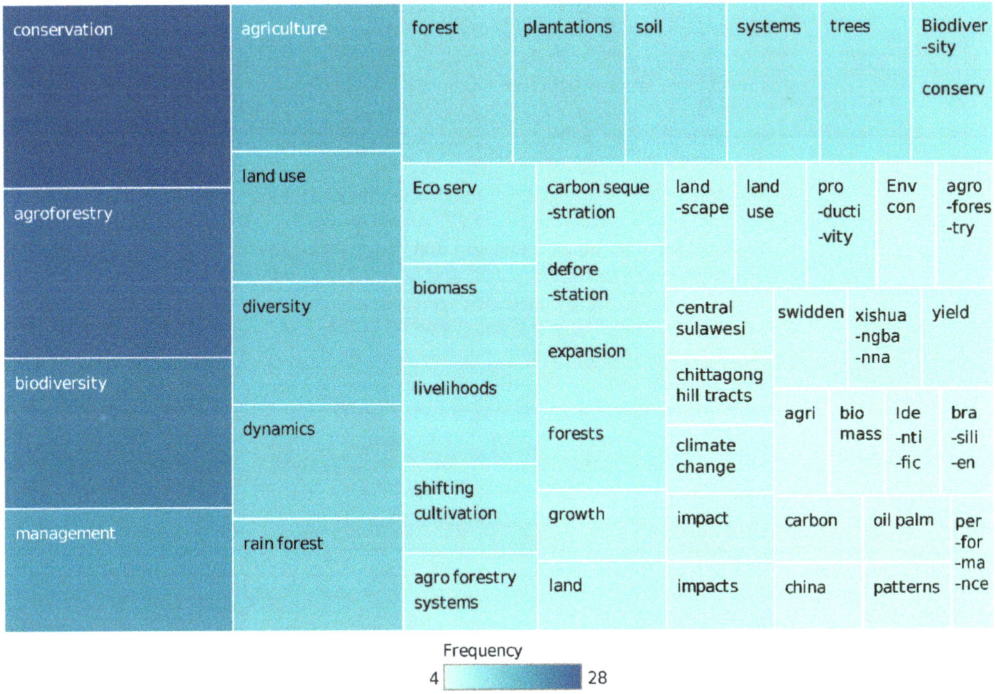

Figure 1. Word tree-map of most frequently used keywords with reference to agroforestry in south Asia.

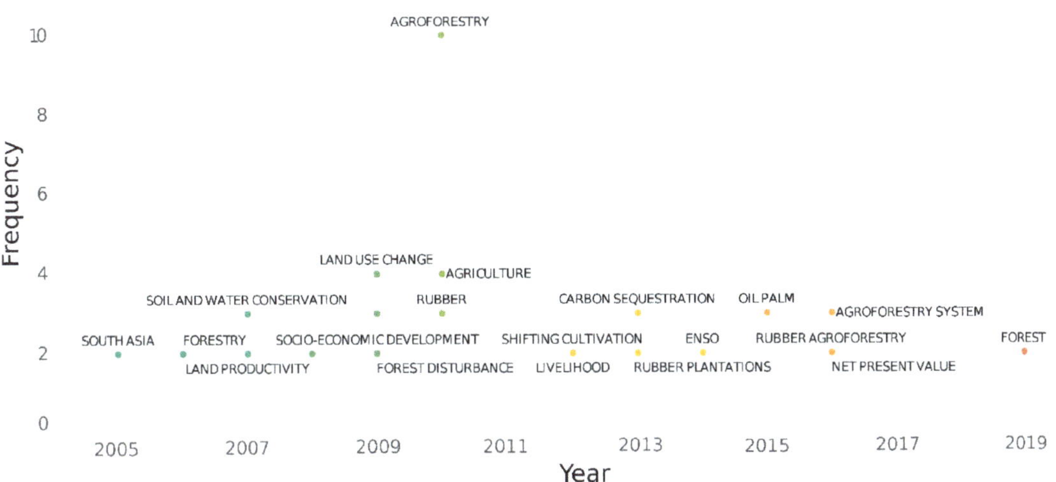

Figure 2. Trends in topics studied from 1991 to 2019 on agroforestry in S. Asia. *Y*-axis represents the frequency of appearance of the author keywords, and *X*-axis represents the year.

Co-word analysis was performed to capture the conceptual structure of research themes by analyzing co-occurrence of author keywords in the bibliometric collection. A bipartite matrix between author keywords and documents has been developed by the *biliometrix* library for analysis. Information on the group of keywords that appear together can be identified and made into clusters based on the k-mean clustering algorithm using R. In order to plot the clusters in a 2D plane, the multiple correspondence analysis dimensionality reduction method was used. The author keywords are grouped into clusters based on proximity in the 2D space, and the keywords that appear in a cluster share same substance of research. The keywords that are placed apart have appeared sparsely together in the collection. Based on our review, the clusters formed according to the analysis is shown in Figure 3. The clusters can be identified as Cluster-1 (land use, forest disturbances, and carbon sequestration), Cluster-2 (AFS), Cluster-3 (land use change, shifting agriculture, rubber plantation, oil palms, livelihood, land use productivity, S. Asia, etc.), and Cluster-4 (bioengineering technology, soil and water conservation, and socio-economic aspects). As the centroid of the Cluster-3 is positioned according to the positive values of X and Y in the 2D space, these themes are known as motor themes, and are central and highly developed themes in the agroforestry research. Cluster-1 and 2 centroids have negative X values and positive Y values, and are known as niche themes (or isolated themes) in the research landscape, focusing on land use change and AFS, respectively. Conversely, Cluster-4 indicates the themes that are central to the research area, but are less dense or transversal in nature. Overall Clusters 1 to 3 are close to each other and the themes also agree with literature discussed AFS, carbon sequestration, climate change, land use change, forest disturbance, livelihoods, and biodiversity.

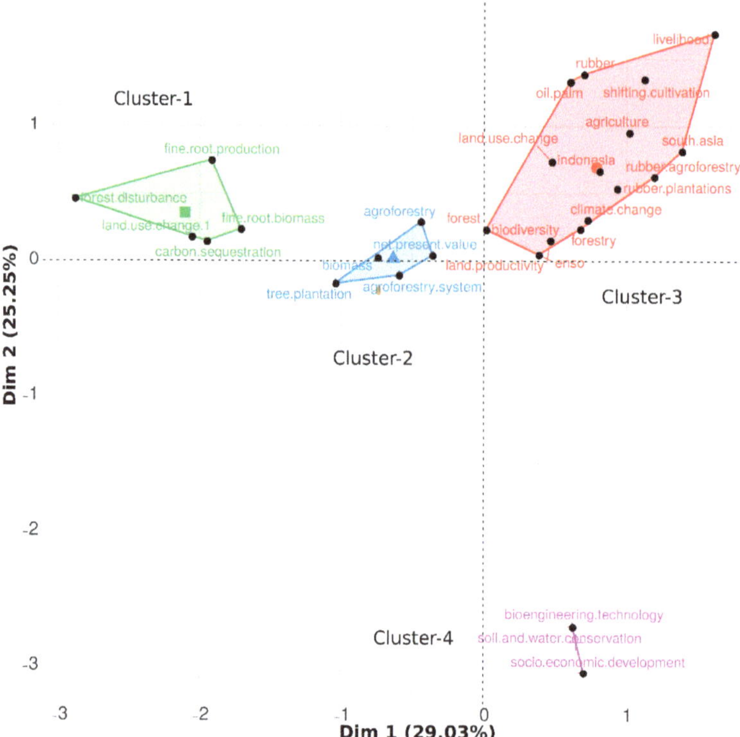

Figure 3. Conceptual structure map of research topics related to agroforestry in S. Asia, prepared using Multiple Correspondence Analysis of author keywords. (Dimension 1 and 2 on X and Y axes, respectively).

The bibliometric analysis was not able to capture the increasing concerns and interest of AFS in the climate dialogue. In general, most of the available information was very fragmentary and isolated in a few case studies. There is a need to further explore the literature to capture and synthesize the available information. Effort to consolidate the information and present it in this paper will be of significant interest to academicians, policymakers, and researchers working on AFS and for mainstreaming AFS in climate dialogues.

3. Global Climate Dialogue around Agroforestry Systems

The United Nations Framework Convention on Climate Change (UNFCCC) along with other prominent international environmental and scientific organizations have stressed the growing need for mainstreaming and implementation of sustainable land management approaches that specifically includes AFS [69–71]. AFS have received substantial recognition from international organizations like the UNFCCC, the Food and Agriculture Organization (FAO), the Convention on Biological Diversity (CBD), and the World Bank [72] (https://agroforestrynetwork.org/, accessed on 25 June 2019). Figure 4 presents an overview of major Conventions and reports that have brought AFS into global focus. The Kyoto Protocol was the first international arrangement to acknowledge the importance of AFS in climate mitigation. Since, then global attention for enhancing carbon sequestration using AFS has increased [30,70]. Although, the Kyoto Protocol was rooted in the Clean Development Mechanism (CDM), the addition of AFS into CDM was hindered due to a lack of uniform protocols to estimate carbon sinks, and associated land right concerns [73]. However, REDD+ (Reduced Emissions from Deforestations and Forest Degradation) brought AFS back into focus in 2007, and several countries have made considerable progress to improve their national planning by understanding the importance of agriculture, forestry, and other land-use (AFOLU) sectors in climate change adaptation and mitigation [74]. AFS are known for their potential to contribute to nine out of the 17 SDGs including SDG 15 (life on land), 13 (climate action), 12 (responsible production and consumption), 2 (zero hunger), 1 (no poverty), 3 (good health and well-being), 8 (decent work and economic growth), 5 (gender equality) and 10 (reduce inequalities) [75–77]. AFS are an important climate mitigation tool, and can help both developing and underdeveloped to achieve policy synergy amongst technologies, landscapes, rights and markets [78] while also improving localization of SDGs (especially 2.4; 13.2 and 15.3), restoration of multifunctional landscapes, climate adaptation and mitigation; reforestation targets in line with the Bonn challenge, UN decade on restoration (2021–2030); and improving food and water security [79–81].

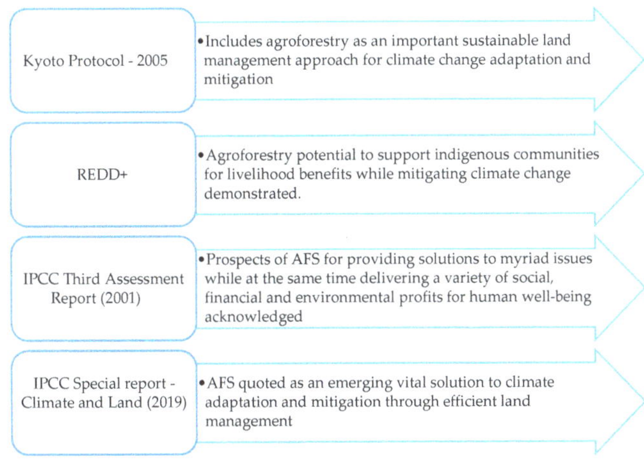

Figure 4. Agroforestry System in key agreements and reports (Source: [82–84]).

3.1. Agroforestry: Role in Climate Change Mitigation and Adaptation

Despite agroforestry being acknowledged for its carbon sequestration potential among all land uses considered in the IPCC (2000), the understanding of carbon sink in different AFS in the region is still very elementary because of insufficient authentic data on carbon stocks of AF interventions, in comparison to agriculture and forestry [85]. While agriculture along with forestry results in large amounts of emissions and also accounts for nearly 21% of the total emissions [86], AFS have significant mitigation potential that has not been scientifically evaluated in global carbon financial plans or national carbon accounts [30]. Limited studies at the global, national and zonal scale have reported carbon stocks in AFS (Table 4). However, for S. Asia, these studies and reports are mostly at the local level. In most of the studies, there is a lack of comprehensive information on both tree and Soil organic carbon (SOC) trends in carbon stocks [82,87,88]. It has been very challenging to gain an understanding of how diverse agroforestry practices can become potential carbon sinks [14,85,89,90].

In farmland biodiversity, the scattered trees in agroforests are the 'keystone species' that expedite and support the movement of wildlife through the landscape [91]. This role of AFS as wildlife corridors is significant under projected climate change as it allows species to adapt in response to unstable climatic conditions by providing the necessary migration paths [90]. In order to optimize the use of AFS in climate adaptation and mitigation, strategic integrated efforts to enhance benefits and reduce negative impacts on climate are needed. Mbow et al. [90] provided an overview of both positive and negative impacts of AFS on the adaptation and mitigation potential. Since most countries in the region are predominantly agrarian, S. Asia region has tremendous potential to promote agroforestry as a tool for climate adaptation and mitigation. A recent study claimed that 69% of the total geographical area of S. Asia retains 55% or even higher suitability for agroforestry [92].

Table 4. Reported carbon stocks in agroforestry systems. Source: [93–97].

Location	Carbon Stock (Mg C ha^{-1})
Global	Biomass—0.29–15.21 Soil—30–300
Global	0.7–1.6
Global	6.3
Pakistan	29.7
India	25.4
Semi-arid	9
Sub-humid	21
Humid	50
Temperate	63

3.2. Nationally Determined Contributions and Agroforestry

Under the Paris Agreement, countries submitted their Intended Nationally Determined Contributions (INDCs) under the Paris Agreement. INDCs, once submitted to UNFCCC, are known as Nationally Determined Contributions (NDCs) and they are the key mechanism towards reducing emissions as per national urgencies, competencies and accountabilities. According to Duguma et al. [98], within the purview of NDCs, agroforestry can provide multi-dimensional benefits by supporting climate adaptation and mitigation actions [98,99]. Nearly 40% of the Non-Annex I countries (developing countries recognized by the UNFCCC as vulnerable to the adverse climate impacts, including areas threatened from sea level rise, desertification and drought) have explicitly proposed agroforestry in their NDCs. A total of 21% of Asian countries have proposed AFS in their NDCs, a ratio that is less than Africa (71%) and the Americas (34%) but higher than Oceania (7%) [20,58].

The S. Asian countries list adaptation actions both at the farm and landscape level. Bangladesh, Nepal, Sri Lanka and Bhutan have proposed "ecosystem-based adaptation" [100], which includes landscape-level actions, spanning management of water resources, crop management by crop rotation, agroforestry and management of natural vegetation. As the sum of carbon flux fundamentally depends on the composition of trees, there needs to be more understanding on it during the implementation phase [101–103].

It is evident from Table 5 that, although countries have not explicitly included agroforestry in their NDCs (Bhutan and Nepal), the existing traditional systems and supporting policies in these countries indicate potential inclusion of AFS as part of a larger mitigation strategy. For example, in Bangladesh, the need to reduce emission from agriculture and further development of the forestry sector is indicated. In line with this, [20] the TOF (croplands, homestead and horticulture based agroforestry) provides significant opportunities in Bangladesh, as it already spreads over 4.1 million hectares or 27.7% of the total land area [20].

Table 5. Nationally determined contributions committed by S. Asian countries and role of agroforestry.

Country	NDC Commitment	Elements of Agroforestry in NDC
Bangladesh *	Emissions reduction from agriculture and development of forest sector. Unconditional contribution to reduce GHG emissions by 5% by 2030 in the power, transport and industry sectors, based on existing resources. Conditional 15% reduction in GHG emissions by 2030 in the power, transport, and industry sectors, subject to appropriate international support	- No mention of agroforestry in the NDC Ecosystem based adaptation (incl. forestry co-management) - Community based conservation of wetlands and coastal areas - Green belt Afforestation and reforestation of mangroves
Bhutan	"No NDC Available"	- Potential of climate-smart agriculture, particularly the development of agro-forestry, agri-silvi-pastoral systems for fodder production, organic agriculture and conservation agriculture are included as mitigation measures [104]
India	- Decrease emissions by 33–35% from the 2005 levels by the year 2030—to be achieved through increase in the segment of non-fossil fuel by 40%, along with sequestering an additional 2.5–3 billion tonnes of carbon through added tree cover by 2030 [105,106]	- Despite India's INDC not mentioning agroforestry specifically, it is believed to play a critical, if not pivotal role in national carbon mitigation targets, given agroforestry is of the sub-missions of the Green India Mission—one of the eight missions under the National Action Plan on Climate Change (NAPCC) [107].
Nepal	- Decrease the dependency on fossil fuels by 2050 and further aim to bring at least 40% of the area of the country under forest cover.	- Ameliorative forest practices including agroforestry as a means to achieve the NDC targets included [107]
Pakistan *	- Mitigation target of 20% of the projected 2030 emissions-subject to international financial support	- Agroforestry implementation included among mitigation strategies.
Sri Lanka *	- Increase forest cover from 29% to 32% by 2030, reduce emissions by 20% in the energy sector and by 10% in other sectors including forest, transport, industry, etc.	- No mention of agroforestry in the NDC.

* Details taken from https://www4.unfccc.int/ (accessed on 25 June 2019), [59,104,106,108,109].

South Asian Association for Regional Cooperation (SAARC) Member States (Afghanistan, Bangladesh, Bhutan, India, the Maldives, Nepal, Pakistan and Sri Lanka) developed the SAARC Regional Coordinated Programme on Agroforestry (SARCOPA) in 2016 that has received active facilitation and technical support from the World Agroforestry Center

(ICRAF) and SAARC Agriculture Centre (SAC). The programme has been divided into two-phases, the first 6-year phase focused on establishing the mechanism and delivery systems and the second 6-year phase focused on upscaling and out scaling the AFS benefits to larger beneficiaries. SARCOPA's first phase is focusing on generating awareness and developing guidelines, policy, and databases of existing information on AFS. India and Nepal already have National Agroforestry Policy in place clearly showing their intent to promote AFS while, Bhutan and Bangladesh are working to develop a National Policy to endorse and recognize the benefits of AFS. In fact, a mere 30% increase in area under AFS is projected to significantly reduce India's total emissions by 2050 [110]. Under SARCOPA there has been support provided for institutional and individual level capacity building and identifying and re-designing specific AFS, and sharing information on successful AFS. The Government of Nepal is implementing a Local Adaptation Plan of Actions through 90 Village Development Committees and seven municipalities. Additionally, about 375 local adaptation plans and approximately 2200 Community Adaptation Plans of Action for community forests have been enacted that will also include the benefits of natural forests, community conservation efforts and traditional AFS [97]. Agroforestry policy put in place by India in 2014 was the first in the region and was seen as a low hanging fruit to not only ensure the benefits from a successful land-use system, but also to harness its economic potential for locals as well for the country [111].

Sri Lanka also committed to supporting climate resilient human settlements, minimizing climate change impacts by ensuring food security, improving climate resilience for key economic support and protection of natural resources and biodiversity. Here again, although agroforestry is not explicitly mentioned, the country has a significant area of land under home gardens (13% of its current land area) that has historically helped in addressing drought and storms disasters, by supporting climate adaptation and so this, by default, will be part of the programme. The Government of Pakistan has initiated a 5-year plantation programme of 100 million trees under the Green Pakistan Programme or Plantation Tsunami to achieve Bonn Targets [108]. Here again, AFS is not explicitly a part of the NDC, but could be included.

The review and synthesis of existing information makes it clear that in S. Asia, there is already a process and approach in place to harness the benefits of AFS in all countries in the region and they are collaborating to share experience and technical support to make implementation a reality across the region.

3.3. Agroforestry in REDD+ and Nationally Appropriate Mitigation Actions (NAMAs)

Trading carbon sinks could be a potential livelihood opportunity for marginalized communities of underdeveloped and developing countries who practice agroforestry [86]. In S. Asian countries, the demand for firewood and timber results in rapid loss of forests and fragmentation, and AFS can help conserve natural forests. REDD+ has been a key feature of climate negotiations in the UNFCCC since 2007. Through REDD+, countries have made considerable progress in national planning to include AFOLU sectors for mitigating extreme climate impacts [74]. The REDD+ policies propose to economically reward countries for improving forest health by conservation and management that reduces GHG emissions [73]. The REDD+ initiative has supported eco-agricultural practices, that help produce surplus food while safeguarding native biodiversity and includes AFS [109]. Co-benefits from AFS are significant to the Koronivia Joint Work on Agriculture (KJWA) of the UNFCCC that addresses resilience building, enhancing soil carbon stocks, soil health, biodiversity and fertility, by supporting sustainable livestock management as well as providing varied nutritional benefits and livelihood diversification [20,58]. However, AFS are not explicitly mentioned in the KJWA. There are also encouraging and substantial evidence to showcase the successful support of AFS by indigenous and local communities [110]. Under the premise of REDD+, activities that lead to improving the capacity of forests to sequester carbon, reduce pressure on forests, and advance diversified livelihood approaches are included. A review of REDD+ strategies in S. Asia show that REDD+ strategies in S. Asian countries are at different stages of development (Table 6).

Table 6. Reduced Emissions from Deforestation and Degradation+ policies and strategies in S. Asian countries.

Country	Status	Scope of REDD+
India	Execution of REDD+ with reference to significant resolutions of COP-16, Warsaw Framework for REDD+, Paris Agreement, and national statutory and policy agenda for conservation and enhancement of forests	Covers forests and TOFs, which potentially includes AFS. The activities of REDD+ contribute to the objective of improving forest and tree cover, thereby ensuring alignment with the National Forest Policy.
Nepal	First draft of REDD+ strategy prepared in 2014, facilitating further consultations and drafting of Version 2 of REDD+ strategy.	- The REDD+ strategy statement established in line with the principles of sustainable development objectives that includes national forestry sector vision of forests for people's prosperity. Scope of the policy is limited to various forest classes including forests under Protected Areas as per Forest Act (1993), the National Parks and Wildlife Conservation Act (1973), and Forest Policy (2015). - Likelihood of inclusion of leasehold forests, sacred forests, forests on public lands and private forests at an advanced stage, to broaden the scope of REDD+ defined.
Pakistan	REDD+ initiated in 2010, envisages forest ecosystems as public goods, a source of multiple benefits required for development and with potential to mitigate climate change, while, building community and ecosystem resilience.	Has key policies that support conservation of forests and ecosystems, viz. National Forest Policy 2015, Climate Change Policy 2012 and Environment Policy 2005-foundation of REDD+ strategy.
Sri Lanka	- 5-year strategy (2018–2022)—National REDD+ Investment Framework and Action Plan prepared with support from UN-REDD Programme. - Is at an advanced stage of REDD+ readiness and includes technical essentials of REDD+ as per the Warsaw Framework (i.e., UNFCCC Decisions 9-15/CP.19).	- 13 policies to address the identified drivers of forest cover change identified. - Policy measure, that cover other forested lands supports agroforestry models for addressing forest degradation, with an objective "to create enabling conditions for making existing agroforestry arrangements financially viable for adoption and implementation".

4. Constraints in Using Agroforestry for Meeting Global Climate Targets

There is a noteworthy gap in country-specific targets and their technical capabilities to measure agroforestry carbon stocks and report to the UNFCCC. SARCOPA will be a great support to bridge this gap in the coming years, but it will take time to develop capacities with reference to carbon stocks stores in AFS. Insufficient data on carbon stocks before land use change along with non-existent reporting on soil carbon stocks is one of the crucial limitations of the AFS database existing in the region [5]. Monitoring, Reporting and Verification (MRV) is a prerequisite for achieving climate adaptation and economic growth aims of countries [112]. Developing robust MRV for AFS in S. Asia is a crucial first stage to facilitate access to national and international funding sources and further backing. Despite the, mounting importance of AFS and TOF in global climate change dialogues, it has been difficult to integrate agroforestry in MRV systems, as proposed by the UNFCCC. MRV protocols developed by one country may not always work for another country. For example, Nepal has comparatively low forest threshold (0.5 ha, 10% tree cover) that supports the addition of AFS in MRV; whereas, in Bangladesh, TOF (also AFS) are omitted from the forest definition in the policies [20]. Local carbon stock change factors are mainly used, which is a limitation. Lack of continued financial support, deviations in government directives, along with the concerns and capacity for data gathering and analysis are projected as other potential constraints in realizing the benefits of AFS in the region. Limited investment in agroforestry sector compared to intensive agriculture adds a key structural restriction for adoption of AFS [18,90].

Institutional constraints have been the most common limiting factor in the majority of countries in S. Asia. Expectations of high agricultural production per hectare followed by non-existent markets, land rights, and technical support are other challenges that impede realization of benefits of AFS in climate policies and implementation. Small landholdings are key limitation for AFS adoption in the region. Livestock size, distance of forest from villages, and a lack of awareness among farmers meanwhile, are other local reasons that limit adoption of AFS. However, poor and marginalized famers show interest in adopting AFS [25]. Shortage of water is another major constraint for promotion and adoption of AFS [108]. In India, the Forest Conservation Amendment Act of 1988 banned wood felling from state forests, amplifying wood prices and providing financial motivation to adopt AFS [113].

Despite widespread environmental and economic benefits, there is still low adoption of AF is largely because of legal and policy constraints including insecure land tenure, complex transit rules, taxes on agriculture based commodities, and socio-economic marginalization of local farmers [61]. Certainly, some key requirements for adoption include a growing need in the regional countries to fulfill market requirements, and formulation of policies that provide clear information on land and tree rights and ownership to enable REDD+ and NAMA contributions. However, farmers in the region are hesitant to plant trees because they do not have the rights to fell the tree for economic benefits. Further, harvesting and transporting of the tree wood from cropland to market is not permissible without prior approval from the forest department, which again deters adoption and promotion of AFS [108]. Farmers in Nepal stress their inability to get financial benefits from AFS because of unsupportive regulations surrounding harvesting and marketing of trees [59]. Farmers and experts in Bangladesh support the need for regulations and guideless for effective implementation of AFS to harness its ecological, economic and climate benefits. In Pakistan, too few trained forest personnel, lack of technical support to farmers, insufficient understanding of tree species, and poor market access along with wood price emerged as major limiting factors [108]. The failure of agroforestry related extension services across S. Asian countries has severely limited the opportunity for AFS to improve land use systems and promote its adoption to address global climate dialogues.

Policy Concerns

The advantage of promoting AFS is the familiarity of small and medium holder farmers, thereby making it a potential low hanging fruit for achieving the NDCs, and contributing to climate mitigation and adaptation. Hence, promotion of AFS alone will not be enough to address the larger concern of using the practice to provide a solution to global climate change. Promotion of AFS in region needs to be backed with an enabling and effective legal policy environment and strategic implementation to achieve the NDCs. Such policy backing would guarantee rights and ownership to communities, and bring incentives and investments, thereby creating a market-based infrastructure. Given the multiple benefits of AFS, countries should consider giving AFS a special place in REDD+ and NAMAs. However, the multiple challenges stressed in the previous sections should be appropriately discussed and addressed for agroforestry to reach its full potential. The following approaches are recommended:

- National and state policies should encourage ways to identify, classify and report on AFS, and expand the finance flow to AFS by increasing knowledge and cooperation among key stakeholders (Table 7).
- National policies addressing agriculture, forest conservation and management practices are required to take stock of both efficient mitigation and adaptation approaches to position agriculture and forestry practices for worldwide sharing of pioneering technologies and improve efficient use of land resources (Table 7).

- Financial incentives and regulatory approaches, are presently being used; however, effective enactment requires recognition of how land-use choices and emerging social-political and economic powers have the capacity to guide this practice in future [89].
- Policy framework to address climate risks need to be comprehensive enough to internalize the negative impacts of climate change, while promoting income from AFS [5].

Table 7. Overview of policies and programmes for promoting Agroforestry system in South Asia

Country	Policies/Programmes	Details
Bhutan	Study of AFS and practices in Bhutan	To contribute to the development of an agroforestry strategy and a national agroforestry programme.
India	Green India Mission	One of the eight missions under the National Action Plan on Climate Change with a target of AFS on 10 Mha of irrigated land, and 18 Mha of rainfed land.
	National Agroforestry Policy	Highlights the environmental implications of AFS including averting deforestation, and stimulating carbon stocks, biodiversity conservation, along with soil and water conservation.
	National REDD+ Policy	Includes broad values for evolving and realizing REDD+ programmes to receive benefits of the international REDD+ mechanism and produce financial enticements for local people who are involved in conservation of forest ecosystems
Nepal	National Agroforestry Policy	Drafted by Climate Technology Centre and Network and World Agroforestry Centre
Pakistan	Green Pakistan Programme	Tree planting efforts striving to realize the Bonn Challenge and address global climate concerns

Source: [44,108].

While AFS in India, through the Agroforestry Policy, aims to contribute to the goal of enhancing forest cover from the existing 23% of geographical area to 33%, the REDD+ strategy aims to slow down forest degradation and halt deforestation. Another programme that is working in this direction, the Green India Mission is another programme that supports AFS in rural parts of the country [45]. The National Agroforestry Policy of Nepal follows up its Nationally Determined Contributions (2016) and the Climate Change Policy (2011) that recognizes forests and trees including AFS to promote climate adaptation and mitigation. A study in Bhutan initiated in June 2020 is facilitated by an EU funded project on Technical Assistance for Renewable Natural Resources and Climate Change Response and Local Governments and Decentralization-Bhutan (EU-TACS). Such agroforestry relevant policies are already being drafted and developed in other smaller countries like Bangladesh and Bhutan, and more efforts will be required under the larger umbrella of SARCOPA for Pakistan, Sri Lanka and the Maldives to draft agroforestry policies relevant for these countries and agro-climatic zones.

5. Recommendations to Improve Mainstreaming of AFS in Climate Change Dialogues

SARCOPA, with support from the World Agroforestry Center (ICRAF), SAARC Agriculture Centre (SAC) and all national governments is a landmark effort in the region to acknowledge and mainstream the benefits of AFS with a special focus on country specific climate action. The UNFCCC encourages countries to produce data from field-based local investigations and carry out reporting under MRV to help create country-specific factors for robust assessment of biomass and SOC stocks [114,115]. Two-phase sampling approaches using laser scanning followed by field-based surveys is an effective method for assessing TOF resources. The region requires more country-specific research on improving TOF models for biomass calculation, that are amended to AFS tree resources [21]. As a first step, it is important to standardize protocols for carbon stock estimation following national REDD+ strategy. India is one of the few countries in the region to pioneer regular basis satellite-based surveys involving RS-GIS tools and has been doing this since the 1980s to assess forest cover changes. India's NDC target could be met by TOF, so its National Agroforestry Policy formulated in 2014 and its National REDD+ strategy, 2018, will benefit the entire process. Incentives for AFS across the region will need more external financial support to strengthen the existing systems. Developing agroforestry pilots for REDD+ can be the next step to building capacity of foresters and local communities, and to generate awareness on mainstreaming AFS for increased benefits. Conflicts with reference to AFS could be avoided by adopting a cautious, site-specific, and participatory approach to project development [18,116]. Skill development and capacity building as per the first phase plan of action SARCOPA by creating model agroforestry farms are already underway across the SAARC region. Discussions on similar issues are becoming common at national and sub-national levels especially in India, Nepal, Bhutan and Bangladesh. Forthcoming research in the region on AFS will requires more mechanistic and process-based surveys followed by models linking AFS and crop development with soil water, carbon and biogeochemical cycles [117].

6. Conclusions

The synthesis presented in this paper clearly supports the importance and potential of AFS in securing human well-being for marginalized and impoverished people that can also help the countries in S. Asia to meet their NDCs and contribute to mitigation of climate change. Although, there are already benefits from AFS that are considerable but they have not been sufficiently harnessed at the local or national level. One key enabling condition for mainstreaming AFS is a regional consensus at the country level and this has already begun as countries work on facilitating and extending support to each other under the larger umbrella of SARCOPA. It is important to mention here that national commitments to acknowledge benefits from AFS and recognize them under national agroforestry policies is the next important step. The phase-wise implementation as per SAARC Resolution on Agroforestry has been initiated and will continue for the next 12 years. These are promising commitments by regional countries and their governments. Countries like India and Nepal have proactively developed agroforestry policies considering AFS is a low hanging fruit that should be appropriately used. Recently, Bhutan, Bangladesh and the Maldives have also initiated their efforts in developing national agroforestry policies. It is certainly relevant for the mountain country of Bhutan, the coastal nation Bangladesh and island countries of Sri Lanka and the Maldives to proactively work in this direction to promote synergy for climate change mitigation and adaptation in the region. Around 21% of the agriculture land area in S. Asia is under trees which is less than other parts of Asia, except for Central Asia (Table 3). Countries across the region need to take steps to set an achievable target to restore degraded AFS and improve the systems by at least 50% in the coming five years as a first step. With years of experience and a traditional knowledge base of AFS across the region, this knowledge could be used to improve the conditions and address the NDCs. Moving beyond awareness and technical cooperation to realize the benefits, fulfilling local livelihood demands and creating more opportunities, is urgently needed to strengthen the ongoing momentum on AFS in the region. Important mechanisms to enhance agricultural productivity of forest dependent marginalized communities and farmers by using enhanced inputs, innovative technologies, and incentives to improve agricultural intensification, and livelihood diversification can help in achieving NDC targets and make headway on several SDGs.

Author Contributions: Conceptualization—S.D. and I.K.M.; methodology—S.D.; validation—S.D. and I.K.M.; formal analysis—S.D. and I.K.M.; investigation S.D., I.K.M., R.K., R.D., M.K.; resources—S.D., I.K.M., R.K., R.D., M.K.; data curation S.D, I.K.M., R.K., R.D., M.K.; writing—original draft preparation S.D, I.K.M., R.K., R.D., M.K., K.A.G.; writing—review and editing—S.D. and I.K.M.; visualization—S.D.; supervision S.D. and I.K.M.; project administration—S.D.; funding acquisition NA. All authors have read and agreed to the published version of the manuscript.

Funding: This research received no external funding.

Acknowledgments: Authors are thankful to Director CSIR-NEERI, Nagpur and Center for Study of Science, Technology and Policy, Bangalore. Knowledge Resource Center, CSIR-NEERI is acknowledged for checking plagiarism under the recorded number CSIR-NEERI/KRC/2021/FEB/WTMD-CTMD/1. Authors are grateful to two anonymous reviewers who have immensely helped in substantially revising and improving the manuscript and Emma Fushimi, IGES Japan for her kind support in proof reading and language corrections. Authors specially thank the special issue editors for their support and CGIAR research program on Forests, Trees and Agroforestry for the partial support to publish the work.

Conflicts of Interest: Authors declare no conflict of interest.

References

1. Ripple, W.J.; Wolf, C.; Newsome, T.M.; Barnard, P.; Moomaw, W.R. World Scientists' Warning of a Climate Emergency. *BioScience* **2020**, *70*, 8–12. [CrossRef]
2. Climate Change. Mitigation of Climate Change. In *Working Group III Contribution to the Fifth Assessment Report of the Intergovernmental Panel on Climate Change*; Intergovernmental Panel on Climate Change; Edenhofer, O., Ed.; Cambridge University Press: New York, NY, USA, 2014; ISBN 978-1-107-05821-7.
3. de Sherbinin, A. Climate Change Hotspots Mapping: What Have We Learned? *Clim. Chang.* **2014**, *123*, 23–37. [CrossRef]
4. Tucker, J.; Daoud, M.; Oates, N.; Few, R.; Conway, D.; Mtisi, S.; Matheson, S. Social Vulnerability in Three High-Poverty Climate Change Hot Spots: What Does the Climate Change Literature Tell Us? *Reg. Env. Chang.* **2015**, *15*, 783–800. [CrossRef]
5. Feliciano, D.; Ledo, A.; Hillier, J.; Nayak, D. Which Agroforestry Options Give the Greatest Soil and above Ground Carbon Benefits in Different World Regions. *Agric. Ecosyst. Environ.* **2018**, *254*, 117–129. [CrossRef]
6. de Coninck, H.; Puig, D. Assessing Climate Change Mitigation Technology Interventions by International Institutions. *Clim. Chang.* **2015**, *131*, 417–433. [CrossRef]
7. Turral, H.; Burke, J.; Faurès, J.M. *Climate Change, Water and Food Security*; Water Report; Food and Agriculture Organization of the United Nations (FAO): Rome, Italy, 2011; p. 200. ISBN 978-92-5-106795-6.
8. Noorudin, I. South Asia: The Road Ahead in 2020. Atlantic Council, 2020. Available online: https://www.atlanticcouncil.org/commentary/feature/south-asia-the-road-ahead-in-2020/ (accessed on 3 March 2020).
9. Sivakumar, M.V.K.; Stefanski, R. Climate Change in South Asia. In *Climate Change and Food Security in South Asia*; Lal, R., Sivakumar, M.V.K., Faiz, S.M.A., Mustafizur Rahman, A.H.M., Islam, K.R., Eds.; Springer: Dordrecht, The Netherlands, 2011; pp. 13–30. ISBN 978-90-481-9516-9.
10. Bandara, J.S.; Cai, Y. The Impact of Climate Change on Food Crop Productivity, Food Prices and Food Security in South Asia. *Econ. Anal. Policy* **2014**, *44*, 451–465. [CrossRef]
11. Maharjan, A.; de Campos, R.S.; Singh, C.; Das, S.; Srinivas, A.; Bhuiyan, M.R.A.; Ishaq, S.; Umar, M.A.; Dilshad, T.; Shrestha, K.; et al. Migration and Household Adaptation in Climate-Sensitive Hotspots in South Asia. *Curr. Clim. Chang. Rep.* **2020**, *6*, 1–16. [CrossRef]
12. Sharma, L.N.; Vetaas, O.R. Does Agroforestry Conserve Trees? A Comparison of Tree Species Diversity between Farmland and Forest in Mid-Hills of Central Himalaya. *Biodivers. Conserv.* **2015**, *24*, 2047–2061. [CrossRef]
13. Laurance, W.F.; Sayer, J.; Cassman, K.G. Agricultural Expansion and Its Impacts on Tropical Nature. *Trends Ecol. Evol.* **2014**, *29*, 107–116. [CrossRef] [PubMed]
14. Westholm, L.; Ostwald, M. Food Production and Gender Relations in Multifunctional Landscapes: A Literature Review. *Agroforest Syst.* **2020**, *94*, 359–374. [CrossRef]
15. Yang, Y.; Liu, B.; Wang, P.; Chen, W.-Q.; Smith, T.M. Toward Sustainable Climate Change Adaptation. *J. Ind. Ecol.* **2020**, *24*, 318–330. [CrossRef]
16. Jat, M.L.; Chakraborty, D.; Ladha, J.K.; Rana, D.S.; Gathala, M.K.; McDonald, A.; Gerard, B. Conservation Agriculture for Sustainable Intensification in South Asia. *Nat. Sustain.* **2020**, *3*, 336–343. [CrossRef]
17. Barros, V.R. (Ed.) *Regional Aspects*; Climate Change 2014; Cambridge University Press: New York, NY, USA, 2014; ISBN 978-1-107-05816-3.
18. Matocha, J.; Schroth, G.; Hills, T.; Hole, D. Integrating Climate Change Adaptation and Mitigation through Agroforestry and Ecosystem Conservation. In *Agroforestry—The Future of Global Land Use*; Nair, P.K.R., Garrity, D., Eds.; Advances in Agroforestry; Springer: Dordrecht, The Netherlands, 2012; pp. 105–126, ISBN 978-94-007-4676-3.

19. Duguma, L.A.; Nzyoka, J.; Minang, P.A.; Bernard, F. How Agroforestry Propels Achievement of Nationally Determined Contributions. *Icraf Policy Brief.* **2017**, *34*, 1–8.
20. Rosenstock, T.S.; Wilkes, A.; Jallo, C.; Namoi, N.; Bulusu, M.; Suber, M.; Mboi, D.; Mulia, R.; Simelton, E.; Richards, M.; et al. Making Trees Count: Measurement and Reporting of Agroforestry in UNFCCC National Communications of Non-Annex I Countries. *Agric. Ecosyst. Environ.* **2019**, *284*, 106569. [CrossRef]
21. Schnell, S.; Kleinn, C.; Ståhl, G. Monitoring Trees Outside Forests: A Review. *Environ. Monit. Assess* **2015**, *187*, 600. [CrossRef] [PubMed]
22. Kumar, V. Multifunctional Agroforestry Systems in Tropics Region. *Nat. Environ. Pollut. Technol.* **2016**, *15*, 365–376.
23. Dhyani, S.; Thummarukuddy, M. Ecological Engineering for Disaster Risk Reduction and Climate Change Adaptation. *Environ. Sci. Pollut. Res.* **2016**, *23*, 20049–20052. [CrossRef] [PubMed]
24. Griscom, B.W.; Adams, J.; Ellis, P.W.; Houghton, R.A.; Lomax, G.; Miteva, D.A.; Schlesinger, W.H.; Shoch, D.; Siikamäki, J.V.; Smith, P.; et al. Natural Climate Solutions. *Proc. Natl. Acad. Sci. USA* **2017**, *114*, 11645–11650. [CrossRef] [PubMed]
25. Dhyani, S.; Karki, M.; Gupta, A.K. Opportunities and Advances to Mainstream Nature-Based Solutions in Disaster Risk Management and Climate Strategy. In *Nature-Based Solutions for Resilient Ecosystems and Societies*; Dhyani, S., Gupta, A.K., Karki, M., Eds.; Disaster Resilience and Green Growth; Springer: Singapore, 2020; pp. 1–26, ISBN 9789811547126.
26. Dhyani, S.; Bartlett, D.; Kadaverugu, R.; Dasgupta, R.; Pujari, P.; Verma, P. Integrated Climate Sensitive Restoration Framework for Transformative Changes to Sustainable Land Restoration. *Restor. Ecol.* **2020**, *28*, 1026–1031. [CrossRef]
27. Dhakal, A.; Rai, R.K. Who Adopts Agroforestry in a Subsistence Economy? Lessons from the Terai of Nepal. *Forests* **2020**, *11*, 565. [CrossRef]
28. Zomer, R.J.; Trabucco, A.; Bossio, D.A.; Verchot, L.V. Climate Change Mitigation: A Spatial Analysis of Global Land Suitability for Clean Development Mechanism Afforestation and Reforestation. *Agric. Ecosyst. Environ.* **2008**, *126*, 67–80. [CrossRef]
29. Zomer, R.; Trabucco, A.; Coe, R.; Place, F.; Van Noordwijk, M.; Xu, J. *Trees on Farms: An Update and Reanalysis of Agroforestry's Global Extent and Socio-Ecological Characteristics*; Volume Working Paper 179; World Agroforestry Center: Bogor, Indonesia, 2014; pp. 1–33.
30. Zomer, R.J.; Neufeldt, H.; Xu, J.; Ahrends, A.; Bossio, D.; Trabucco, A.; van Noordwijk, M.; Wang, M. Global Tree Cover and Biomass Carbon on Agricultural Land: The Contribution of Agroforestry to Global and National Carbon Budgets. *Sci. Rep.* **2016**, *6*, 29987. [CrossRef]
31. Ishii-Eiteman, M.; Li Ching, L. The IAASTD Report. *Development* **2008**, *51*, 570–573. [CrossRef]
32. Nair, P.K.R.; Kumar, B.M.; Nair, V.D. Agroforestry as a Strategy for Carbon Sequestration. *J. Plant Nutr. Soil Sci.* **2009**, *172*, 10–23. [CrossRef]
33. Zomer, R.; Trabucco, A.; Coe, R.; Place, F. Trees on Farm: Analysis of Global Extent and Geographical Patterns of Agroforestry. *Icraf Work. Pap. World Agrofor. Cent.* **2009**, *89*, 63.
34. Mohan Kumar, B.; Singh, A.K.; Dhyani, S. South Asian Agroforestry: Traditions, Transformations, and Prospects. In *Agroforestry-The Future of Global Land Use*; Springer: Berlin/Heidelberg, Germany, 2012; pp. 359–389. ISBN 978-94-007-4675-6.
35. Shin, S.; Soe, K.T.; Lee, H.; Kim, T.H.; Lee, S.; Park, M.S. A Systematic Map of Agroforestry Research Focusing on Ecosystem Services in the Asia-Pacific Region. *Forests* **2020**, *11*, 368. [CrossRef]
36. Nair, P.K.R.; Viswanath, S.; Lubina, P.A. Cinderella Agroforestry Systems. *Agroforest Syst.* **2017**, *91*, 901–917. [CrossRef]
37. Dhyani, S.; Maikhuri, R.K.; Dhyani, D. Energy Budget of Fodder Harvesting Pattern along the Altitudinal Gradient in Garhwal Himalaya, India. *Biomass Bioenergy* **2011**, *35*, 1823–1832. [CrossRef]
38. Dhyani, S.; Maikhuri, R.K.; Dhyani, D. Utility of Fodder Banks for Reducing Women Drudgery and Anthropogenic Pressure from Forests of Western Himalaya. *Natl. Acad. Sci. Lett.* **2013**, *36*, 453–460. [CrossRef]
39. Dagar, J.C.; Singh, A.K.; Arunachalam, A. (Eds.) *Agroforestry Systems in India: Livelihood Security & Ecosystem Services*; Advances in Agroforestry; Springer: Berlin/Heidelberg, Germany, 2014; ISBN 978-81-322-1661-2.
40. Puri, S.; Nair, P.K.R. Agroforestry Research for Development in India: 25 Years of Experiences of a National Program. *Agrofor. Syst.* **2004**, *61*, 437–452. [CrossRef]
41. Chavan, S.; Dhyani, S.K.; Handa, A.K.; Newaj, R. National Agroforestry Policy in India: A Low Hanging Fruit. Available online: https://www.researchgate.net/publication/281363272_National_Agroforestry_Policy_in_India_A_low_hanging_fruit (accessed on 21 December 2020).
42. Oli, B.N.; Treue, T.; Larsen, H.O. Socio-Economic Determinants of Growing Trees on Farms in the Middle Hills of Nepal. *Agroforest Syst.* **2015**, *89*, 765–777. [CrossRef]
43. Chakraborty, M.; Haider, M.; Rahaman, M.M. Socio-Economic Impact of Cropland Agroforestry: Evidence from Jessore District of Bangladesh. *Int. J. Res. Agric. For.* **2015**, *2*, 11–20.
44. Inder, D.; Ram, A.; Bhaskar, S.; Chaturvedi, O.P. Role of Agroforestry in current scenario. In *Agroforestry for Climate Resilience and Rural Livelihood*; Scientific Publishers: Jodhpur, India, 2018; pp. 1–10.
45. Basu, J.P. Agroforestry, Climate Change Mitigation and Livelihood Security in India. *N. Z. J. Sci.* **2014**, *44*, S11. [CrossRef]
46. Murthy, I.K.; Gupta, M.; Tomar, S.; Munsi, M.; Hegde, R.T.G.; Nh, R. Carbon Sequestration Potential of Agroforestry Systems in India. *J. Earth Sci. Clim. Chang.* **2013**, *4*, 1–7. [CrossRef]
47. Dhyani, S.; Kadaverugu, R. Food Security and Cultural Benefits from Urban Green Spaces: Exploring Urban Foraging as a Silently Growing Global Movement. *Clim. Chang. Environ. Sustain.* **2020**, *8*, 219–225. [CrossRef]

48. Perrings, C.; Naeem, S.; Ahrestani, F.S.; Bunker, D.E.; Burkill, P.; Canziani, G.; Elmqvist, T.; Fuhrman, J.A.; Jaksic, F.M.; Kawabata, Z.; et al. Ecosystem Services, Targets, and Indicators for the Conservation and Sustainable Use of Biodiversity. *Front. Ecol. Environ.* **2011**, *9*, 512–520. [CrossRef]
49. Clark, D.A. Defining and Measuring Human Well-Being. In *Global Environmental Change*; Freedman, B., Ed.; Handbook of Global Environmental Pollution; Springer: Dordrecht, The Netherlands, 2014; Volume 1, pp. 833–855, ISBN 978-94-007-5784-4. [CrossRef]
50. Potschin-Young, M.; Haines-Young, R.; Görg, C.; Heink, U.; Jax, K.; Schleyer, C. Understanding the Role of Conceptual Frameworks: Reading the Ecosystem Service Cascade. *Ecosyst. Serv.* **2018**, *29*, 428–440. [CrossRef] [PubMed]
51. Tschora, H.; Cherubini, F. Co-Benefits and Trade-Offs of Agroforestry for Climate Change Mitigation and Other Sustainability Goals in West Africa. *Glob. Ecol. Conserv.* **2020**, *22*, e00919. [CrossRef]
52. Murthy, K.I.; Dutta, S.; Varghese, V.; Joshi, P.P.; Kumar, P. Impact of Agroforestry Systems on Ecological and Socio- Economic Systems: A Review. *Glob. J. Sci. Front Res. H Environ. Earth. Sci.* **2016**, *16*, 15–27.
53. Udawatta, P.R.; Rankoth, L.; Jose, S. Agroforestry and Biodiversity. *Sustainability* **2019**, *11*, 2879. [CrossRef]
54. Rahman, M.S.; Roy, P.R.; Ali, M.M.; Bari, M.S.; Sarmin, I.J.; Rahman, M.A. Cost-Benefit Analysis of Different Agroforestry Systems and Practices of Kaharole Upazila of Dinajpur District, Bangladesh. *South Asian J. Soc. Stud. Econ.* **2020**, 87–97. [CrossRef]
55. Montagnini, F. Conclusions: Lessons Learned and Pending Challenges. In *Integrating Landscapes: Agroforestry for Biodiversity Conservation and Food Sovereignty*; Montagnini, F., Ed.; Advances in Agroforestry; Springer International Publishing: Cham, Switzerland, 2017; pp. 479–494, ISBN 978-3-319-69371-2.
56. Dhungana, S.P.; Bhattarai, R.C. Exploring Economic and Market Dimensions of Forestry Sector in Nepal. *J. For. Livelihood* **2008**, *7*, 58–69.
57. Kumar, B.M.; Nair, P.K.R. (Eds.) *Carbon Sequestration Potential of Agroforestry Systems: Opportunities and Challenges*; Advances in Agroforestry; Springer: Dordrecht, The Netherlands, 2011; ISBN 978-94-007-1629-2.
58. Rosenstock, T.S.; Dawson, I.K.; Aynekulu, E.; Chomba, S.; Degrande, A.; Fornace, K.; Jamnadass, R.; Kimaro, A.; Kindt, R.; Lamanna, C.; et al. A Planetary Health Perspective on Agroforestry in Sub-Saharan Africa. *One Earth* **2019**, *1*, 330–344. [CrossRef]
59. Cedamon, E.D.; Nuberg, I.; Mulia, R.; Lusiana, B.; Subedi, Y.R.; Shrestha, K.K. Contribution of Integrated Forest-Farm System on Household Food Security in the Mid-Hills of Nepal: Assessment with EnLiFT Model. *Aust. For.* **2019**, *82*, 32–44. [CrossRef]
60. Nkonya, E.; Srinivasan, R.; Anderson, W.; Kato, E. Economics of Land Degradation and Improvement in Bhutan. In *Economics of Land Degradation and Improvement–A Global Assessment for Sustainable Development*; Nkonya, E., Mirzabaev, A., von Braun, J., Eds.; Springer International Publishing: Cham, Switzerland, 2016; pp. 327–383, ISBN 978-3-319-19168-3.
61. Rasul, G.; Thapa, G.B. Financial and Economic Suitability of Agroforestry as an Alternative to Shifting Cultivation: The Case of the Chittagong Hill Tracts, Bangladesh. *Agric. Syst.* **2006**, *91*, 29–50. [CrossRef]
62. Pandey, D.N. Carbon Sequestration in Agroforestry Systems. *Clim. Policy* **2002**, *2*, 367–377. [CrossRef]
63. Muschler, R.G. Agroforestry: Essential for Sustainable and Climate-Smart Land Use? In *Tropical Forestry Handbook*; Pancel, L., Köhl, M., Eds.; Springer: Berlin/Heidelberg, Germany, 2014; pp. 1–104. ISBN 978-3-642-41554-8.
64. IPCC. Annex II Glossary. In *Climate Change 2014: Synthesis Report. Contribution of Working Groups I, II and III to the Fifth Assessment Report of the Intergovernmental Panel on Climate Change*; Pachauri, R.K., Meyer, L.A., Eds.; IPCC: Geneva, Switzerland, 2014; pp. 117–130.
65. Castro, P.; Azul, A.M.; Leal Filho, W.; Azeiteiro, U.M. Climate Change-Resilient Agriculture and Agroforestry-Ecosystem Services and Sustainability. Available online: https://www.springer.com/gp/book/9783319750033 (accessed on 21 December 2020).
66. Khamzina, A.; Lamers, J.; Vlek, P. Conversion of Degraded Cropland to Tree Plantations for Ecosystem and Livelihood Benefits. In *Cotton, Water, Salts and Soums: Economic and Ecological Restructuring in Khorezm, Uzbekistan*; Springer: Dordrecht, The Netherlands, 2012; pp. 235–248. ISBN 978-94-007-1962-0. [CrossRef]
67. Djanibekov, U.; Djanibekov, N.; Khamzina, A.; Bhaduri, A.; Lamers, J.P.A.; Berg, E. Impacts of Innovative Forestry Land Use on Rural Livelihood in a Bimodal Agricultural System in Irrigated Drylands. *Land Use Policy* **2013**, *35*, 95–106. [CrossRef]
68. Kattumuri, R.; Ravindranath, D.; Esteves, T. Local Adaptation Strategies in Semi-Arid Regions: Study of Two Villages in Karnataka, India. *Clim. Dev.* **2017**, *9*, 36–49. [CrossRef]
69. Bélanger, J.; Pilling, D. *The State of the World's Biodiversity for Food and Agriculture*; FAO: Rome, Italy, 2019; ISBN 978-92-5-131270-4.
70. IPCC. Summary for Policymakers. In *Climate Change and Land: An Ipcc Special Report on Climate Change, Desertification, Land Degradation, Sustainable Land Management, Food Security, and Greenhouse Gas Fluxes in Terrestrial Ecosystems*; Shukla, P.R., Skea, J., Calvo Buendia, E., Masson-Delmotte, V., Pörtner, H.-O., Roberts, D.C., Zhai, P., Slade, R., Connors, S., van Diemen, R., et al., Eds.; IPCC Press Office: Geneva, Switzerland, 2019; p. 36.
71. Bongaarts, J. IPBES, 2019. Summary for Policymakers of the Global Assessment Report on Biodiversity and Ecosystem Services of the Intergovernmental Science-Policy Platform on Biodiversity and Ecosystem Services. *Popul. Dev. Rev.* **2019**, *45*, 680–681. [CrossRef]
72. Buttoud, G. *Advancing Agroforestry on the Policy Agenda: A Guide for Decision-Makers*; Agroforestry Working Paper; FAO: Rome, Italy, 2013; ISBN 978-92-5-107470-1.
73. Atangana, A.; Khasa, D.; Chang, S.; Degrande, A. Agroforestry and the Carbon Market in the Tropics. In *Tropical Agroforestry*; Atangana, A., Khasa, D., Chang, S., Degrande, A., Eds.; Springer: Dordrecht, The Netherlands, 2014; pp. 353–365, ISBN 978-94-007-7723-1.

74. Fortuna, S.; Tjarvar, A.; Borelli, S.; Simelton, E. Agroforestry in REDD+ and NDCs Ways to Fulfill the Paris Agreement and Reduce Deforestation. In Proceedings of the 4th World Congress on Agroforestry, Montpellier, France, 20–22 May 2019.
75. Van Noordwijk, M. Agroforestry as Nexus of Sustainable Development Goals. *Iop Conf. Ser. Earth Environ. Sci.* **2020**, *449*, 012001. [CrossRef]
76. Leimona, B.; Noordwijk, M.V. Smallholder Agroforestry for Sustainable Development Goals: Ecosystem Services and Food Security | CAPSA. *Palawija Newslett.* **2017**, *34*, 1–6.
77. van Noordwijk, M.; Duguma, L.A.; Dewi, S.; Leimona, B.; Catacutan, D.C.; Lusiana, B.; Öborn, I.; Hairiah, K.; Minang, P.A. SDG Synergy between Agriculture and Forestry in the Food, Energy, Water and Income Nexus: Reinventing Agroforestry? *Curr. Opin. Environ. Sustain.* **2018**, *34*, 33–42. [CrossRef]
78. Van Noordwijk, M.; Duguma, L.; Dewi, S.; Leimona, B.; Catacutan, D.; Lusiana, B.; Oborn, I.; Hairiah, K.; Minang, P.; Ekadinata, A.; et al. Agroforestry into its fifth decade: Local responses to global challenges and goals in the Anthropocene. In *Sustainable Development through Trees on Farms: Agroforestry in Its Fifth Decade*; van Noordwijk, M., Ed.; World Agroforestry (ICRAF): Bogor, Indonesia, 2019; pp. 397–418.
79. Waldron, A.; Garrity, D.; Malhi, Y.; Girardin, C.; Miller, D.C.; Seddon, N. Agroforestry Can Enhance Food Security While Meeting Other Sustainable Development Goals. *Trop. Conserv. Sci.* **2017**, *10*, 1940082917720667. [CrossRef]
80. Borah, B.; Bhattarcharjee, A.; Ishwar, N.M. *Bonn Challenge and India: Progress on Restoration Efforts across States and Landscapes*, 1st ed.; IUCN, International Union for Conservation of Nature: Gland, Switzerland, 2018; ISBN 978-2-8317-1912-2.
81. Fagan, M.E.; Reid, J.L.; Holland, M.B.; Drew, J.G.; Zahawi, R.A. How Feasible Are Global Forest Restoration Commitments? *Conserv. Lett.* **2020**, *13*, e12700. [CrossRef]
82. Yasin, G.; Nawaz, M.F.; Martin, T.A.; Niazi, N.K.; Gul, S.; Yousaf, M.T.B. Evaluation of Agroforestry Carbon Storage Status and Potential in Irrigated Plains of Pakistan. *Forests* **2019**, *10*, 640. [CrossRef]
83. Mccarthy, J.; Canziani, O.; Leary, N.; Dokken, D.; White, K. *Climate Change 2001: Impacts, Adaptation, and Vulnerability*; Contribution of Working Group II to the Fourth Assessment Report of the Intergovernmental Panel on Climate Change; Cambridge University Press: Cambridge, UK, 2001; 1032p.
84. Shukla, P.R.; Skea, J.; Calvo Buendia, E.; Masson-Delmotte, V.; Pörtner, H.-O.; Roberts, D.C.; Zhai, P.; Slade, R.; Connors, S.; Van Diemen, R.; et al. *IPCC, 2019: Climate Change and Land: An IPCC Special Report on Climate Change, Desertification, Land Degradation, Sustainable Land Management, Food Security, and Greenhouse Gas Fluxes in Terrestrial Ecosystems*; Intergovernmental Panel on Climate Change (IPCC): Geneva, Switzerland, 2019.
85. Jose, S.; Bardhan, S. Agroforestry for Biomass Production and Carbon Sequestration: An Overview. *Agroforest Syst.* **2012**, *86*, 105–111. [CrossRef]
86. FAO. *Forest Landscape Restoration for Asia-Pacific Forests*; FAO: Rome, Italy, 2016; ISBN 978-92-5-109094-7.
87. Sharma, P.; Tiwari, P.; Verma, K.; Singh, M. Agroforestry Systems: Opportunities and Challenges in India. *J. Pharmacogn. Phtochem.* **2017**, 953–957.
88. Manjaiah, K.M.; Sandeep, S.; Ramesh, T.; Mayadevi, M.R. Soil Organic Carbon Stocks Under Different Agroforestry Systems of North-Eastern Regions of India. In *Agroforestry: Anecdotal to Modern Science*; Dagar, J.C., Tewari, V.P., Eds.; Springer: Singapore, 2017; pp. 299–316. ISBN 978-981-10-7650-3.
89. Bustamante, M.; Robledo-Abad, C.; Harper, R.; Mbow, C.; Ravindranat, N.H.; Sperling, F.; Haberl, H.; de Pinto, A.S.; Smith, P. Co-Benefits, Trade-Offs, Barriers and Policies for Greenhouse Gas Mitigation in the Agriculture, Forestry and Other Land Use (AFOLU) Sector. *Glob. Chang. Biol.* **2014**, *20*, 3270–3290. [CrossRef] [PubMed]
90. Mbow, C.; Smith, P.; Skole, D.; Duguma, L.; Bustamante, M. Achieving Mitigation and Adaptation to Climate Change through Sustainable Agroforestry Practices in Africa. *Curr. Opin. Environ. Sustain.* **2014**, *6*, 8–14. [CrossRef]
91. Manning, A.D.; Gibbons, P.; Lindenmayer, D.B. Scattered Trees: A Complementary Strategy for Facilitating Adaptive Responses to Climate Change in Modified Landscapes? Manning-2009-Journal of Applied Ecology-Wiley Online Library. Available online: https://besjournals.onlinelibrary.wiley.com/doi/10.1111/j.1365-2664.2009.01657.x (accessed on 21 December 2020).
92. Ahmad, F.; Uddin, M.M.; Goparaju, L.; Rizvi, J.; Biradar, C. Quantification of the Land Potential for Scaling Agroforestry in South Asia. *KN J. Cart. Geogr. Inf.* **2020**, *70*, 71–89. [CrossRef]
93. Trexler, M.; Haugen, C. Keeping It Green: Tropical Forestry Opportunities for Mitigating Climate Change. Available online: /paper/Keeping-it-green-%3A-tropical-forestry-opportunities-Trexler-Haugen/1179b3456cfe6374791923e4b001546d8dac76ab (accessed on 24 October 2020).
94. Brown, S.; Sathaye, J.; Cannell, M.; Kauppi, P.E. Mitigation of Carbon Emissions to the Atmosphere by Forest Management. *Commonw. For. Rev.* **1996**, *75*, 80–91.
95. Montagnini, F.; Nair, P.K.R. Carbon sequestration: An underexploited environmental benefit of agroforestry systems. In *New Vistas in Agroforestry: A Compendium for 1st World Congress of Agroforestry, 2004*; Nair, P.K.R., Rao, M.R., Buck, L.E., Eds.; Advances in Agroforestry; Springer: Dordrecht, The Netherlands, 2004; pp. 281–295, ISBN 978-94-017-2424-1.
96. Sathaye, J.A.; Ravindranath, N.H. Climate Change Mitigation in the Energy and Forestry Sectors of Developing Countries. *Annu. Rev. Energy Environ.* **1998**, *23*, 387–437. [CrossRef]
97. Nair, P.; Nair, V.; Mohan Kumar, B.; Showalter, J. Carbon Sequestration in Agroforestry Systems. *Adv. Agron.* **2010**, *108*, 237–307. [CrossRef]

98. Duguma, L.; Nzyoka, J.; Minang, P.; Bernard, F. *How Agroforestry Propels Achievement of Nationally Determined Contributions*; World Agroforestry Centre: Nairobi, Kenya, 2017.
99. Chavan, S.B.; Newaj, R.; Rizvi, R.H.; Ajit; Prasad, R.; Alam, B.; Handa, A.K.; Dhyani, S.K.; Jain, A.; Tripathi, D. Reduction of Global Warming Potential Vis-à-Vis Greenhouse Gases through Traditional Agroforestry Systems in Rajasthan, India. *Environ. Dev. Sustain.* **2021**, *23*, 4573–4593. [CrossRef]
100. Munang, R.; Thiaw, I.; Alverson, K.; Liu, J.; Han, Z. The Role of Ecosystem Services in Climate Change Adaptation and Disaster Risk Reduction. *Curr. Opin. Environ. Sustain.* **2013**, *5*, 47–52. [CrossRef]
101. Newaj, R.; Chavan, S.; Alam, B.; Dhyani, S. Biomass and Carbon Storage in Trees Grown under Different Agroforestry Systems in Semi Arid Region of Central India. *Indian For.* **2016**, *142*, 642–648.
102. Gosain, B.G.; Negi, G.C.; Dhyani, P.P.; Bargali, S.S.; Saxena, R. Ecosystem Services of Forests: Carbon Stock in Vegetation and Soil Components in a Watershed of Kumaun Himalaya, India. *Int. J. Ecol. Environ. Sci.* **2015**, *41*, 177–188.
103. *Carbon Management in Tropical and Sub-Tropical Terrestrial Systems*; Ghosh, P.K.; Mahanta, S.K.; Mandal, D.; Mandal, B.; Ramakrishnan, S. (Eds.) Springer: Singapore, 2020; ISBN 9789811396274.
104. Tenzin, K.; Norbu, L. Leveraging Conservation Benefits through Ecosystem-Based Services Approach and Community Engagement in Wetland and Riparian Ecosystems: The Case of Conserving Black-Necked Crane and White-Bellied Heron in Bhutan. In *Nature-Based Solutions for Resilient Ecosystems and Societies*; Springer: Berlin/Heidelberg, Germany, 2020; pp. 163–183.
105. Nath, A.J.; Sileshi, G.W.; Laskar, S.Y.; Pathak, K.; Reang, D.; Nath, A.; Das, A.K. Quantifying Carbon Stocks and Sequestration Potential in Agroforestry Systems under Divergent Management Scenarios Relevant to India's Nationally Determined Contribution. *J. Clean. Prod.* **2020**, 124831. [CrossRef]
106. Sahoo, G.R.; Wani, A.M.; Satpathy, B. Greening Wastelands for Environmental Security through Agroforestry. *Iint. J. Adv. Res. Sci. Technol. (IJARST)* **2020**, *7*, 2581–9429.
107. Cedamon, E.; Nuberg, I.; Pandit, B.H.; Shrestha, K.K. Adaptation Factors and Futures of Agroforestry Systems in Nepal. *Agrofor. Syst.* **2018**, *92*, 1437–1453. [CrossRef]
108. Baig, M.B.; Burgess, P.J.; Fike, J.H. Agroforestry for Healthy Ecosystems: Constraints, Improvement Strategies and Extension in Pakistan. *Agroforest Syst.* **2020**. [CrossRef]
109. Namirembe, S.; McFatridge, S.; Sassen, M.; van Soesbergen, A.; Duguma, L.; Bernard, F.; Minang, P.; Akalu, E. Agroforestry: An Attractive REDD+ Policy Option. 2015. Available online: http://img.teebweb.org/wp-content/uploads/2016/05/FeederStudy_Agroforestry_Report.pdf (accessed on 25 October 2012).
110. Holmes, I.; Kirby, K.R.; Potvin, C. Agroforestry within REDD+: Experiences of an Indigenous Emberá Community in Panama. *Agroforest Syst.* **2017**, *91*, 1181–1197. [CrossRef]
111. Chavan, S.; Keerthika, A.; Dhyani, S.K.; Handa, A.K.; Newaj, R.; Rajarajan, K. National Agroforestry Policy in India: A Low Hanging Fruit. *Curr. Sci.* **2015**. [CrossRef]
112. Olander, L.P.; Wollenberg, E.; Tubiello, F.N.; Herold, M. Synthesis and Review: Advancing Agricultural Greenhouse Gas Quantification. *Environ. Res. Lett.* **2014**, *9*, 075003. [CrossRef]
113. UNEP Emissions Gap Report 2013. Available online: http://www.unenvironment.org/resources/emissions-gap-report-2013 (accessed on 22 December 2020).
114. Pandit, B.H.; Neupane, R.P.; Sitaula, B.K.; Bajracharya, R.M. Contribution of Small-Scale Agroforestry Systems to Carbon Pools and Fluxes: A Case Study from Middle Hills of Nepal. *Small Scale For.* **2013**, *12*, 475–487. [CrossRef]
115. Cardinael, R.; Umulisa, V.; Toudert, A.; Olivier, A.; Bockel, L.; Bernoux, M. Revisiting IPCC Tier 1 Coefficients for Soil Organic and Biomass Carbon Storage in Agroforestry Systems. *Environ. Res. Lett.* **2018**, *13*, 124020. [CrossRef]
116. Sharma, V.; Chaudhry, S. An Overview of Indian Forestry Sector with REDD+ Approach. Available online: https://www.hindawi.com/journals/isrn/2013/298735/ (accessed on 5 November 2020).
117. Shi, L.; Feng, W.; Xu, J.; Kuzyakov, Y. Agroforestry Systems: Meta-Analysis of Soil Carbon Stocks, Sequestration Processes, and Future Potentials. *Land Degrad. Dev.* **2018**, *29*, 3886–3897. [CrossRef]

 *forests*

Article

Who Adopts Agroforestry in a Subsistence Economy?—Lessons from the Terai of Nepal

Arun Dhakal [1] and **Rajesh Kumar Rai** [2,*]

1. Nepal Agroforestry Foundation (NAF), Kathmandu 44600, Nepal; arun_dhakal2004@yahoo.com
2. The Center for People and Forests (RECOFTC), Bangkok 10903, Thailand
* Correspondence: rjerung@gmail.com

Received: 7 May 2020; Accepted: 15 May 2020; Published: 18 May 2020

Abstract: Agroforestry is recognized as a sustainable land use practice. However, the uptake of such a promising land use practice is slow. Through this research, carried out in a Terai district of Nepal, we thoroughly examine what influences farmers' choice of agroforestry adoption and what discourages the adoption. For this, a total of 288 households were surveyed using a structured questionnaire. Two agroforestry practices were compared with conventional agriculture with the help of the Multinomial Logistic Regression (MNL) model. The likelihood of adoption was found to be influenced by gender: the male-headed households were more likely to adopt the tree-based farming practice. Having a source of off-farm income was positively associated with the adoption decision of farmers. Area of farmland was found as the major constraint to agroforestry adoption for smallholder farmers. Some other variables that affected positively included livestock herd size, provision of extension service, home-to- forest distance, farmers' group membership and awareness of farmers about environmental benefits of agroforestry. Irrigation was another adoption constraint that the study area farmers were faced with. The households with a means of transport and with a larger family (household) size were found to be reluctant regarding agroforestry adoption. A collective farming practice could be a strategy to engage the smallholder farmers in agroforestry.

Keywords: agroforestry practices; adoption determinants; smallholder farmers; Nepal

1. Introduction

Land degradation, a persistent decline in soil quality and its productivity caused by natural or anthropogenic factors, has adversely affected food production, the supply of ecosystem services and livelihoods globally [1]. Even though it occurs throughout the world, the extent and degree of degradation vary with region. For instance, dryland areas of African countries and Australia, mountain ranges of the Himalayas, and densely populated areas of South Asia are more vulnerable [2,3]. The consequences of land degradation are severe as they impact adversely on farm-productivity, and hence on food security [4]. By 2030, the demand for food is expected to increase by at least 50%, which requires conservation and restoration of the productivity of agricultural land. It is estimated that a 60% increase in agricultural productivity, will be necessary by 2050 in order to overcome hunger and food insecurity [5].

Many factors are responsible for the global spread of agricultural land degradation. The spread and growth of populations, inappropriate land-use practices, excessive use of chemicals, mechanized agriculture and natural phenomena such as erosion, floods and drought are the proximate causes of degradation [4]. In countries like Nepal, where the demographic pattern is changing substantially due to the outmigration of the economically active population, agriculture land degradation is becoming a serious issue [6]. In a subsistence economy, farmers are forced to cultivate marginal lands, use agrochemicals, and follow intensive farming and mechanized agriculture to sustain their livelihoods. All these activities have supported a gradual decline in soil fertility [7,8].

Since the underlying causes of land degradation are multifaceted, it requires an integrated approach of land management [4]. A single strategy may be counterproductive, for instance, a reduction in chemical fertilizer application may result in decreased crop yield and, hence, food insecurity. In this context, agroforestry, which is an integrated tree-based farming system, has come into the forefront given its potential to address land degradation with additional environmental and social benefits [9,10]. Agroforestry supports biodiversity conservation [11–14]. Similarly, it has higher financial returns compared to that of conventional agriculture [15]. It also provides biosafety, as the crop failure is less likely compared to the treeless system [16]. This may be because the agroforestry system restores soil fertility [15,17] and rehabilitates degraded agricultural land [18]. The scientific evidence clearly indicates that agroforestry can be a viable land-use option with its potential to address various issues ranging from household-level issues such as food insecurity to global issues including climate change and biodiversity. It is of the utmost urgency to make such a promising land-use reach a wider geographic coverage and motivate farmers to adopt it.

Having so many economic and environmental benefits, agroforestry should be a widely adopted practice. However, the status of agroforestry adoption is not encouraging and not widespread as expected, even though several national and international organizations are involved in its promotion and extension. There might be disincentives to establishing trees including lack of knowledge, upfront costs, length of time until there is a return and a short-to-medium-term reduction in cash flow and/or household food production [19]. Nonetheless, there has been a wealth of research works on agroforestry adoption and the factors associated with it [20]. The existing agroforestry literature documents four broad categories of factors/determinants influencing farmers' adoption decisions: farmers' preferences, resource endowments, institutional impediments and risk/uncertainty [21]. However, the influence of these determinants/factors on the adoption decision of farmers differs from one place to another. For example, in some African countries (Sudan and Uganda), factors such as gender of household head, household family size, level of education, farmer's experience, membership within farmers' associations, contact with extension workers, land tenure security, agroecological zone, distance of the village from the nearest town, village accessibility and income were the major factors that determined the adoption of agroforestry systems by the smallholder farmers [22,23]. On the contrary, a study by Beyene et al. [24] in Ethiopia reveals that gender has no role in the agroforestry adoption decision of farmers. A similar result was documented by Oli et al. [25] from a mid-hills district of Nepal, that agroforestry practice is a gender-neutral activity. In another study carried out in Vietnam by Catacutan and Naz [26], female-headed households were found to be less likely to adopt agroforestry practices. Likewise, the issue of land tenure security is the prominent one influencing the adoption decision of African farmers, while this has no impact in the Nepalese context [22–24,27]. Therefore, understanding the region-specific determinants of agroforestry adoption is crucial for the successful uptake and diffusion of agroforestry practices in that region.

Against the above backdrop, this study attempts to assesses the determining factors of adoption of two agroforestry practices, agroforest/woodlot system (AFS) and alley cropping system (ACS), in Nepal. The findings of the study are useful for policymakers, development agencies and academicians.

2. Materials and Methods

2.1. Study Area and Descriptions

Dhanusha district, which is in the southern part of Nepal and shares a border with India, was selected for this study. About 60% of the land comes under agriculture out of the total area of the district, 119,000 ha. Data were collected from May through August 2014. Like other parts of Nepal, agriculture is the major economy of the district, where about 90% of people are actively engaged in the cultivation of rice, maize, wheat and sugarcane [28]. After the state federalization, the district now falls in province no. 2. Located approximately 95 m above the sea level, the district is one of the hottest districts of Nepal with the average annual rainfall being 2199 mm. The meteorological data show that April is the warmest month, with the average temperature being 39.6 °C, while January is the coldest, with the average temperature of 21.4 °C [29].

Administratively, the district consists of one sub-metropolitan city, eleven urban municipalities and six rural municipalities. The Terai Private Forest Development Association (TPFDA), a local NGO, has worked to promote a tree-based farming practice in then nine Village Development Committees (VDCs), covering 10,500 hectares (Figure 1). Therefore, these nine VDCs were selected as the study site. After the state is restructured, some parts of the study site fall in the urban municipality while most parts are still VDCs, now known as rural municipalities.

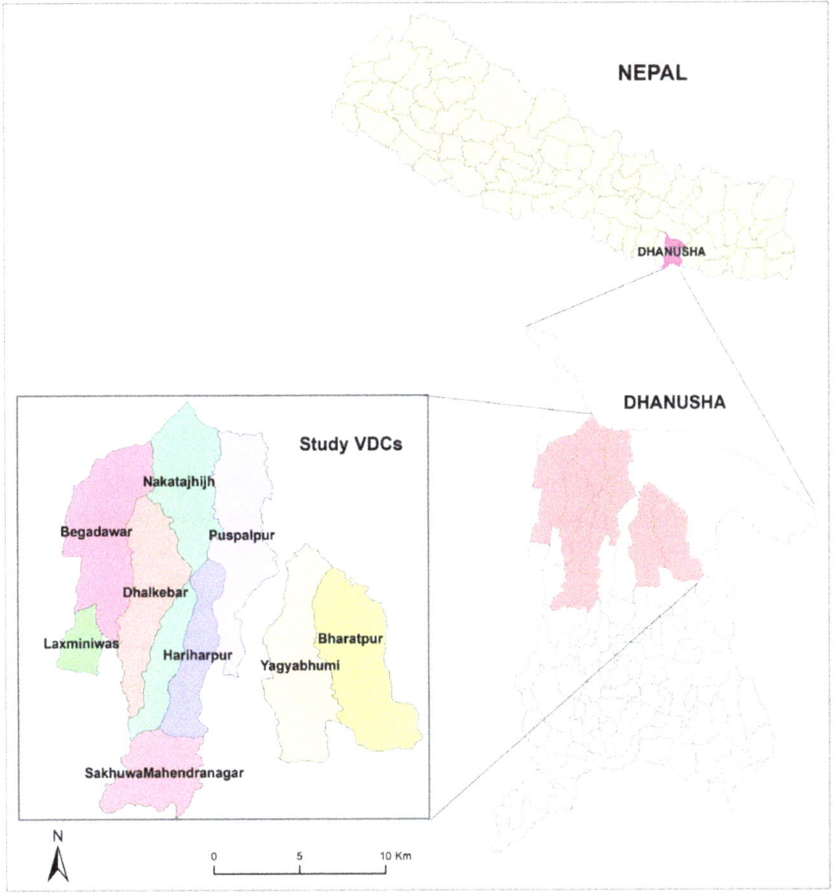

Figure 1. Study area.

2.2. Overview of Farming Systems in the Study Area

Dhakal et al. [11] documented two forms of farming systems in the study area: conventional agriculture and agroforestry. Under agroforestry, they identified three variants depending on the arrangement of trees on the farmland: alley cropping, agroforest/woodlot and a combination of the two variants. Dhakal [16] added one more variant to the list, boundary plantation, however, this practice is very scanty. Since the trees grown along the farm boundary have an adverse impact on the field crops of adjoining farmland of fellow farmers due to shading, boundary plantation is not an acceptable practice in the study area.

Conventional agriculture is a cereal-based farming system in the study area that has evolved over the years from a simple mono-cropping to a complex and intensive multi-cropping system [11] (Figure 2a). This practice includes rice, wheat/mustard and maize as major crops, and lentil, beans, groundnuts, pea and millet as inter-crops. The mono-cropping of sugarcane is also a part of conventional agriculture in the area [28].

Alley cropping is an agroforestry system in which trees are grown on the farm bunds as an alley. *Eucalyptus camaldulensis, Bauhinia variegata, Leucaena leucocephala* and *Melia azedarach* are most preferred by farmers for alley cropping [16] (Figure 2b). Agroforest/woodlot is a kind of agroforestry, which is grown as a plantation for a commercial purpose, requiring a separate patch of farmland. Farmers prefer multipurpose tree species such as *Eucalyptus camaldulensis, Gmelia arborea, Tectona grandis, Melia azedarach* and *Anthocephalus chinensis* as agroforest species [30] (Figure 2c). Both agroforestry systems are spread in the study district as spillover effects of the Sagarnath Forestry Project, which promoted production forestry and Taungya cultivation of *Eucalyptus camaldulensis* and *Dalbergia sissoo* [16]. The further spread of these systems was supported by Nepal Agroforestry Foundation (NAF) by introducing a private forestry project in the district. However, most farmers ceased continuing to grow trees on their farmlands upon completion of the project [31].

(a)

(b)

(c)

Figure 2. (**a**). Conventional agriculture (Paddy), (**b**) Alley cropping, (**c**) Agroforest/woodlot.

2.3. Selection of Farmers for Questionnaire Survey

A two-stage sampling approach was chosen for this study. This approach is widely used in agroforestry adoption studies [32–34]. At the first stage, one representative ward (Ward is the lowest administrative unit) from each VDC was selected, thus making altogether nine wards [33]. These nine wards were selected for two reasons: 1) the community in these wards is composed of both the native

(Madhesi ethnic group) and the migrants, coming from hilly regions of Nepal and northern India and; 2) agroforestry and conventional agriculture, a cereal-based farming practice, are the two most dominant forms of agriculture in the wards [11]. In the second stage, we considered the total number of households in the wards while determining the sample size for this study. The VDC records showed the total households in each ward falling in the range from 309 to 338 while the average number of households was 320. Since there is no vast difference in household numbers across the sample wards, we took the average as a reference and assumed a 10% sample size would serve our purpose. Hence, there were thirty-two households from each ward, and they were selected randomly. This means 288 sample households were selected. In-person interviews were conducted with the head of the sample households using a structured questionnaire. A total of 18 households were dropped out of the analysis since these households were practicing a combination of two or more agroforestry practices, agroforest/woodlot, boundary plantation and alley cropping.

2.4. Analytical Model

Multinomial probit and logit models are the two commonly used regression models when there are more than two dependent variables and the dependent variables are unordered and categorical [35,36]. We chose the Multinomial Logit (MNL) model over the probit model for this study because: (1) It gives more precise parameter estimation [37]; (2) It estimates the likelihood of adoption of non-reference categories against a reference (base) category in terms of relative risk ratio (RRR) [36]; (3) The model has been more commonly used in recent studies [36,38–40]; (4) The Probit model is not usually used, largely because of the practical difficulty involved in its estimation process [35]; and (5) The model is the best choice when the data are not normally distributed [35].

With three farming practices in place, farmers can choose the one they prefer the most from the three alternatives. That means their choice is mutually exclusive. We assumed farmers follow the random utility theory, while making the choice out of the three farming practices available. Therefore, we used a random utility model while determining the farmers' choice of farming practices, as given by Greene [41]

$$U_{ij} = \beta_j X_{ij} + \varepsilon_{ij} \tag{1}$$

where U_{ij} denotes the utility of farmer i obtained from farming choice j, X_{ij} denotes all the factors affecting farmers' decision to adopt a farming practice j and β_j is the parameter that reflects changes in U_{ij} due to changes in X_{ij}. We assumed the error terms to have an independent and identical distribution (iid) [42]. According to profit maximization, farmer i will, thus, only choose a specific alternative j if $U_{ij} > U_{ik}$ for all $k \neq j$., which means when each farming practice is considered as a possible adoption decision, it is expected that farmers will choose the alternative that maximizes utility given the three farming practices available. This choice of j depends on a number of independent variables, as denoted by X_{ij} in the above equation. If Y_i is a random choice that a farmer can make, the MNL model can be expressed as

$$\text{Prob}\ (Yi\ =\ j) = \frac{e^{\beta_j x_i}}{\sum_{j=1}^{j} e^{\beta_j x_i}} \tag{2}$$

the above equation estimates probabilities for j+1 farming choices, i.e., three practices for farmers with a number of independent variables, X_{ij}. Here, we estimate the probabilities of two non-reference farming practices, agroforest system and Alley cropping system against the reference category, i.e., conventional agriculture. Therefore, the probabilities can be estimated by the following equation.

$$\text{Prob}\ (Yi\ =\ j) = \frac{e^{\beta_j x_i}}{1 + \sum_{j=1}^{j} e^{\beta_j x_i}} \tag{3}$$

2.5. Variables Defined

We extensively reviewed the contemporary literature on adoption to identify and determine independent variables. The independent variables included socio-economic, biophysical and institutional characteristics. We hypothesized these variables as adoption determinants and constraints (Table 1). However, some variables were excluded in the model. The variable 'farmers' perception on agroforestry' was dropped off the model because studies suggest it had no relationship with adoption [43–46] and the methodological challenge we faced to precisely measure the perception made us drop this variable off the model [47]. The 'slope gradient' is another variable we ignored because of little altitudinal variation across the sampled households. The third variable 'access to credit facility' was also excluded because of no financial guarantee from the financial institutions for agroforestry promotion in the study area.

Table 1. Description of the independent variables specified in the multinomial logistic model.

Variables	Description	Type of Measure	Expected Sign
Education	Years of formal education of household head	Years	+
Age	Age of the household head	Years	–
Sex	Sex of the household head	1 if male, 0 otherwise	+
Household size	Number of family members between 15 to 60 years	Years	–
Off-farm income	Farmer has any off-farm source of income	1 if yes, 0 otherwise	+
Landholding size	Total cultivated area	Katha [1]	+
Livestock herd size	Total livestock standard units (LSU) [2] kept by a surveyed household	Numbers	+
Extension service	Total number of training received and visits by extension workers in the last five years	Numbers	+
Home to forest distance	Distance from home to nearest government forest	Kilometers	+
Transport	Means of transport possessed by the surveyed household	1 if a farmer has own means of transport, 0 otherwise	+, -
Irrigation facility	Farm has any source of irrigation	1 if yes, 0 otherwise	+
Membership	Member of farmers' group and organization	1 if yes, 0 otherwise	+
Origin	Farmer is native	1 if yes, 0 otherwise	+
Risk taking attitude	Farmer is risk-averse, risk-neutral and risk loving	1 if risk loving, 2 if risk-neutral and 3 if risk-averse	+
Awareness	Farmer is aware of environmental benefits of an agroforestry practice	1 if yes, 0 otherwise	+

[1] Katha is a unit of area. 30 Katha= 1 hector. [2] adult buffalo = 1 LSU, 2 young buffalos (1.5 to 3 years) = 1 LSU, 2 cattle = 1 LSU and 3 young cattle = 1 LSU.

The dependent variable is the choice of farming practices by farmers as denoted by Yi. For MNL model, the dependent variable was denoted as:

Yi = 0 if a household adopts conventional agriculture system (CAS) -reference category- (j = 0);

Yi = 1 if a household adopts agroforest system (AFS)- non-reference category- (j = 1);

Yi = 2 if a household adopts alley cropping system (ACS)- non-reference category- ((j = 2).

Before the model is run, all the hypothesized independent variables were tested for multicollinearity using the variance inflation factor (VIF). We found the VIFs of the independent variables were below 10 (1.09–2.03), indicating that there is no issue of multicollinearity [35].

The estimation of the MNL model for this study was undertaken by selecting CAS as the base category. The odds of two other farming systems, namely AFS and ACS, against the CAS were estimated in this study. Since the CAS was the base category, it was hypothesized that most predictor variables will positively impact the adoption of the tree-based farming practices, i.e., a one-unit increase in an independent variable will increase the likelihood of AFS and ACS adoption.

2.6. Method of Data Analysis

The household survey data were analyzed by descriptive statistics and multinomial logit model using STATA (version 14).

3. Results

3.1. Socio-Economic, Biophysical and Institutional Characteristics of Sample Households

Out of 270 sample households surveyed, 60% were involved in conventional agriculture. The average age of the sampled household heads was 44 years with a minimum of 26 years and a maximum of 75 years. AFS farmers were younger than the other two (Table 2). The survey results showed that 57% were males, while the remainder (43%) were females. The family size of the sample household ranged from two to 14, with a mean family size of seven, which is nearly 1.5 times larger than the national average national, i.e., 4.9 people per family [29]. If only the economically active family members (year 15 to year 60) are considered, the average household size was 4.5 with the lowest household size in the AFS group (Table 2). The survey results indicated that out of the total sample households, the majority (57%) had no source of off-farm income, while 75% of AFS households had off-farm income sources, which is the highest among the three farming groups (Table 2). In terms of access to irrigation, the survey reveals that 56% of the sample farm households had no access to any kind of irrigation facility. However, if we see specifically, the majority of the AFS households had access to irrigation. About 15% of sample household heads had no formal education, of which 94% were females. On average, the household head had 6 years of schooling. Among the three farming groups, the AFS household head had the highest education.

Table 2. Characteristics of sample households in the study area.

Variables.	Mean Values of the Variables		
	CAS (*n* = 162)	ACS (*n* = 60)	AFS (*n* = 48)
Education (Years of schooling)	5.0 (3.6) [a]	6.3 (3.7) [b]	9.6 (4.0) [c]
Age of household head	46.6 (13.2) [a]	43.6 (9.9)	39.4 (10.0) [b]
Sex of household head	0.55 (0.50)	0.56 (0.50)	0.64 (0.48)
Household size	4.7 (2.1) [a]	4.4 (1.9)	3.9 (1.3) [b]
Off-farm income	0.32 (0.50)	0.49 (0.50)	0.75 (0.43)
Landholding size	23.8 (21.1) [a]	34.7 (25.4) [b]	74.3 (36.7) [c]
Livestock herd size (LSU) *	2.9 (1.9) [a]	3.7 (2.6) [b]	6.7 (2.8) [c]
Extension service	0.80 (1.1) [a]	3.2 (2.2) [b]	5.5 (1.7) [c]
Distance from home to nearest government forest	4.2 (2.7) [a]	9.0 (5.6) [b]	9.3 (5.5) [b]
Transport (tractor, bullock cart)	0.6 (0.51)	0.4 (0.51)	0.3 (0.48)
Irrigation	0.35 (0.48)	0.46 (0.50)	0.63 (0.49)
Membership	0.25 (0.43)	0.51 (0.50)	0.73 (0.45)
Origin	0.41(0.49)	0.40 (0.49)	0.58 (0.50)
Risk taking attitude	2.4 (0.80)	1.71 (0.77)	1.52 (0.74)
Awareness	0.28 (0.45)	0.51 (0.50)	0.69 (0.47)

Note: Figure in the parenthesis is the standard deviation. Means in a row with different superscripts are significant at 0.05 level, and with similar superscripts are insignificant. CAS: Conventional agricultural system; ACS: Alley cropping system and AFS: Agroforest system. *: adult buffalo = 1 LSU.

Farmland is the primary livelihood asset of Nepalese farmers. The survey results indicate that the average landholding size of the sample farm households was 1.16 hectares (ha), slightly above the national average, which is 0.8 ha [48]. However, the group-wise distribution of landholding was different: the AFS farmers had the highest average. The livestock herd size was measured in terms of livestock standard unit (LSU). Only the young and adult buffaloes and cattle were considered while estimating the LSU. The average LSU of sample farm households was 3.8 with the highest average LSU in the AFS group. The study area community is composed of both native and migrant people. Out of the total sample households, 56% were migrants. The migrants included people coming from both the hilly regions of Nepal and northern India.

3.2. Determinants of AFS and ACS Adoption

The determinants of agroforestry adoption were examined using the multinomial logit (MNL) model. Since conventional agriculture is the base category, two models were estimated: one is for agroforest/woodlot adoption relative to conventional agriculture and the other is for alley-cropping adoption relative to conventional agriculture. The relative risk ratio (RRR), coefficients and significance levels are presented in Table 3. The model is good-fit, as it was significant at the 1% level. The RRR shows the relative risks/likelihood/chances of AFS and ACS adoption relative to CAS.

Table 3. Parameter estimates and RRR of a multinomial logistic model for AFS and ACS.

Independent Variables	AFS ($n = 48$)			ACS ($n = 60$)		
	Coefficient	RRR	p Value	Coefficient	RRR	p Value
Years of schooling (education)	0.159	1.172	0.247	0.114	1.121	0.194
Age of household head	−0.048	0.953	0.315	−0.008	1.008	0.753
Sex of household head	0.280	1.323 **	0.044	0.202	0.823	0.714
Household size	−0.618	0.539 **	0.041	−0.078	0.925	0.580
Off-farm income	1.083	2.954 **	0.023	0.148	1.159	0.262
Landholding size	0.123	3.130 ***	0.000	0.095	1.099 ***	0.003
Livestock herd size	0.555	1.742 ***	0.003	0.178	1.195	0.179
Extension service	1.064	2.910 ***	0.000	0.529	1.697 ***	0.003
Distance from home to government forest	0.376	1.457 ***	0.001	0.322	1.380 ***	0.000
Transport	−0.682	0.506 ***	0.005	−0.172	0.842 *	0.086
Irrigation	0.549	1.732 **	0.042	0.302	0.352	0.571
Membership	0.217	1.242 **	0.038	0.115	1.122 **	0.019
Origin	1.215	3.371	0.188	−0.336	0.714	0.551
Risk averse [a]	−2.134	0.118 **	0.041	−1.208	0.299	0.123
Risk neutral [a]	−1.049	0.350	0.326	−0.384	0.681	0.577
Awareness	0.189	1.208 *	0.058	0.821	2.273	0.122
Constant	−10.110	0.00004 ***	0.004	−5.213	0.0054 ***	0.002
Diagnostics						
Base category	CAS ($n = 162$)					
Number of observations	270					
LR chi-square	373.13 ***					
Log likelihood	−93.45					
Pseudo R^2	0.67					

[a] risk loving is the reference category. AFS: Agroforest system, ACS: Alley cropping system, CAS: Conventional agricultural system. RRR: Relative risk ratio. * $p < 0.10$, ** $p < 0.05$ and *** $p < 0.01$.

The analysis of the MNL model showed that the adoption of AFS and ACS was influenced by several factors. AFS adoption was influenced by eleven variables including the sex of household head, household size, off-farm income, landholding size, livestock size, extension service, distance from home to government forest, transport, irrigation, membership and risk-taking. Out of eleven, the influence of three variables, household size, transport and risk-taking was negative. The adoption of ACS was influenced by five variables only. They included landholding size, extension service, distance from home to government forest, transport and membership (Table 3).

The sex of household head had a positive and significant effect on the adoption of AFS. This implies that the relative risk/chance of adopting this practice would be 1.32 times more likely when the household head were males. In other words, if the household head was a male, we would expect him to be more likely to prefer AFS over conventional agriculture.

The negative and significant sign of household size indicates that larger families were less likely to adopt agroforest/woodlot. In other words, if household size increased, we would expect farmers to be more likely to prefer conventional agriculture over agroforest/woodlot. Landholding size positively and significantly influenced the adoption of AFS and ACS. In other words, if farmers held larger landholdings, we would expect them to be more likely to prefer AFS and ACS over conventional agriculture.

Livestock herd size (expressed in terms of LSU) is positively and significantly associated with the adoption of AFS, which means if the herd size is increased by one unit, the relative risk of AFS adoption relative to conventional agriculture would be expected to increase by a factor of 1.742. The positive association and the significance of extension service revealed that training for farmers and visits by extension officials are important for the adoption of both practices.

The negative and significant sign of transport indicated that when a farmer had a means of transport, the farmer would be expected to be less likely to adopt agroforest/woodlot and alley cropping. Farmers' association with farmer groups and agricultural organizations positively and significantly affected the adoption of these agroforestry practices. The risk-taking farmers and those living farther from the government forest were more likely to adopt AFS. The distant farmers also preferred alley cropping to conventional agriculture.

4. Discussion

The cereal-based farming practice (conventional agriculture) is the most dominant in the study area. However, the continuation of the practice is uncertain given the shortage of labor/workforce. Farmers are forced to grow one or two field crops only and even some farmlands are left all barren. A large section of the workforce is now in gulf countries for jobs, which has dropped farming activities considerably in Nepal [49]. A tree-based farming practice, which could be a viable alternative to conventional agriculture, is slow-growing in the study area. Although it holds the potential of enhancing the household economy and contributing to climate change mitigation and biodiversity conservation, the uptake of the practice by farmers is at a snail's pace. We attempted to address the slow-uptake issue through this study by analyzing the adoption factors using the MNL model.

The role of gender in agroforestry adoption is vividly discussed in the literature. Both men and women have influenced the adoption decisions depending on their circumstances. For example, in Malawi, male-headed households are more likely to adopt agroforestry in patrilocal societies, while in matrilocal societies, it is the female-headed households who are more interested in the adoption [21]. In another study from the Rulindo district of Rwanda, men were found to be reluctant regarding agroforestry adoption. The reason for this is attributed to the agroforestry trees, which lack commercial values such as timber and only have subsistence uses such as fodder, firewood and soil fertility improvement. However, many other studies claimed that agroforestry adoption has been the male-headed households' preference [50–52]. A study by Catacutan and Naz [26] in Vietnam highlights the reasons for women's reluctance towards the adoption being a lack of knowledge, low education level and poor access to extension. In line with the above studies, our finding also reinforces that the adoption of agroforest/woodlot is the male-headed households' affair. The reasons for this can be attributed to the commercial values of agroforest/woodlot in the study area, and lower education level of female heads, which might have limited their access to extension officials. In the study area, the agroforest is composed of commercially important multipurpose tree species while fuelwood and fodder species are preferred for alley cropping [11,30].

Access to land and land tenure security are considered two important determinants of agroforestry adoption [53,54]. However, for the kind of agroforestry we have in the study area, more important is landholding size. Our result suggests that the adoption of AFS and ACS is dependent on farm size: the larger the farm size is, the greater the chance of adoption is, and the result was as expected. To better understand why the large farmers are likely to favor agroforest/woodlot, we need to know the very nature of these practices. AFS is different from ACS. Farmers are required to have allocated parts of their farmland and wait for at least 10 years for returns if they want to grow trees as an agroforest [11]. The reduction in farmland after land-sparing decreases annual food production, which might fall short of fulfilling the annual food demand of the family and livestock. Since large landholding guarantees food security, farmers are willing to allocate parts of their farmlands for long-term investments such as AFS [55]. Ahmed et al. [56] argue that farmers with more farmland are less risk-averse, and therefore tend to and are more willing to try new technologies. In the case of ACS, land allocation is not required since the trees are grown on farm bunds. However, there exists competition between tree crops and agricultural crops for light, water and nutrients, thus increasing the risk of a decrease in food production. Therefore, smallholder farmers are less likely to shift from conventional agriculture to any of the two agroforestry practices. Our finding is supported by previous studies [24,28,57,58] which found that farm size is the significant factor positively influencing the adoption of agroforestry.

Agroforest/woodlot and alley cropping are new practices in the study area. Early adopters of such new practices tend to be the better-off households [11,24]. In rural Nepal, land, livestock and off-farm income are the measure of wealth. Just as more farmland, we hypothesized off-farm income of farm households positively influences the adoption of both practices. The influence of the variable was found to be positive but not significant for ACS adoption. Alley-cropping is in practice in the study area, mainly for subsistence uses such as fodder and firewood. This might be the reason why farm households with a good source of off-farm income are not interested in ACS adoption. A study by Kassie [59] carried out in northwest Ethiopia revealed that agroforestry adopters tend to have more off-farm income diversification than non-adopters. Off-farm source of income acts as a safety net and helps solve the cash constraints of the farm households, thus inducing them to perform long-term investments, which are expected to yield higher returns in the future [59]. Financial security backs them up to take risks and they tend to try technologies such as agroforest/woodlot [56]. Studies from Swaziland [60] and Indonesia [61] are some examples supporting the hypothesis that off-farm sources of income positively influence agroforestry adoption. Our finding that risk-averse farmers are less likely to adopt AFS also reinforces the argument that agroforestry adopters are less risk-averse.

Irrigation and livestock are two important endowments (inputs) for agriculture. These inputs contribute to enhancing farm productivity. Our result reveals that these endowments are positively associated with AFS adoption. Studies by Sood and Mitchell [62] in Himachal, India and Pingale et al. [63] in Pantnagar, India report that having a source of irrigation favors agroforestry adoption. These are interesting results because farmers generally use irrigated farmlands for field crop and cash crop production. Our finding agrees with these studies too. To understand why the farmers of the study area prefer agroforest/woodlot to conventional agriculture when irrigation is available, we need to see the physical properties of soil of the study area. The study area falls in the Bhabar zone of Nepal. The Bhabar is characterized by the low water holding capacity and high rate of infiltration and percolation [64]. These characteristics favor tree plantations. In the study area, *Eucalyptus camaldulensis* is the most preferred agroforestry species because of its high-value poles that are used as utility poles [11]. Farmers have experienced faster growth of the species when grown in the irrigated fields. The harvest cycle of the species for pole production is considered 10 years. However, Dhakal [16] reported the harvest cycle to be seven years in the irrigated farmlands. A similar result is reported by Pingale et al. [63] from India, that farmers preferred woodlots of *Populus deltoides* and *Eucalyptus camaldulensis* in the irrigated fields for their high industrial values. Likewise, livestock is a good source of farmyard manure to improve farm production and an agroforest is a good source of feed to livestock as it provides fodder. Therefore, our finding that there exists a positive association between livestock herd size and AFS adoption was as expected. However, this is not true for ACS adoption. This is because farmers' choice of tree species for farm bunds are mostly fodder species (medium-sized trees with minimum shading effects), which cannot fulfill both needs: commercial (timber and pole) as well as subsistence (fodder and firewood) like the agroforest does. Similar results are reported by Neupane et al. [27] and Oli et al. [25] from the studies carried out in the mid-hills districts of Nepal.

Labor is one of the major factors of a production system. In recent years, Nepalese farmers have witnessed a shortage of workforce. Family labor is the main source of the workforce in Nepal. The shortage of labor has resulted in low-intensity farming. Since cereal-based farming is a labor-intensive activity, many farmers are forced to leave their farmlands barren [65]. Our finding that an agroforestry practice such as agroforest/woodlot is favored when the workforce is not enough holds great significance in the present Nepalese farming context. However, there is no consensus on whether an agroforestry practice is less labor-intensive. Depending on the types and objectives of agroforestry, it can be either less or more labor-intensive. For example, coffee-based agroforestry and cocoa-based agroforestry are more labor-intensive than conventional agriculture [26,66] while timber/fuelwood-based agroforestry such as agroforest is less labor-intensive [16,19]. A study by Kassie [59] reveals that farmers are shifting to timber-based agroforestry when they found food crop farming is more labor-intensive. In a timber/fuelwood-based agroforestry such as agroforest, no labor is required after the second year of establishment until the harvest year. A recent study by Cedamon et al. [67] from Nepal's mid-hills also reinforces our findings. They argue that the emerging remittance economy of the country has increased the outmigration of Nepalese youths, resulting in a short supply of labor force, which made the Nepalese farmers adopt less labor-intensive cultivation practices such as agroforestry with multipurpose tree species.

Training and farmers' field visits are two important extension services, widely used to transfer knowledge and skills about agricultural innovations such as agroforestry to farmers [68]. These services not only assist farmers in gaining skills on nursery techniques, tree planting and raising and tree harvesting but also provide opportunities to establish a good rapport with extension workers and extension offices/agents, which may increase their access to information and keep them abreast of the latest developments in agroforestry [69]. Against this backdrop, our finding that extension services positively influence the adoption of AFS, and ACS was as expected. Our result is supported by previous studies which proved that provision of training and contact with extension workers are the significant factors positively affecting the uptake of agroforestry [28,55,68–71].

As hypothesized, membership in farmer groups and local agricultural organizations had a positive and significant sign, which implies that the farmers, who are affiliated to a group/organization, were more likely to prefer AFS and ACS to CAS. This is because being in the group provides farmers with opportunities to share information, knowledge, and experiences about the new technologies and learn from one another, which positively influences the adoption behavior of individual farmers [70,71]. Our finding is supported by previous studies, which documented the significant and positive influence of group membership on the adoption behavior of farmers [55,67,70–72].

The influence of forest distance from home was positive and significant. This implies that the chance of adopting AFS and ACS increases when farmers live at a distant location from the nearest forest. Our finding was as expected. This is because the distant farmers may find it difficult and time-consuming to go to the forest very often for grazing their livestock and collecting fodder and fuelwood. Having a private source of fodder and fuelwood such as AFS and ACS would save time and labor, which farmers could utilize in other farming activities. Our finding corroborates with previous studies [28,31,73].

There is a wealth of literature that describes the environmental benefits of agroforestry including biodiversity conservation, climate change mitigation and carbon sequestration. We attempted to examine whether Nepalese farmers are aware of these benefits and their awareness positively influences the adoption of AFS and ACS. Our finding that awareness increased farmers' willingness to adopt AFS was expected. This is because, in recent years, peoples' awareness of environmental issues such as climate change and the role of trees in climate change mitigation has increased [74]. Last but not least, our finding that having a private transport (bullock-cart) decreases farmers' willingness to adopt AFS and ACS seems to be unexpected, as we see from the result of a study [60] that documented that bullock-carts are used to carry timber and fuelwood to the proximate market centers and there exists a strong and positive relationship between transport means and timber/fuelwood-based agroforestry adoption. In the study area, however, bullock-carts are used mainly to carry food crops (food grains) and sell them at the farmer markets. Even though AFS is a timber/fuelwood-based agroforestry, carts are not needed; the sale of agroforestry products (timber, poles and fuelwood) is managed by the local contractors who transport the products to the market centers by using their transport means [16]. ACS is mainly a fodder-based agroforestry practice and there exist no formal markets for fodder. Based on the current practice in the study area, our result was as expected.

5. Conclusions

The study indicates that landholding size, extension services, distance from home to forest, and membership in farmer groups have positive impact on selecting both agroforestry systems over conventional farming. This clearly suggests that agroforestry can be promoted with less effort in the communities, that are distantly located. In addition, well-off households (i.e. having more farmlands) can be the entry point of agroforestry promotion program compared to their smallholder neighbors as the former are less risk averse. However, extension services and the formation of farmer groups are essential conditions for information sharing and learning about these agroforestry systems.

The results also show that male-headed households having large livestock herd and small working family size with irrigated land preferred AFS over conventional farming system. In the context of growing labor shortage for farming activities in rural areas, there is a huge scope and potential for farmers to utilize agroforest/woodlot as a viable strategy to address the 'land fallow' issue. While the labor constraint is a favorable condition for AFS promotion, farm size is the major challenge to the wider uptake of this practice.

These results clearly suggest that the agroforestry program should not be considered as a poverty reduction strategy. This is because smallholders may not be able to afford the initial production loss due to a shift from conventional farming to agroforestry. For this, a policy intervention is imperative to involve smallholders in agroforestry promotion. The interventions may include provisioning public land to smallholder farmers under a legal framework and organizing them to initiate collective farming through a cooperative approach both in private as well as public land [75]. However, these interventions are to be supported by some other programs such as extension services and off-farm income-generating activities.

Nepal has recently adopted an agroforestry policy, the impact of which has yet to be realized at farmers' level. The policy might bring changes in the perception and adoption behavior of farmers, which could be the future agenda of research in the field of agroforestry adoption in Nepal.

Author Contributions: A.D., Conceptualization, formal analysis, and original draft preparation; R.K.R., review and editing, and inputs in methodological section. All authors have read and agreed to the published version of the manuscript.

Funding: This research received no external funding.

Acknowledgments: We acknowledge the financial contribution from the Nepal Agroforestry Foundation (NAF) to carry out this research. The staff of Terai Private Forest Development Association (TPFDA) are also acknowledged for their support during data collection. The authors extend thanks to Nab Raj Subedi who helped produce the study area map. We appreciate the reviewers' comments and suggestions.

Conflicts of Interest: There are no conflicts of interest.

References

1. Kotiaho, J.S.; Halme, P. *The IPBES Assessment Report on Land Degradation and Restoration*; IPBES Secretariat, UN Campus: Bonn, Germany, 2018.

2. Bai, Z.G.; Dent, D.L.; Olsson, L.; Schaepman, M.E. Proxy global assessment of land degradation. *Soil Use Manag.* **2008**, *24*, 223–234. [CrossRef]

3. Nachtergaele, F.; Biancalani, R.; Petri, M. *Land Degradation: SOLAW Background Thematic Report 3*; Food and Agriculture Organization of the United Nations: Rome, Italy, 2011.

4. Conacher, A. Land degradation: A global perspective. *N. Z. Geog.* **2009**, *65*, 91–94. [CrossRef]

5. Alexandratos, N.; Bruinsma, J. *World Agriculture towards 2030/2050: The 2012 Revision*; Food and Agriculture Organization of the United Nations: Rome, Italy, 2012.

6. Jaquet, S.; Schwilch, G.; Hartung-Hofmann, F.; Adhikari, A.; Sudmeier-Rieux, K.; Shrestha, G.; Kohler, T. Does outmigration lead to land degradation? Labor shortage and land management in a western Nepal watershed. *Appl. Geogr.* **2015**, *62*, 157–170. [CrossRef]

7. Rasul, G.; Thapa, G.B. Shifting cultivation in the mountains of South and Southeast Asia: Regional patterns and factors influencing the change. *Land Degrad. Dev.* **2003**, *14*, 495–508. [CrossRef]

8. Westarp, S.V.; Sandra Brown, H.S.; Shah, P.B. Agricultural intensification and the impacts on soil fertility in the Middle Mountains of Nepal. *Can. J. Soil Sci.* **2004**, *84*, 323–332. [CrossRef]

9. Jose, S. Agroforestry for ecosystem services and environmental benefits: An overview. *Agrofor. Syst.* **2009**, *76*, 1–10. [CrossRef]

10. Nair, P.K.; Mohan Kumar, B.; Nair, V.D. Agroforestry as a strategy for carbon sequestration. *J. Plant Nutr. Soil Sci.* **2009**, *172*, 10–23. [CrossRef]

11. Dhakal, A.; Cockfield, G.; Maraseni, T.N. Evolution of agroforestry-based farming systems: A study of Dhanusha District, Nepal. *Agrofor. Syst.* **2012**, *86*, 17–33. [CrossRef]

12. Harvey, C.A.; Gonzalez, J.; Somarriba, E. Dung beetle and terrestrial mammal diversity in forests, indigenous agroforestry systems and plantain monocultures in Talamanca, Costa Rica. *Biodivers. Conserv.* **2006**, *15*, 555–585. [CrossRef]

13. Kabir, M.E.; Webb, E.L. Can home gardens conserve biodiversity in Bangladesh? *Biotropica* **2008**, *40*, 95–103. [CrossRef]

14. Moguel, P.; Toledo, V.M. Biodiversity conservation in traditional coffee systems of Mexico. *Conserv. Biol.* **1999**, *13*, 11–21. [CrossRef]

15. Neupane, R.P.; Thapa, G.B. Impact of agroforestry intervention on soil fertility and farm income under the subsistence farming system of the middle hills, Nepal. *Agric. Ecosyst. Environ.* **2001**, *84*, 157–167. [CrossRef]

16. Dhakal, A. Evolution, Adoption and Economic Evaluation of an Agroforestry-based Farming System with and without Carbon Values: The Case of Nepal. Ph.D. Thesis, University of Southern Queensland, Toowoomba, Australia, 2013.

17. Schwab, N.; Schickhoff, U.; Fischer, E. Transition to agroforestry significantly improves soil quality: A case study in the central mid-hills of Nepal. *Agric. Ecosyst. Environ.* **2015**, *205*, 57–69. [CrossRef]

18. Acharya, A.K.; Kafle, N. Land degradation issues in Nepal and its management through agroforestry. *J. Agric. Environ.* **2009**, *10*, 133–143. [CrossRef]

19. Cockfield, G.J. Evaluating a Markets-based Incentive Scheme for Farm Forestry: A Case Study. Ph.D. Thesis, University of Queensland, Brisbane, Australia, 2005.

20. Mercer, D.E. Adoption of agroforestry innovations in the tropics: A review. *Agrofor. Syst.* **2004**, *61*, 311–328. [CrossRef]

21. Toth, G.G.; Nair, P.R.; Duffy, C.P.; Franzel, S.C. Constraints to the adoption of fodder tree technology in Malawi. *Sustain. Sci.* **2017**, *12*, 641–656. [CrossRef]

22. Mfitumukiza, D.; Barasa, B.; Ingrid, A. Determinants of agroforestry adoption as an adaptation means to drought among smallholder farmers in Nakasongola District, Central Uganda. *Afr. J. Agric. Res.* **2017**, *12*, 2024–2035. [CrossRef]

23. Sanou, L.; Savadogo, P.; Ezebilo, E.E.; Thiombiano, A. Drivers of farmers' decisions to adopt agroforestry: Evidence from the Sudanian savanna zone, Burkina Faso. *Renew. Agric. Food Syst.* **2019**, *34*, 116–133. [CrossRef]

24. Beyene, A.D.; Mekonnen, A.; Randall, B.; Deribe, R. Household level determinants of agroforestry practices adoption in rural Ethiopia. *For. Trees Livelihoods* **2019**, *28*, 194–213. [CrossRef]

25. Oli, B.N.; Treue, T.; Larsen, H.O. Socio-economic determinants of growing trees on farms in the middle hills of Nepal. *Agrofor. Syst.* **2015**, *89*, 765–777. [CrossRef]

26. Catacutan, D.; Naz, F. Gender roles, decision-making and challenges to agroforestry adoption in Northwest Vietnam. *Int. For. Rev.* **2015**, *17*, 22–32. [CrossRef]

27. Neupane, R.P.; Sharma, K.R.; Thapa, G.B. Adoption of agroforestry in the hills of Nepal: A logistic regression analysis. *Agric. Syst.* **2002**, *72*, 177–196. [CrossRef]

28. Dhakal, A.; Cockfield, G.; Maraseni, T.N. Deriving an index of adoption rate and assessing factors affecting adoption of an agroforestry-based farming system in Dhanusha District, Nepal. *Agrofor. Syst.* **2015**, *89*, 645–661. [CrossRef]

29. Central Bureau of Statistics. *Environment Statistics of Nepal*; Central Bureau of Statistics: Kathmandu, Nepal, 2012.

30. Dhakal, A. *Silviculture and Productivity of Five Economically Important Timber Species of Central Terai of Nepal*; Nepal Agroforestry Foundation (NAF): Kathmandu, Nepal, 2008.

31. Nepal Agroforestry Foundation (NAF). *Terai Project Report*; Nepal Agroforestry Foundation: Kathmandu, Nepal, 2018.

32. Borremans, L.; Reubens, B.; Van Gils, B.; Baeyens, D.; Vandevelde, C.; Wauters, E. A sociopsychological analysis of agroforestry adoption in Flanders: Understanding the discrepancy between conceptual opportunities and actual implementation. *Agroecol. Sustain. Food Syst.* **2016**, *40*, 1008–1036. [CrossRef]

33. Idumah, F.O.; Owombo, P.T. Determinants of yam production and resource use efficiency under agroforestry system in Edo State, Nigeria. *Tan. J. Agric. Sci.* **2019**, *18*, 35–42.

34. Tadele, M.; Birhane, E.; Kidu, G.; G-Wahid, H.; Rannestad, M.M. Contribution of parkland agroforestry in meeting fuel wood demand in the dry lands of Tigray, Ethiopia. *J. Sustain. For.* **2020**, *39*, 1–13. [CrossRef]

35. Sarker, M.A.R.; Alam, K.; Gow, J. Assessing the determinants of rice farmers' adaptation strategies to climate change in Bangladesh. *Inter. J. Clim. Chang. Strat. Mgmt.* **2013**, *5*, 382–403. [CrossRef]

36. Miheretu, B.A.; Yimer, A.A. Determinants of farmers' adoption of land management practices in Gelana sub-watershed of Northern highlands of Ethiopia. *Ecol. Process.* **2017**, *6*, 19–30. [CrossRef]

37. Kropko, J. Choosing between Multinomial Logit and Multinomial Probit Models for Analysis of Unordered Choice Data. Master's Thesis, University of North Carolina, Chapel Hill, NC, USA, 2008. [CrossRef]

38. Lin, Y.; Deng, X.; Li, X.; Ma, E. Comparison of multinomial logistic regression and logistic regression: Which is more efficient in allocating land use? *Front. Earth Sci.* **2014**, *8*, 512–523. [CrossRef]

39. Luus, F.P.; Salmon, B.P.; Van den Bergh, F.; Maharaj, B.T.J. Multiview deep learning for land-use classification. *IEEE Geosci. Remote Sens. Lett.* **2015**, *12*, 2448–2452. [CrossRef]

40. Paton, L.; Troffaes, M.C.; Boatman, N.; Hussein, M.; Hart, A. Multinomial logistic regression on Markov chains for crop rotation modelling. In *Information Processing and Management of Uncertainty in Knowledge-Based Systems*; Springer International Publishing: Cham, Switzerland, 2014. [CrossRef]

41. Greene, W.H. *Econometric Analysis*, 5th Ed. ed; Pearson Education: New Delhi, India, 2003; pp. 67–185.

42. Cheng, S.; Long, J.S. Testing for IIA in the multinomial logit model. *Sociol. Methods Res.* **2007**, *35*, 583–600. [CrossRef]

43. Alavalapati, J.R.R.; Luckert, M.K.; Gill, D.S. Adoption of agroforestry practices: A case study from Andhra Pradesh, India. *Agrofor. Syst.* **1995**, *32*, 1–14. [CrossRef]

44. Anley, Y.; Bogale, A.; Haile-Gabriel, A. Adoption decision and use intensity of soil and water conservation measures by smallholder subsistence farmers in Dedo district, Western Ethiopia. *Land Degrad. Dev* **2007**, *18*, 289–302. [CrossRef]

45. Carlson, J.E.; Schnabel, B.; Beus, C.E.; Dillman, D.A. Changes in the soil conservation attitudes and behaviors of farmers in the Palouse and Camas prairies: 1976–1990. *J. Soil Water Conserv.* **1994**, *49*, 493–500.

46. Thangata, P.H.; Alavalapati, J.R. Agroforestry adoption in southern Malawi: The case of mixed intercropping of *Gliricidia sepium* and maize. *Agric. Syst.* **2003**, *78*, 57–71. [CrossRef]

47. Roberts, J.S.; Laughlin, J.E.; Wedell, D.H. Validity issues in the Likert and Thurstone approaches to attitude measurement. *Educ. Psychol. Meas.* **1999**, *59*, 211–233. [CrossRef]

48. Food and Agriculture Organization. *The Economic Lives of Smallholder Farmers: An Analysis Based on Household Data from Nine Countries*; FAO: Rome, Italy, 2015.

49. Khanal, U. Why are farmers keeping cultivatable lands fallow even though there is food scarcity in Nepal? *Food Sec.* **2018**, *10*, 603–614. [CrossRef]

50. Adesina, A.A.; Mbila, D.; Nkamleu, G.B.; Endamana, D. Econometric analysis of the determinants of adoption of alley farming by farmers in the forest zone of southwest Cameroon. *Agric. Ecosyst. Environ.* **2000**, *80*, 255–265. [CrossRef]

51. Adesina, A.A.; Chianu, J. Determinants of farmers' adoption and adaptation of alley farming technology in Nigeria. *Agrofor. Syst.* **2002**, *55*, 99–112. [CrossRef]

52. Fabiyi, Y.L.; Idowu, E.O.; Oguntade, A.E. Land tenure and management constraints to the adoption of alley farming by women in Oyo State of Nigeria. *Niger. J. Agric. Ext.* **1991**, *6*, 40–46.

53. Djalilov, B.M.; Khamzina, A.; Hornidge, A.K.; Lamers, J.P. Exploring constraints and incentives for the adoption of agroforestry practices on degraded cropland in Uzbekistan. *J. Environ. Plan. Manag.* **2016**, *59*, 142–162. [CrossRef]

54. Nkomoki, W.; Bavorová, M.; Banout, J. Adoption of sustainable agricultural practices and food security threats: Effects of land tenure in Zambia. *Land Use Policy* **2018**, *78*, 532–538. [CrossRef]

55. Meijer, S.S.; Catacutan, D.; Ajayi, O.C.; Sileshi, G.W.; Nieuwenhuis, M. The role of knowledge, attitudes and perceptions in the uptake of agricultural and agroforestry innovations among smallholder farmers in sub-Saharan Africa. *Int. J. Agr. Sustain.* **2015**, *13*, 40–54. [CrossRef]

56. Ahmed, A.U.; Hernandez, R.; Naher, F. Adoption of Stress-Tolerant Rice Varieties in Bangladesh. In *Technological and Institutional Innovations for Marginalized Smallholders in Agricultural Development*; Gatzweiler, F.W., vun Brawn, J., Eds.; Springer: Cham, Switzerland; New York, NY, USA, 2016; pp. 241–255.

57. Mwase, W.; Sefasi, A.; Njoloma, J.; Nyoka, B.I.; Manduwa, D.; Nyaika, J. Factors affecting adoption of agroforestry and evergreen agriculture in Southern Africa. *Environ. Nat. Resour. Res.* **2015**, *5*, 148–157. [CrossRef]

58. Coulibaly, J.Y.; Chiputwa, B.; Nakelse, T.; Kundhlande, G. Adoption of agroforestry and the impact on household food security among farmers in Malawi. *Agric. Syst.* **2017**, *155*, 52–69. [CrossRef]

59. Kassie, G.W. Agroforestry and farm income diversification: Synergy or trade-off? The case of Ethiopia. *Environ. Syst. Res.* **2017**, *6*, 8–21. [CrossRef]

60. Mabuza, M.L.; Sithole, M.M.; Wale, E.; Ortmann, G.F.; Darroch, M.A.G. Factors influencing the use of alternative land cultivation technologies in Swaziland: Implications for smallholder farming on customary Swazi Nation Land. *Land Use Policy* **2013**, *33*, 71–80. [CrossRef]

61. Sabastian, G.; Kanowski, P.; Race, D.; Williams, E.; Roshetko, J.M. Household and farm attributes affecting adoption of smallholder timber management practices by tree growers in Gunungkidul region, Indonesia. *Agrofor. Syst.* **2014**, *88*, 257–268. [CrossRef]

62. Sood, K.K.; Mitchell, C.P. Identifying important biophysical and social determinants of on-farm tree growing in subsistence-based traditional agroforestry systems. *Agrofor. Syst.* **2009**, *75*, 175–187. [CrossRef]

63. Pingale, B.; Bana, O.P.S.; Banga, A.; Chaturvedi, S.; Kaushal, R.; Tewari, S.; Neema, S. Accounting biomass and carbon dynamics in *Populus deltoides* plantation under varying density in terai of central Himalaya. *J. Tree Sci.* **2014**, *33*, 1–6.

64. Pathak, D. Water availability and hydrogeological condition in the Siwalik foothill of east Nepal. *Nepal J. Sci. Technol.* **2016**, *17*, 31–38. [CrossRef]

65. Rai, R.K.; Bhatta, L.D.; Acharya, U.; Bhatta, A.P. Assessing climate-resilient agriculture for smallholders. *Environ. Dev.* **2018**, *27*, 26–33. [CrossRef]

66. Andres, C.; Comoé, H.; Beerli, A.; Schneider, M.; Rist, S.; Jacobi, J. Cocoa in Monoculture and Dynamic Agroforestry. In *Sustainable Agriculture Reviews*; Lichtfouse, E., Ed.; Springer: Cham, Switzerland; New York, NY, USA, 2016; Volume 19, pp. 121–153. [CrossRef]

67. Cedamon, E.; Nuberg, I.; Pandit, B.H.; Shrestha, K.K. Adaptation factors and futures of agroforestry systems in Nepal. *Agrofor. Syst.* **2018**, *92*, 1437–1453. [CrossRef]

68. Islam, M.A.; Sofi, P.A.; Bhat, G.M.; Wani, A.A.; Gatoo, A.A.; Singh, A.; Malik, A.R. Prediction of agroforestry adoption among farming communities of Kashmir valley, India: A logistic regression approach. *J. Nat. Appl. Sci.* **2016**, *8*, 2133–2140. [CrossRef]

69. Simelton, E.S.; Catacutan, D.C.; Dao, T.C.; Dam, B.V.; Le, T.D. Factors constraining and enabling agroforestry adoption in Viet Nam: A multi-level policy analysis. *Agrofor. Syst.* **2017**, *91*, 51–67. [CrossRef]

70. Basamba, T.A.; Mayanja, C.; Kiiza, B.; Nakileza, B.; Matsiko, F.; Nyende, P.; Ssekabira, K. Enhancing Adoption of Agroforestry in the Eastern Agro Ecological Zone of Uganda. *Int. J. Ecol. Sci. Environ. Eng.* **2016**, *3*, 20–31.

71. Etshekape, P.G.; Atangana, A.R.; Khasa, D.P. Tree planting in urban and peri-urban of Kinshasa: Survey of factors facilitating agroforestry adoption. *Urban For. Urban Green.* **2018**, *30*, 12–23. [CrossRef]

72. Haider, H.; Smale, M.; Theriault, V. Intensification and intrahousehold decisions: Fertilizer adoption in Burkina Faso. *World Dev.* **2018**, *105*, 310–320. [CrossRef]

73. Rai, R.K.; Dhakal, A.; Khadayat, M.S.; Ranabhat, S. Is collaborative forest management in Nepal able to provide benefits to distantly located users? *For. Policy Econ.* **2017**, *83*, 156–161. [CrossRef]

74. Pradhan, N.S.; Sijapati, S.; Bajracharya, S.R. Farmers' responses to climate change impact on water availability: Insights from the Indrawati Basin in Nepal. *Int. J. Water Resour. Dev.* **2015**, *31*, 269–283. [CrossRef]

75. Bhattarai, S.; Pant, B.; Laudari, H.K.; Timalsina, N.; Rai, R.K. Restoring Landscapes through Trees Outside Forests: A Case from Nepal's Terai Region. *Int. For. Rev.* **2020**, *22*, 33–48. [CrossRef]

Article

Assessing Sustainable Bamboo-Based Income Generation Using a Value Chain Approach: Case Study of Nongboua Village in Lao PDR

Bohwi Lee [1], Hakjun Rhee [1,*], Sebin Kim [1], Joon-Woo Lee [1], Seungmo Koo [2], Sang-Jin Lee [1], Phayvanh Alounsavath [3] and Yeon-Su Kim [4]

1   Department of Environment and Forest Resources, Chungnam National University, Daejeon 34134, Korea; bohwi00@gmail.com (B.L.); sbkim@cnu.ac.kr (S.K.); jwlee@cnu.ac.kr (J.-W.L.); sangjin78@gmail.com (S.-J.L.)
2   Department of Agricultural Economics, Chungnam National University, Daejeon 34134, Korea; koosm@cnu.ac.kr
3   Department of Forestry, Ministry of Agriculture and Forestry, Vientiane Capital 0100, Laos; vanhlasv@gmail.com
4   School of Forestry, Northern Arizona University, Flagstaff, AZ 86011, USA; ysk@nau.edu
*   Correspondence: hakjun.rhee@gmail.com; Tel.: +82-10-4899-8751

Abstract: Many bamboo species are well suited for agroforestry as they are more versatile and rapidly renewable than trees. Bamboo is an important income source for rural villagers around the world, especially in tropical developing countries, such as Lao PDR (Lao People's Democratic Republic). This study applied a value chain approach to compare potential incomes from different bamboo utilization models: (1) existing model of selling semi-processed raw materials (bamboo splits), and (2) new model of producing handcraft products locally. Using a rural village in eastern Lao PDR (Nongboua village in Vientiane Capital province) as a case study, we provided empirical assessments of two bamboo value chains. Based on interviews with the villagers and stakeholders and government statistical data from 2017 to 2019, existing and new bamboo production chains were evaluated. In the existing value chain, the final products, bamboo chopsticks, are worth \$6.74/kg. The value chain starts with bamboo harvesting, collection, and management, which are done by villagers in Lao PDR and taxed by the Lao PDR government. Bamboo splits are then transported to Vietnam to make the final products to sell. Local villagers received only 4.9% of the total value. The new bamboo handicraft model could produce 9 bamboo cups and 60 medals from one bamboo stem worth \$52.6–61.7 and \$343.8. In this value chain, bamboo harvesting, management, and processing to final products are done by villagers. The handcrafts were collected by traders to be sold at souvenir shops. Local villagers could capture 29.4%–42.3% of the total values. Producing bamboo cup and medal could generate 1.12–2.17 and 234.8–244.6 times higher income for villagers per labor hour and per bamboo stem, respectively, and allow them to use more bamboo resource than producing bamboo splits to export to Vietnam. If applied to other rural areas in Lao PDR, the new bamboo product model for handicrafts can be a better income source for local villagers in Lao PDR with sustainable use of bamboo resources than the existing model. However, it requires extensive bamboo handicrafts training over a year. Although alternative uses of bamboo would be different depending on social, economic, and market contexts, the value chain analysis demonstrated in this study can be applied elsewhere to increase local retention of economic values generated from agroforestry.

Keywords: agroforestry; bamboo handicraft; bamboo; income generation; Laos; sustainability; value chain approach

Citation: Lee, B.; Rhee, H.; Kim, S.; Lee, J.-W.; Koo, S.; Lee, S.-J.; Alounsavath, P.; Kim, Y.-S. Assessing Sustainable Bamboo-Based Income Generation Using a Value Chain Approach: Case Study of Nongboua Village in Lao PDR. *Forests* **2021**, *12*, 153. https://doi.org/10.3390/f12020153

Academic Editors: Mi Sun Park and Himlal Baral

Received: 31 December 2020
Accepted: 23 January 2021
Published: 28 January 2021

Publisher's Note: MDPI stays neutral with regard to jurisdictional claims in published maps and institutional affiliations.

1. Introduction

Agroforestry is a collection of land-use systems and practices where woody plants are integrated with other crops and animals [1]. Agroforestry consists of agrisilviculture, silvopasture, forest farming, and other practices, of which the forest farming plays a key role in offering timber and non-timber forest products (NTFPs) in tropical and subtropical region [2]. Recently, bamboo has gained increasing recognitions as a substitute for timber and a viable resource for rural poverty alleviation, as well as a high potential for carbon sequestration and climate change adaption [3–5]. Bamboo-based agroforestry can be used to improve land productivity, sustainability, and resource conservation, and is well suited for agroforestry [6–8].

Bamboo is regarded as one of the fastest-growing plants on earth [9]. It belongs to Gramineae family and Bambusoideae subfamily that has approximately 121 genera and 1662 species of bamboo, of which 232 (14%) have been found worldwide beyond their native ranges [10]. Bamboo, which has played an important role in rural economies for centuries, became an indispensable part of emerging economies around the world as one of the most traded NTFPs especially in tropical regions [11–14].

Sustainably managing natural resources in tropical developing countries must be balanced with addressing livelihood needs of local communities that would ensure community participation [15–17]. For bamboo to fulfill its potential for conservation and development, we need to better understand the current uses and value chains and develop and introduce new income models that allows local villagers to actively participate and improve their livelihoods while managing their bamboo resource sustainably.

A value chain analysis is a widely used method to assess business activities and recognize which activities are the most valuable and which ones could be improved to create competitive advantage for a business [18]. The value chain consists of a set of activities that a specific industry firm carries out to produce and deliver a product of value, which is either or both good and service, to the market [18]. It helps identify weak points in the chain and actions to add more value. The value chain approach is a method for "decomposing an enterprise into strategically important activities" [19,20], and the overall logic of a value chain can be applied to any industry [18,20]. We applied a value chain approach to compare the existing and potential bamboo product models using Vientiane Capital province in Lao PDR as a case study.

Bamboo is one of the most important natural resources in Lao PDR. Local governments and organizations consider bamboo as "green gold" to alleviate poverty, and promote production, processing, and trading of bamboo by the smallholders [21]. For many rural communities in Lao PDR, bamboo is an abundant and accessible resource. It is commonly used as a house building material, foods such as bamboo shoot, and a raw material for handicraft to make furniture, baskets, chopsticks, and incense sticks. With improved transportation infrastructure in northern part of Lao PDR over the past decades, increasing demand from neighboring countries, such as Vietnam, China, and Thailand, is adding pressure on bamboo resources of Lao PDR [22]. Recently, semi-processed products, such as dried bamboo splits, are exported to Vietnam and Thailand in large quantities. While producing bamboo splits may increase income of local villagers and improve their livelihoods, it may encourage overharvest. Bamboo splits are often sold directly to Vietnamese companies at a low price, and much of value adding economic activities occur outside Lao PDR without contributing much to local economy. When bamboo handicrafts are produced and sold to tourists, it can be an income source that generates multiple economic activities for local villagers [23]. However, little is known about different bamboo-based income models and their local economic impacts.

The objective of this study is to assess bamboo-based income generations for local villagers in Vientiane Capital province in Lao PDR from the existing and new bamboo product models using a value chain approach. We developed value chains for the existing bamboo product, bamboo chopsticks in Vietnam, and the new bamboo products, bamboo handicrafts such as bamboo cups and medals. The two value chains were compared to assess the impacts of the two models on local villagers' livelihoods and sustainable use of bamboo resources. Understanding the value chains for the existing and new bamboo products provides useful information for local communities and policy makers in Lao PDR, as well as future studies in other developing countries.

2. Materials and Methods

2.1. Study Area

The Sangthong district is one of the poorest districts in Vientiane Capital province. Total area of the district is 75,980 ha, of which forest area accounts for 55%, agriculture land 33%, and wetland and other land use 10% of the land area. Existing forestlands are highly fragmented and degraded [24].

Four main soil groups (FAO/UNESCO Classification) in Sangthong district are: *Alisols, Acrisols, Luvisols,* and *Regosols.* The Acrisols and Alisols are dominant soils, covering about 80% of the total area. The climate is monsoonal, with a rainy season from May to October and a dry season during the rest of the year. The average annual rainfall is 1558 mm, of which over 90% occurs between May and October. The average mean temperature of the study area is 31.9 °C and maximum mean temperature of 35.9 °C has been recorded in April. The study area receives the average daily sunlight of 6.5 h [24].

The Sangthong district consists of 35 villages of which the Nongboua village was selected for this study. Nongboua village in Sangthong district, about 60 km to the north of Vientiane (Figure 1), represents a low-income rural village in Lao PDR with rich bamboo resources. The Nongboua village consisted of 83 households and had a total population of 377 inhabitants, of which 177 inhabitants were women. The 80% of the total population was engaged in agriculture, and relied on paddy rice cultivation, shifting cultivation, gardening, livestock raising, and collection of non-timber forest products.

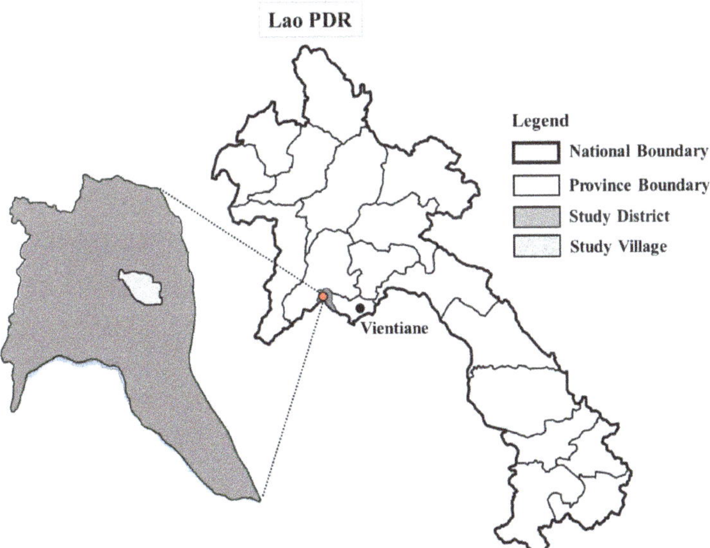

Figure 1. Location of the project site in Sangthong district.

Land area of the Nongboua is 1135.92 ha, of which forest area is 504.40 ha and agricultural land is 599.52 ha (Table 1). These land use data were collected at the village and district level by the Department of Forestry in Lao PDR (hereafter called DOF). There were three major land use types: forest, agriculture, and settlement. The details of each land use are described below.

Table 1. Land use of Nongboua village.

Serial Number	Land Use Types	Area (ha)
I	**Forest area**	**504.40**
1.1	Village conservation forest	80.00
1.2	Village protection forest	345.00
1.3	Village Utility forest	43.00
1.4	Sacred forest	18.00
1.5	Tree plantation	18.40
II	**Agricultural land**	**599.52**
2.1	Rice paddy field	378.00
2.2	Shifting cultivation	24.30
2.3	Garden and other uses	16.50
2.4	Rangeland	180.72
III	**Settlement area**	**32.00**
IV	**Other land use**	
	Total	**1135.92**

(Source: Nongboua, Namiang, and Tauhai villages, investigated by the local expert in Lao PDR in 2015).

The Sangthong district has one of the richest bamboo resources in Vientiane Capital province. The bamboo area of Sangthong district accounted for 1.3% (20,565 ha; 7,094,803 stems) of the total bamboo area of 1,612,000 ha in Lao PDR [11,12,25]. The Nongboua village was estimated to account for 2.6% (530 ha; 182,367 stems) of the bamboo resources in Sangthong district.

In Nongboua village there are a total of nine bamboo species: Mai phai, Mai Phang (small), Mai Phang (big), Mai Sang Phai, Mai Khao Larm, Mai Xord, Mai Bong, Mai Hia, and Mai Ria (Table 2). Usually, villagers harvest bamboo stems of Mai Phang and Mai Sang Phai, which have a wider space between internodes than the others and are therefore preferred by bamboo companies for efficient product processing.

Table 2. List of Bamboo species in Sangthong district, Lao PDR.

No	Lao Name	Scientific Name
1	Mai Phai	*Bambusablumeana Schultes*
2	Mai Phang (small)	*Dendrocalamus longifimbriatus Gamble*
3	Mai Phang (big)	*Schizostachyumgrande Kurz cephalostachyum sp*
4	Mai Sang Phai	*Dendrocalamusbrandissi (Munro) Kurk*
5	Mai Khao Larm	*Cephalostachuympergracile Munro*
6	Mai Xord	*Vietnamosasaciliata (A, Camus) Nguyen*
7	Mai Bong	*Bambusatulda Roxb.*
8	Mai Hia	*Schizostachyumblumei (C, Virgatum Munro & Kurz)*
9	Mai Rai	*Cephalostachyum sp*

2.2. Methods

To identify the underlying structure of value chains associated with bamboo, first we mapped existing and new bamboo value chains in Nongboua village to assess various barriers, weak points, solutions, and the position of each actor in the value chain. We compared two value chains to analyze a way for local villagers to improve added value using their bamboo resources. We collected data through survey of selected villagers, field observation of the research team, as well as stakeholder interview and Focus Group Interview (F.G.I.).

2.2.1. Data Collection of Existing Bamboo Products Value Chain

Survey Questionnaires

To find out the status of bamboo-based income, the survey was conducted on 50 households out of 83 households in Nongboua village for a week in mid-October of 2019. The questions in the survey were as follows:

- Whether there was bamboo-based income or not
- Bamboo harvesting or not
- Annual working period for bamboo harvesting
- Average monthly and daily working hours for bamboo harvesting
- Labor type, self or hired
- Daily bamboo sales (kg) and income (kip)
- Total income related to bamboo
- Other bamboo income such as handicraft

Field Observation

In mid-October 2019, a total of six villagers participated in demonstrating bamboo harvesting and trading in Nongboua village. We observed and measured how much local villagers harvested and sold bamboos from forest to a Vietnamese company.

Stakeholder Interview

We conducted stakeholder interviews to examine bamboo value-chain for Nongboua village and Sangthong district in 2019. The interviews were conducted on key stakeholders of the bamboo value chain, including two local villagers from Nongboua village, two national and two local government officials in Lao PDR, and one staff of a Vietnamese bamboo company from Sangthong district. To each stakeholder group, the following questions were asked.

- Local villagers: Status and problems of the existing bamboo production system.
- Government officials of Sangthong District Agriculture and Forestry Office (DAFO) and DOF: Bamboo import and export data, annual bamboo harvesting quota, and regional bamboo tax and royalty of Sangthong district in 2017–2019. Especially, in Lao PDR, commercial uses of NTFPs are regulated with a quota system to protect bamboo forests [26]. Harvesting bamboos in Sangthong district also followed the quota system. Each company needs to obtain their business and quota licenses from the Provincial Agriculture and Forestry Office (PAFO) and Department of Industry and Commerce (DIC) annually. Different government agencies require appropriate licenses for harvesting, collecting, and transporting bamboo resource [27]. The quota was issued at a provincial level [28].

Once provincial annual quota was approved by the Ministry of Agriculture and Forestry (MAF), the PAFO held a meeting with the Provincial Trade Office and District Governor Office where the provincial quota was allocated to the DAFO. These quotas were issued for specific zones within each district and were further divided to traders contracted with the DAFO. These contractors were not allowed to collect NFTPs directly from forests, but to buy them from villagers. The actual harvesting of bamboo stems mostly took place during the dry season (November to April). Lack of information on bamboo resources made it difficult to calculate the reasonable quota [27,28]. Annual bamboo quota and actual harvest from 2017 to 2019 were reported in the Sangthong DAFO (The Sangthong DAFO had not collected official statistics of bamboo quotas prior to 2017). The quota in 2018 was 2,100,000 stems (16,800 ton), but the actual harvest increased to 2,625,000 stems (21,004 ton), harvesting 525,500 stems (4204 ton) more than the quota. In addition, the quota in 2019 was twice of the quota in the previous year.

The quota was allocated by district level, so each village's quota depended on demand from the Vietnamese companies rather than reliable estimates of available bamboo resource.

It is difficult to provide unit basis quota in Nongboua village, thus raising a question about sustainability of bamboo forests in the area.

Vietnamese bamboo companies that sought to buy NTFPs from villagers had to contract the village chief. Each company paid $0.02/stem (200 kip/stem or 25 kip/green kg) as a contribution to village development fund at village level, in accordance with the quota allocated by the Lao PDR government. When bamboo companies transported dried bamboo splits from the collection sites, they had to pay an export tax of $0.0044/kg (40 kip/kg) at district level. In addition, the companies needed to pay one-time $5.6 (50,000 kip) and $2.2/year (20,000 kip/year) for registration and annual fees to the PAFO.

- Staff of Vietnamese bamboo company: In the existing bamboo value chain, the Vietnamese company is an important actor. The company headquarter and factory were located in Vietnam. The company sent and hired workers to collect and manage bamboos from the Sangthong district. It was difficult to access detailed data and visit the bamboo-collecting areas where the harvested bamboos were transported and sold to the company. Therefore, we relied on interviews with Laotian staff who were working for the company and collected the information on the existing bamboo value chain, such as average cost and amount of bamboo materials to produce final bamboo products (e.g., chopsticks and incense stick).

2.2.2. Data Collection of New Bamboo Products Value Chain

We surveyed bamboo handicraft items in the neighboring villages and markets in 2017. Most of bamboo handicrafts were weaved items such as rice basket, bag, mat, chicken cage, fishing trap, and furniture. For the new model we decided to make small bamboo products for foreign tourists (i.e., cup, medal, and hairpin) that required relatively low handicraft skills and could be sold to a trader.

The bamboo quota and tax by imposed by the DOF involves only the existing bamboo chain, e.g., collecting bamboo splits for Vietnamese company. The existing quota and tax do not apply to the new value chain of bamboo handicrafts in this study.

Survey Questionnaires

We used the same field survey that was conducted on 50 households in Nongboua village for the existing bamboo product model.

Field Observation

In 2018, a total of 12 people volunteered for a new bamboo-based income model. We recruited seven male participants in their early-40s to 60s and five female participants in their mid-30s to 40s through the DOF. As of 2019, only six villagers were still participating in the new income model (one male early-40s and five females). A total of five trainings were conducted: two times in 2018, and three times in 2019. Each training averaged about 10 days. Their motives for the participation were to contribute to family livelihoods and to learn advanced bamboo handicraft skills. We observed the participants' engagements during the trainings and changes in their motivation, skill proficiency, and income generation.

Focus Group Interview (F.G.I.)

The F.G.I. is a qualitative data collection method for small groups [29]. A focus group in this study consisted of the volunteered participants. The participants were interviewed in 2018 and 2019 about income generation and contribution to their livelihoods before and after the participation of the new bamboo income model.

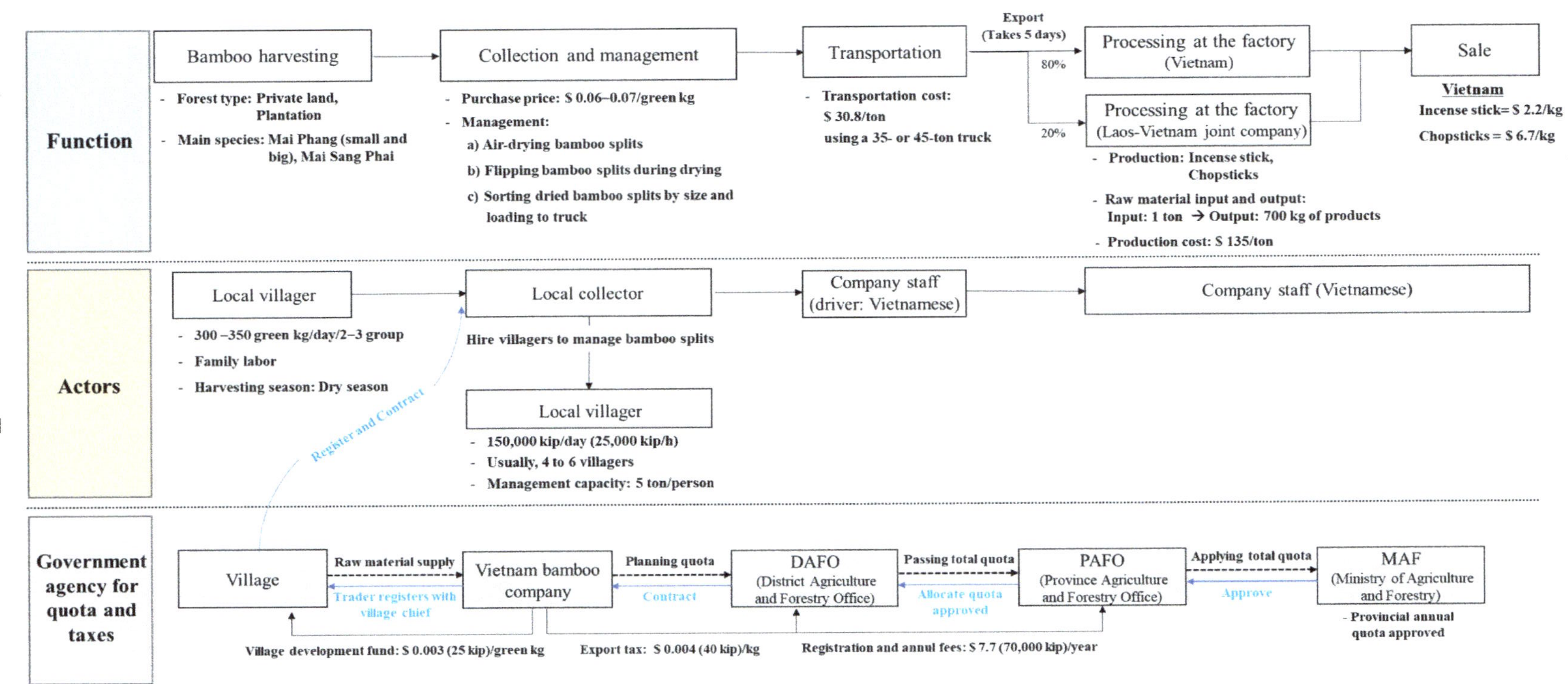

Figure 2. Overview of the existing bamboo products value chain in Nongboua village.

3. Results

3.1. Exisiting Bamboo Products Value Chain in Nongboua Village

The value chain of the existing bamboo products produced mostly bamboo chopsticks and incense stick in Vietnamese markets, which required raw materials (i.e., dried bamboo splits) from local villages in Lao PDR. This value chain had the functions in the following order: bamboo harvesting, collection and management of bamboo splits, transportation, processing at factories, and sales (Figure 2).

3.1.1. Bamboo Harvesting

During the dry season, local villagers harvested bamboo stems from bamboo forest to supply raw materials for chopsticks and incense stick in Vietnam. The procedure for the villagers to harvest bamboo stems was as follows:

- Vietnamese bamboo companies registered at the MAF in Lao PDR to acquire raw bamboo materials from Sangthong district.
- The companies received annual quotas and villages allocated by the MAF, and established collection sites in town.
- The companies usually either hire local villagers to harvest bamboos on a daily basis or purchase harvested split bamboos from local villagers. At the time of our interview, the company in the Nongboua village purchased the bamboos from the villagers.
- Nongboua villagers harvested bamboos at plantation sites near the village. Following the MAF regulation, only bamboos 3 years or older were harvested [26].
- After cutting down bamboo stands, the villagers split the bamboos into seven sizes: 65 cm, 75 cm, 82 cm, 92 cm, 102 cm, 112 cm, and 122 cm (Figure 3a,b).
- Bamboo splits were gathered by the same size and tied up. Finally, bamboo splits were carried by a tractor and sold to the company at a collection site (Figure 3c).

Figure 3. Bamboo harvesting and semi-processing steps: (**a**) villagers harvested whole bamboos in forest, (**b**) villagers cut and split bamboo stems (internodes) into smaller pieces on-site, and (**c**) villagers sold bamboo splits by weight to a company at the collecting area in town.

The interviews and field observations found that more than 60% of Nongboua villagers earned cash income from harvesting and selling bamboos during the dry season (about 5 months). Bamboo harvesting was a highly labor-intensive work and required a minimum of three people to work efficiently. A three-person family could produce 300–350 green kg bamboo splits at a time, which would take an average of 2 h. The interviewed villagers worked 5 times a month to harvest bamboos. The labor and income information by a three-person family is shown in Table 3.

Table 3. Labor and income information on bamboo harvesting for a three-person village family.

Average Annual Labor Period (Month)	Average Month Number of Labor (Time)	Average Sales Volume of Bamboo (Green kg)	Bamboo Split Purchase Price ($/Green kg)	Average Labor Time for a Day (h)	Cash Income by One Family		
					Daily ($)	Monthly ($)	Annual ($)
5	5	300–350	0.06–0.07	2	18–24.5	90–122.5	450–612.5

3.1.2. Collection and Management

The villagers in Sangthong district sold bamboo splits to a Vietnamese company at bamboo collection sites that were distributed to villages nearby, thus it would be easy to sell the bamboo splits. According to a local official in the Sangthong DAFO, there were estimated to be a total number of 20 collection sites in 35 villages of Sangthong district.

The collected bamboo splits were sorted by size and dried until a truck arrived. The villagers mainly worked as follows: (1) positioning and air-drying bamboo splits for at least 3–4 weeks (Figure 4a), (2) flipping top to bottom of bamboo splits during the drying, (3) classifying bamboo splits by the seven sizes, and (4) storing and loading dried bamboo splits to a truck. The company hired local villagers in two ways: On daily basis, $16.9/day (150,000 kip/day) for which a worker could manage 167 kg of bamboo splits (5 tons/month), and on a regular basis, $0.9/ton (8000 kip/ton) for which a worker stayed in a collection site to dry bamboo splits.

Figure 4. Bamboo collection and management: (**a**) air drying bamboo splits and (**b**) loading bamboo splits into a truck owned by a Vietnamese company.

3.1.3. Transportation

Vietnamese bamboo companies transported bamboo splits completely dried to processing factories in Vietnam and near Lao PDR-Vietnam border. Most of the dried bamboo splits (about 80%) were transported to factories in Vietnam and some (20%) were transported to factories near the border in Lao PDR. Over the past several years, road networks in northern and central Lao PDR have been improved (e.g., road No. 1 and 13), and enabled to trade NTFPs with neighboring countries (e.g., Vietnam and China).

Interviews with a local DAFO official and a Vietnamese company staff identified that the main transportation route was Sangthong district → Vientiane capital → No.13 Road south-bound → No. 8 Road east-bound → Nam Phao international checkpoint in Khamkerd district → Vinh in Vietnam. Total traveling distance was about 500 km and it took 5 days to travel. An average transportation cost from the Sangthong district to the factories in Vietnam was about $30.8/ton (265,000 kip/ton), including gas and depreciation. Typically, a company transportation truck (35 or 45 ton) visited a collection site at least every two months, depending on the company's demand of raw materials (Figure 4b).

3.1.4. Processing and Production

Dried bamboo splits were unloaded and processed at bamboo factories in Vietnam or near the border in Lao PDR. The bamboo products made in all factories were consumed in Vietnamese markets.

According to an interview with the company staff, one ton of dried bamboo split could produce an average of 700 kg of bamboo products regardless of items (bamboo chopsticks and incense stick). The average production cost was about $135/ton (1,200,000 kip/ton).

3.1.5. Sales

The bamboo sector in Lao PDR was undeveloped and depended heavily on Vietnamese industries. All bamboo products that used bamboos in Sangthong district, including Nongboua village, were sold in Vietnamese markets. Based on the interview with the company staff, the selling prices of bamboo chopsticks and incense stick were $6.7/kg (60,000 kip/kg) and $2.2/kg (20,000 kip/kg), respectively.

3.1.6. Estimation of the Existing Bamboo Value Chain: Bamboo Chopsticks

Bamboo splits dried in Nongboua village were used to produce mainly chopsticks. Local villagers sold green bamboos to the Vietnamese companies; on the other hand, bamboo chopsticks factories used dried bamboos. We converted green kg to dry kg to estimate resource volume and values to have been used for bamboo chopsticks.

Normally green bamboo has an average of 60–75% moisture content [30]. When bamboos are dried, the moisture content drops to 12–12.5% [31–34], losing 48–63% moisture. This study assumed 50% weight loss from the air drying in the Nongboua village. In addition, the bamboo processing factories in Vietnam used 1 kg dried bamboo material to produce 0.7 kg of bamboo chopsticks. In other words, 1 kg of bamboo chopsticks needs 1.429 kg of dry bamboos or 2.858 kg of green bamboos.

The green bamboo splits were sold at an average price of $0.065/green kg (575 kip/green kg), which was $0.13/kg (1150 kip/kg) and $0.18/kg chopsticks (1643 kip/kg chopsticks). The costs for collection and management, tax, transportation, and processing in Vietnam were $0.14 (1286 kip), $0.01 (129 kip), $0.04 (379 kip), and $0.19 (1715 kip), respectively, to produce 1 kg of bamboo chopsticks that were sold at a price of $6.74 (60,000 kip) in Vietnamese markets. Therefore, the production cost of bamboo chopsticks excluding wages and salaries of drivers and workers in the factories was $0.58/kg chopsticks, which was only 8.6% of the bamboo chopsticks market value.

Local villagers were a key actor in the bamboo chopsticks value chain as part of bamboo harvesting and collection and management, which generated low values ($0.18 and $0.14/kg chopsticks). On the contrary, the Vietnamese bamboo company was a key actor as part of transportation, processing, and sales, which generated a total value of $6.4/kg chopsticks or 94.9% of the total value (Table 4). Although Lao PDR was able to generate added-values of bamboo resource domestically, most added-values were created by the Vietnamese companies out of the villages.

3.2. New Bamboo Products Value Chain in Nongboua Village

This study attempted to strengthen and build capacity of residents and to develop a new bamboo handicraft product model. The new model allowed local villagers to actively participate in the production of bamboo handicrafts and contributed to the socio-economic improvement and sustainable bamboo resource management within the area. A value chain approach shows how much values were added by local villagers (Figure 5) compared with the values added from the existing bamboo products.

Table 4. Bamboo chopsticks value chain in Nongboua village, Sangthong district, in 2018.

Value Chain	Function	Bamboo Harvesting	Collection and Management	Tax (Lao PDR)	Transportation [1]	Processing [2] (Vietnam)	Sales (Vietnam)
	Actor	Local Villager (3-Person Labor)	Local Villager (Hired Labor)	Lao PDR Government	Vietnamese Bamboo Company		
Bamboo Volume	kg	2.858 green kg		1.429 dry kg			1 kg of chopsticks
Price	kip	1643	2929	3058	3437	5152	60,000
	\$ [3]	0.18	0.33	0.34	0.39	0.58	6.74
	kip	1643	1286	129	379	1715	54,848
	\$	0.18	0.14	0.01	0.04	0.19	6.16
Value	Calculation	575 kip/green kg × 2.858 green kg	150,000 kip/day ÷ 5000 dry kg/month × 30 days/month × 1.429 dry kg	200 kip/stem ÷ 8 green kg/stem × 2.858 green kg + 40 kip/dry kg × 1.429 dry kg [4]	265 kip/dry kg × 1.429 dry kg	1200 kip/dry kg × 1.429 dry kg	-
	%	2.7	2.1	0.2	0.6	2.9	91.4

[1] Excluding wages and salaries of drivers. [2] Excluding wages and salaries of workers in the factories. [3] Currency rate 8900 kip/\$ based on 20 March 2020 (https://ko.exchange-rates.org/converter/USD/LAK/1/Y). [4] Total tax/kg chopsticks = village development fund + export tax. One-time registration and annual fees were excluded due to their insignificance.

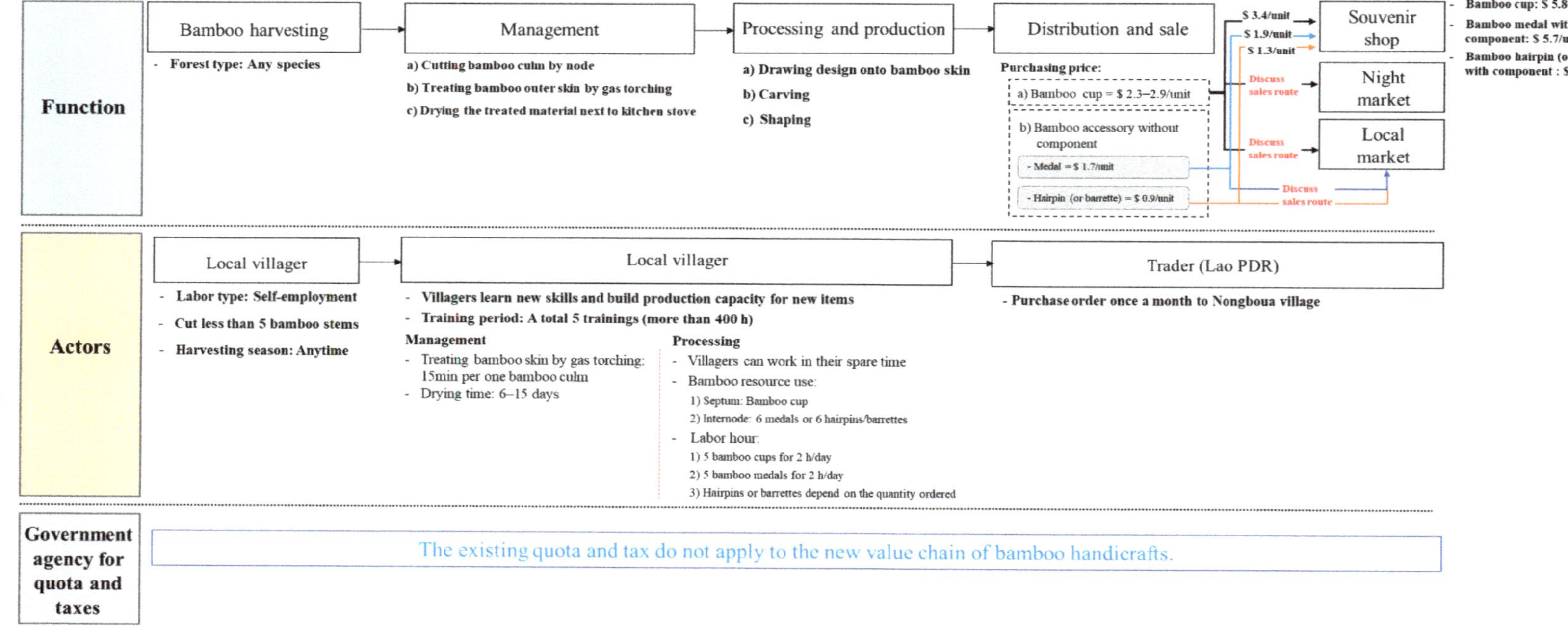

Figure 5. Overview of the new bamboo products value chain in Nongboua village.

3.2.1. Bamboo Harvesting

The new handicraft model was able to use any bamboo species more than 3 years old, unlike the existing model that used only two species (Mai Pahang and Mai Sang Phai). When six participants in the new model needed raw materials from bamboo forest, they harvested green bamboos to meet their demands. During the period from 2018 to 2019, each participant cut an average of five green bamboo stems that were more than 3 years old to make bamboo handicraft products.

The new model does not need a large quantity of bamboos. The villagers harvested a small quantity for dry season (November to April) [35]. On the other hand, villagers harvested bamboos for the existing model even in the rainy season depending on company orders.

3.2.2. Management

After harvesting green bamboos, the participants cut bamboo culm by node (Figure 6a), treated bamboo outer skin by gas torching (Figure 6b), and dried the treated material next to a kitchen stove (Figure 6c). This process was a key step for producing high quality bamboo handicrafts and improving the product life. This drying method is similar to kiln oven drying that is a commonly used method to dry bamboo culms [4]. Advantages of kiln drying is easy to control drying conditions and the required final moisture content in short time. It can also ensure higher level dried bamboo quality than air drying. However, the kiln oven/machine is difficulty to build/purchase and maintain in rural areas and only a small amount of bamboo is needed for handicrafts. A kitchen stove was used instead, which can result in similar effects.

Figure 6. Bamboo harvesting, collection, and management steps: (**a**) cutting bamboo culm by node, (**b**) treating bamboo outer skin with gas torch, and (**c**) drying the treated bamboo material using a kitchen stove.

It took 6–15 days for villagers to dry bamboo material in this way. This drying method did not require a lot of manpower, took short drying time, and prevented a fungal infection inside bamboo.

3.2.3. Processing and Production

The processing and production of the new model required bamboo handicraft skills that local villagers did not have initially. It took time and efforts at this step to build the processing and production capacity of the villagers by training them. This study conducted a total of five training sessions for bamboo processing and handicraft production from 2018 to 2019. A Korean bamboo handicraftsman was invited to train local villagers to have enough bamboo craftsmanship skills for handicrafts. Each technique training took a total of 80 h for 10 days.

Each bamboo item needs a different part from bamboo culm (Figure 7). Bamboo cups need a septum for the cup bottom. The internode of which the septum was taken away for a bamboo cup can be used to produce 6 round-shaped medals, 6 ornamental hairpins, or 6 barrettes (hairclips), thus creating highly added value per bamboo stem.

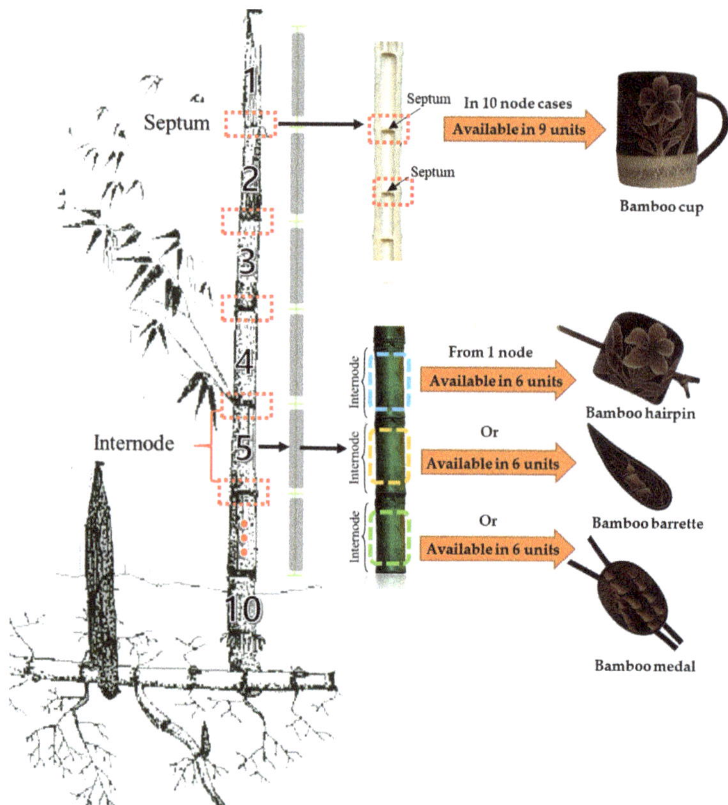

Figure 7. The bamboo parts used for the new bamboo handicraft model (modified after [36]).

3.2.4. Distribution and Sales

Small agriculture producers in developing countries are often linked with buyers through spot market-type transactions. Thus, the new bamboo model also needs to secure sales routes for the products. In 2019, a distributor (middle trader) visited the village and purchased bamboo handicrafts (Figure 8a). The purchase price of bamboo cup was $2.3–2.9 (20,000–25,000 kip) per unit depending on the cup size, which was decided based on the sales trial price. The price of bamboo medal was decided to be $1.7 (15,000 kip), and the hairpin and barrette were $0.9 (8000 kip) per piece. The distributor sold the bamboo cup, medal, hairpin, and barrette to souvenir shops at prices of $3.4 (30,000 kip), $1.9 (17,000 kip), $1.3 (12,000 kip), and $1.3 (12,000 kip). These bamboo products were sold at prices of $5.8–6.9 (52,000–61,000 kip), $5.7 (51,000 kip), $2.2 (20,000 kip), and $2.2 (20,000 kip), respectively, at the souvenir shops (Figure 5).

Figure 8. Distribution and sales of the new bamboo products: (**a**) display and sales of bamboo products in a souvenir shop and (**b**) night market booth in Vientiane, Lao PDR.

Another sales route was direct sales by participating in events such as local festivals or national handcraft festivals (Figure 8b). In this case, local villagers were the only actor in the value chain from bamboo harvesting to sales. For example, the villagers went to the Sangthong District Festival and sold 25 bamboo cups, generating a total revenue of $70.2 (625,000 kip) by six people or $11.7/person (104,000 kip/person) in a day. Lao PDR had a monthly minimum wage of $124 (1,100,000 kip) as of 2018 [37] and a daily wage of $5.4–6.7 (50,000–60,000 kip) in rural areas [38]. Therefore, the new bamboo model could be highly profitable, thus contributing to cash circulation, and improve local livelihoods in Lao PDR.

Despite the high potential income increase, this model lacks a stable sales route. The new model needs to expand market-based sales by strengthening partnerships with local governments and associations that already have sales routes, such as the DAFO, PAFO, LHA (Lao Handicraft Association, The Lao Handicraft Association is a group of over 100 handicrafts producers in Lao PDR. The products include jewelry, textiles, carpets, metal work, wood crafts, and other handicraft items, all of which are made of local materials by local artisans [39]), and GDA (Gender Development Association, The GDA, formerly the Gender and Development Group (GDG) and Women in Development (WID) network, was formed in 1991. This association plays a role as a platform for knowledge sharing on gender issues. Especially, the GDA in Sangthong district has implemented building women's capacity on bamboo weaving with the Lao Women's Union (LWU). Therefore, it is one of the Sangthong bamboo value chain producer groups [40]).

3.2.5. Estimation of New Bamboo Value Chain: Bamboo Cup and Medal

Table 5 shows the new bamboo value chain. Local villagers were restricted to providing raw material to the supplier in the existing bamboo value chain. However, being able to process the final products, they became a key actor who played important roles as producers in the new value chain and negotiated the price of their bamboo products with traders.

Local villagers started acting on their own to engage in new activities, such as showing their skills and products in domestic festivals and other events, and to teach bamboo handicraft skills to other villagers as a Trainer of Trainers (TOT). However, this study did not include the opportunity cost of participants such as villagers' time and efforts to learn bamboo handicraft skills. The villagers received a total of five trainings and spent more than 400 h (10 day/training × 8 h/day × 5 trainings) to acquire the new skills.

Table 5. New bamboo products value chain in Nongboua village, Sangthong district.

Value Chain	Function	Bamboo Harvesting	Management	Processing		Distribution		Sales	
				Bamboo Cup [1]	Bamboo Medal [1]	Bamboo Cup	Bamboo Medal	Bamboo Cup	Bamboo Medal
	Actor		Local Villager			Trader from Lao PDR		Souvenir Shop (End User)	
Bamboo Volume	Stem			1 (10 nodes and 9 septa per stem)					
	Green kg			8					
Price	kip/stem	0	0	180,000–225,000	900,000	270,000	1,020,000	468,000–549,000	3,060,000
	$/stem [2]	0	0	20.2–25.3	101.1	30.3	114.6	52.6–61.7	343.8
	Calculation		Self	20,000–25,000 kip/cup × 9 cups/stem	15,000 kip/medal × 6 medals/node × 10 nodes/stem	30,000 kip/cup × 9 cups/stem	17,000 kip/medal × 6 medals/node × 10 nodes/stem	52,000–61,000 kip/cup × 9 cups/stem	51,000 kip/medal [3] × 6 medals/node × 10 nodes/stem
Value	kip	0	0	180,000–225,000	900,000	45,000–90,000	120,000	198,000–279,000	2,040,000
	$	0	0	20.2–25.3	101.1	5.1–10.1	13.5	22.2–31.3	229.2
	%	0	0	37.9–42.6	29.4	10.6–15.2	3.9	46.8–47.0	66.7

[1] We selected two bamboo products (i.e., cup and medal) due to high demands by traders. [2] The price included medal components, such as strap and strap adjusting link. [3] Currency rate 8900 kip/$ based on 20 March 2020 (https://ko.exchange-rates.org/converter/USD/LAK/1/Y).

3.3. Comparison of the Exisiting and New Bamboo Products Value Chains for Nongboua Villagers

It is necessary to compare income generation and use of bamboo resource between the existing and new models to assess their impacts on local villagers' livelihood and sustainability of bamboo forest. However, the existing and new value chains were assessed using different units: unit weight (1 kg) of the final bamboo products and one bamboo stem. This study compares the income generations from the two models for Nongboua villagers in two ways: labor hour and use of bamboo resource.

3.3.1. Labor Hour

The bamboo harvesting in the existing value chain required a minimum of three people, which would take an average of 2 h. A three-person family was assumed to work 5 times a month and could produce 300–350 green kg bamboo splits at a time, which was purchased at an average price of $0.06/green kg (575 kip/green kg). Thus, the hourly income of one person for bamboo splits was $3.23–3.77 (28,750–33,542 kip) (Table 6).

Table 6. Comparison of Nongboua villager's incomes from the existing and new bamboo product models by labor hour.

Product Type	No. of Workers	Daily Production [1]	Unit Price	Hourly Income for One Person
Bamboo splits	3 people	300–350 green kg/day	$0.06 (575 kip)/green kg	$3.23–3.77 (28,750–33,542 kip)
Bamboo cup	1 person	5 units/day [2]	$2.25–2.81 (20,000–25,000 kip)/unit	$5.62–7.02 (50,000–62,500 kip)
Bamboo medal	1 person	5 units/day [3]	$1.69 (15,000 kip)/unit	$4.21 (37,500 kip)

[1] Based on 2 h/day. [2] Villagers could make 4–6 bamboo cups for 2 h. The average value, 5 bamboo cups, was used. [3] Villagers could make 4–6 bamboo medals for 2 h. The average value, 5 bamboo medals, was used.

On the other hand, a villager could make about five bamboo cups for two hours a day. The purchase price was $2.25–2.81/unit (20,000–25,000 kip/unit), and therefore the hourly income for one person was $5.62–7.02 (50,000–62,500 kip). Alternatively, a villager could make about five bamboo medals for two hours a day. The purchase price was $1.69/unit (15,000 kip/unit), and the hourly income for one person was $4.21 (37,500 kip) (Table 6). The new models for bamboo cup and medal can generate 1.49–2.17 and 1.12–1.30 times higher income than the existing model for bamboo splits.

Labor hour of bamboo splits included harvesting and splitting bamboos; however, that of bamboo cups and medals included only processing products. The bamboo harvesting and management in the new model were conducted as needed, which took over 6–15 days, and was therefore excluded in the comparison. Also, the existing model required highly labor-intensive outdoor work during the daytime, while the new model did not require heavy labor so villagers can work on bamboo handicrafts indoor regardless of outside weather conditions. However, the new model required a certain level of bamboo handicraft skills that took 5 sessions of bamboo skill trainings, equivalent to 400 h of training.

3.3.2. Use of Bamboo Resource

A fully-grown bamboo stem could produce 8 green kg of bamboo splits, of which the purchase price was $0.06/green kg (575 kip/green kg). Therefore, one bamboo stem could generate a total income of $0.52 (4600 kip) for the villagers by making bamboo splits. For the new model, one bamboo stem could produce 9 cups and 60 medals of which the purchase prices were $2.25–2.81/unit (20,000–25,000 kip/unit) and $1.69/unit (15,000 kip/unit), respectively. Therefore, one bamboo stem could generate a total income of $121.35–126.40 (1,080,000–1,125,000 kip) for the villagers by making bamboo cups and medals. From one bamboo stem, the new model for bamboo cup and medal can generate 234.8–244.6 times higher income than the existing model for bamboo splits. The new bamboo product model has more benefits than the existing model considering higher income generation for local villagers and much less use of local bamboo resource.

4. Discussion

The existing value chain for bamboo chopsticks proceeded as follows: bamboo harvesting, collection and management by villagers in Lao PDR, taxed by Lao PDR government, transportation to Vietnam, processing, and sales. Each function of the value chain produced $0.18 (2.7%), $0.14 (2.1%), $0.01 (0.2%), $0.04 (0.6%), $0.19 (2.9%), and $6.16 (91.4%) from 1 kg of bamboo chopsticks that was worth $6.74 (100%). Our analysis indicated that local villagers received only 4.9% of the total product value and Lao PDR, i.e., local villagers and Lao PDR government, received 5.1%. Lao PDR provided a raw material, bamboo, but most of the value (94.9%) was created by Vietnamese bamboo companies. Providing bamboo splits for bamboo chopsticks contributes little to livelihoods of local villagers.

The new model value chain for bamboo handicrafts proceeded as follows: bamboo harvesting, management, processing by villagers, distribution by traders, and sales by souvenir shops. One bamboo stem was estimated to produce 9 bamboo cups and 60 medals that were worth $52.6–61.7 and $343.8, respectively, totaling $396.4–405.5. From the 9 bamboo cups, bamboo harvesting, management, and processing produced $20.2–25.3 (37.9%–42.6%); distribution, $5.1–10.1 (10.6–15.2%); and sales, $22.2–31.3 (46.8–47.0%). From the 60 bamboo medals, bamboo harvesting, management, and processing produced $101.1 (29.4%); distribution, $13.5 (3.9%); and sales, $229.2 (66.7%). Most of the values (46.8%–66.7%) were created by souvenir shops, but 29.4%–42.6% of the values was retained by local villagers. Thus, producing bamboo handicrafts can potentially increase income and expand livelihood options of local villagers.

A comparison of the value chains of the existing and new bamboo products by labor hour showed that producing bamboo cup and medal could generate 1.49–2.17 and 1.12–1.30 times higher income for local villagers than producing bamboo splits to be exported to Vietnam. Also, a comparison of the two value chains by use of bamboo resource (i.e., bamboo stem) showed producing bamboo cup and medal could generate 234.8–244.6 times higher income than producing bamboo splits. Therefore, this study concludes that the new bamboo product model for handicrafts is better for local villagers in Lao PDR than the existing model for chopsticks made in Vietnam, considering both higher income generation and less use of bamboo resource. The new model enables local villagers to increase their income while sustainably using the bamboo resource. Similar results can be found in the previous study in India [41] where bamboo handicrafts produced much higher income (2700 USD) than selling raw bamboo (800 USD) annually, and boosted farmers' income.

The new model, however, required a certain level of bamboo handicraft skills that needed 5 sessions of bamboo skill trainings, which took 400 h of training. To apply the new bamboo handicraft model to other rural areas in Lao PDR and potentially other developing countries, an intensive bamboo handicraft skill training would be required, which is a difficult task.

One solution is TOT model that educates a group of local villagers to acquire enough bamboo crafts skills, which allows them to train others with less skills, spreading bamboo handicrafts skill and expanding the new bamboo handicrafts model throughout a region. TOT for the new bamboo income model can be designed based on the lessons learned from the strategy of *Saemaul Undong* (New Village Movement), which was a community-driven development program and a government-initiated movement in the 1970s in the Korea [42]. Recently, this movement have been promoted by Korea in more than 30 developing countries across Africa, Asia, and Latin America as an integrated rural development model to increase income of rural poor through small-scale and self-help projects [43]. In particular, TOT has been one of the main activities in *Saemaul Undong* of Vietnam, where the government leaders organized TOT for poverty reduction and rural development. To develop the new bamboo model in Lao PDR would need various social, economic, and technical supports by the Lao PDR government who can develop TOT that can systematically facilitate spreading bamboo handicrafts skill and expanding the new bamboo handicrafts model throughout a region. Already Na Lad villagers adjacent to the Nongboua village requested bamboo handicraft trainings to the DOF, which shows potential applicability of the new model.

The new model needs more diverse and stable sales routes that this study did not explore. Without stable market demand for bamboo handicrafts, local villagers cannot continue producing the handicraft products. Also, understanding market trends of bamboo handicrafts will ensure local villagers to produce and supply the bamboo handicrafts that the market demands. More efforts should be made to bridge the supplier, local villagers, and the market of the new bamboo handicraft model. The needs to link bamboo resource and consumers, and to expand extension education for producers are found in other developing country [44].

In the Mekong basin region such as Cambodia, Laos, Myanmar, Thailand, and Vietnam, more than 400,000 people rely on their livelihoods from bamboo, and it could expect to increase 800,000 people [45] who are mostly lower income group in need of economic security. Processing raw bamboo and producing value-added products can generate local employment opportunities by enhancing the local bamboo value chain [46]. Although alternative uses of bamboo would be different depending on social, economic, and market contexts, the value chain analysis demonstrated in this study can be applied elsewhere. The first step is to evaluate an alternative approach for utilizing local resources that increases local value capture. Then, assessing available labor force characteristics with necessary job skills and providing targeted job trainings can increase livelihoods of local villagers and improve local economies while sustainably using bamboo resources.

5. Conclusions

This study assessed bamboo-based income generations from the existing and new bamboo product models for local villagers using a value chain approach and provided empirical assessments of local economic impacts that can be enhanced by the new bamboo utilization approach. About 60% of the Nongboua households relied on bamboo-based income during the dry season. Most of them engaged in simply cutting and selling green bamboos. A new income model should encourage local villagers to increase their roles within the value chain by selling more value-added products with local bamboo resources. The new income model made it possible for villagers to build their capacities to make and sell their own products from local bamboo resources.

This study was conducted at a village level and did not provide complete information on the whole value chains, especially for bamboo chopsticks made in Vietnam, which would have provided a better view on the existing and new bamboo product models. Further study is needed to demonstrate the potential increase in income from the new model with respect to sales route, bamboo handcraft design, and skill level as well as the economic and ecological impacts at different scale, such as regional and provincial levels. The value chain assessment provides the government and policymakers useful information so that they can find a way to improve livelihoods of local villagers and to manage bamboo forests in a sustainable way. As this model can utilize any bamboo species, it can be applied to other local villagers in Lao PDR, as well as neighboring countries.

Author Contributions: B.L., H.R., S.K. (Sebin Kim), J.-W.L. and S.K. (Seungmo Koo) conceived and designed the study; B.L., H.R., S.-J.L. and P.A. collected the data; B.L. and H.R. analyzed the data; and B.L., H.R. and Y.-S.K. wrote the paper. All authors have read and agreed to the published version of the manuscript.

Funding: This study was carried out with the support of "R & D Program for Forest Science Technology (Project No. 2019126A00-1920-0001)" provided by Korea Forest Service (Korea Forestry Promotion Institute).

Institutional Review Board Statement: Not applicable.

Informed Consent Statement: Informed consent was obtained from all subjects involved in the study.

Data Availability Statement: The data presented in this study are available on request from the corresponding author. The data are not publicly available due to personal information protection.

Acknowledgments: The authors would like to thank bamboo craftsman, Dongsuk Park, from the Korea, for his help in bamboo technique trainings and Oupakone Alounsavath at Department of Forestry, Ministry of Agriculture and Forestry, Lao People's Democratic Republic, for his help in data collection.

Conflicts of Interest: The authors declare no conflict of interest.

References

1. Leakey, R.R.B. Definition of agroforestry revisited. *Agrofor. Today* **1996**, *8*, 5–7.
2. Shin, S.; Soe, K.T.; Lee, H.; Kim, T.H.; Lee, S.; Park, M.S. A systematic map of agroforestry research focusing on ecosystem services in the Asia-Pacific Region. *Forests* **2020**, *11*, 368. [CrossRef]
3. Nang'ole, E.M.; Mithöfer, D.; Franzel, S. *Review of Guidelines and Manuals for Value Chain Analysis for Agricultural and Forest Products*; Occasional Paper No. 17; World Agroforestry Centre: Nairobi, Kenya, 2011; pp. 1–30.
4. Liese, W.; Köhl, M. *Bamboo: The Plant and Its Uses*, 1st ed.; Springer International Publishing: Cham, Switzerland, 2015; pp. 5–6. [CrossRef]
5. Partey, S.T.; Sarfo, D.A.; Frith, O.; Kwaku, M.; Thevathasan, N.V. Potentials of bamboo-based agroforestry for sustainable development in Sub-Saharan Africa: A review. *Agric. Res.* **2017**, *6*, 22–32. [CrossRef]
6. Christanty, L.; Kimmins, J.P.; Mailly, D. 'Without bamboo, the land dies': A conceptual model of the biogeochemical role of bamboo in an Indonesian agroforestry system. *For. Ecol. Manag.* **1997**, *91*, 83–91. [CrossRef]
7. Nath, A.J.; Das, A.K. Carbon pool and sequestration potential of village bamboos in the agroforestry system of northeast India. *Trop. Ecol.* **2012**, *53*, 287–293.
8. Tewari, S.; Banik, R.L.; Kaushal, R.; Bhardwaj, D.R.; Chaturvedi, O.P.; Gupta, A. Bamboo based agroforestry systems. In *Bamboos in India*, 1st ed.; Kaushik, S., Singh, Y.P., Kumar, D., Thapliyal, M., Barthwal, S., Eds.; ENVIS Centre on Forestry: Dehradun, India, 2015; pp. 261–284. Available online: http://frienvis.nic.in/WriteReadData/UserFiles/file/Publication/Books/2015-Bamboos-in-India.pdf (accessed on 11 November 2020).
9. Buckingham, K.C.; Wu, L.; Lou, Y. Can't see the (bamboo) forest for the trees: Examining bamboo's fit within international forestry institutions. *Ambio* **2014**, *43*, 770–778. [CrossRef] [PubMed]
10. Canavan, S.; Richardson, D.M.; Visser, V.; Le, J.J.; Vorontsova, M.S.; Wilson, J.R.U. The global distribution of bamboos: Assessing correlates of introduction and invasion. *AoB Plants* **2016**, *9*, 1–18. [CrossRef] [PubMed]
11. Lovovikov, M.; Paudel, S.; Piazza, M.; Ren, H.; Wu, J. *World Bamboo Resources: A Thematic Study Prepared in the Framework of the Global Forest Resources Assessment 2005*; Non-Wood Forest Products 18; FAO: Rome, Italy, 2007; p. 81. Available online: http://www.fao.org/3/a1243e/a1243e00.htm (accessed on 27 October 2020).
12. Lovovikov, M.; Lou, Y.; Schoene, D.; Widenoja, R. *The Poor Man's Carbon Sink: Bamboo in Climate Change and Poverty Alleviation*; Non-Wood Forest Products Working Document No 8; FAO: Rome, Italy, 2009; p. 72. Available online: http://www.fao.org/tempref/docrep/fao/012/k6887e/k6887e00.pdf (accessed on 27 October 2020).
13. Warner, K.; McCall, E.; Garner, S. *The Role of NTFPs in Poverty Alleviation and Biodiversity Conservation*; International Union for Conservation of Nature (IUCN): Hanoi, Vietnam, 2008; p. 260.
14. Paudyal, K.; Adhikari, S.; Sharma, S.; Samsudin, Y.B.; Paudyal, B.R.; Bhandari, A.; Birhane, E.; Darcha, G.; Trinh, T.L.; Baral, H. *Framework for Assessing Ecosystem Services from Bamboo Forests: Lessons from Asia and Africa*; Working paper 255; CIFOR: Bogor, Indonesia, 2019; pp. 1–42.
15. Sakala, W.D.; Moyo, S. Socio-economic benefits of community participation in wildlife management in Zambia. *Sustain. Res. Manag. J.* **2017**, *2*, 1–18.
16. Jabeen, S.; Nisar, M.; Saq, U.I.; Hussain, I. Community participation in socio-economic development through secondary education in one of the remotest regions of Pakistan. *J. Nat. Soc. Sci.* **2018**, *7*, 663–680.
17. Alexander, A.F.; Hezekiah, A.A. Socioeconomic characteristics and community participation in infrastructure provision in Akure. *Cogent Soc. Sci.* **2018**, *4*, 1–13. [CrossRef]
18. Porter, M. *Competitive Advantage: Creating and Sustaining Superior Performance*; Free Press: New York, NY, USA, 1985; p. 592.
19. Stabell, C.B.; Fjeldstad, O.D. Configuring value for competitive advantage: On chains, shops, and networks. *Strateg. Manag. J.* **1998**, *19*, 413–437. [CrossRef]
20. Koc, T.; Bozdag, E. Measuring the degree of novelty of innovation based on Porter's value chain approach. *Eur. J. Oper. Res.* **2017**, *257*, 559–567. [CrossRef]
21. World Wide Fund for Nature (WWF). Bamboo Can Become the Green Gold of Small Holders to Decrease Poverty. 2016. Available online: https://www.wwf.org.la/?271273/Launching-of-a-Lao-Bamboo-Platform-and-National-strategy (accessed on 2 November 2020).
22. Phounvisouk, L.; Zuo, T.; Kiat, N.C. Non-timber forest products marketing: Trading network of trader and market chain in Luang Namtha Province, Lao PDR. *J. Humanit. Soc. Sci.* **2013**, *18*, 48–57. [CrossRef]

23. Yumkella, K.K.; Hai, D.T.; Dinh, D.N. *Greening Value Chains for Sustainable Handicrafts Production in Vietnam*; United Nations Industrial Development Organization (UNIDO): Hanoi, Vietnam, 2013; pp. 1–44.
24. Department of Forestry (DOF). *Work Plan: Village Based Forest Rehabilitation Project in Lao PDR*; DOF: Vientiane, Lao PDR, 2015; p. 78.
25. Lee, B.; Rhee, H.; Lee, S.; Alounsavath, P.; Lee, J.; Koo, S.; Kim, S. Effectiveness analysis on new bamboo-based income system of Lao PDR: Case study of Nongboua mountain village. *J. Korean Soc. Int. Agric.* **2019**, *31*, 312–321. [CrossRef]
26. Greijmans, M.; Oudomvilay, B.; Banzon, J. *Houaphanh Bamboo Value Chain Analysis: Identifying SNV's Potential Advisory Services for the Development of the Bamboo Value Chain*; Stichting Nederlandse Vrijwilligers (SNV) Netherlands Development Organisation: Hague, The Netherland, 2007; p. 19.
27. Phimmavong, B.; Chanthavong, V. *Challenges and Opportunities for Lao DPR's Small and Medium Forest Enterprises (SMFEs)*; Food and Agriculture Organization (FAO): Rome, Italy, 2009; pp. 9–10.
28. Phommasane, S.; Foppes, J.; Wanneng, P.; Sirithirath, S. *Positioning Study of the Bamboo Value Chain in Sang Thong District*; SNV Netherlands Development Organisation: Hague, The Netherlands, 2007; pp. 10–52.
29. McLafferty, I. Focus group interviews as a data collecting strategy. *J. Adv. Nurs.* **2004**, *48*, 187–194. [CrossRef] [PubMed]
30. Wakchaure, M.R.; Kute, S.Y. Effect of moisture content on physical and mechanical properties of bamboo. *Asian J. Civ. Eng.* **2012**, *13*, 753–763.
31. Gandhi, Y. Preliminary study on the drying the bamboo (*Bambusa blumeana*) in a wood waste fired kiln. In Proceedings of the 5th International Bamboo Congress and the 6 International Bamboo Workshop, San Jose, Costa Rica, 2–6 November 1998.
32. Chung, W.F.; Yu, W.K. Mechanical properties of structural bamboo for bamboo scaffoldings. *Eng. Struct.* **2002**, *24*, 429–442. [CrossRef]
33. Montoya-Arango, J.A. Trocknungsverfahren für die Bambusart Guadua Angustifolia unter Tropischen Bedingungen. Ph.D. Dissertation, University Hamburg, Hamburg, Germany, 2006.
34. Qingfeng, X.; Kent, H.; Xiangmin, L.; Qiong, L.; Jennifer, G. Mechanical properties of structural bamboo following immersion in water. *Eng. Struct.* **2014**, *81*, 230–239.
35. Ministry of Agriculture and Forestry (MAF). *Forestry Strategy to the Year 2020 of the Lao PDR*; MAF: Vientiane, Lao PDR, 2005; pp. 18–78.
36. Zhu, Z. Bamboo industry's impact evaluation on rural sustainable development. In *International Training Workshop on Small Bamboo Daily Product Processing Technologies and Machines, Zhejiang, China, September 2004*; Zhu, Z., Jiang, C., Eds.; International Farm Forestry Training Center, Chinese Academy of Forestry (CAF): Beijing, China, 2004.
37. Bissinger, J. *Discussion Paper on Minimal Wage Reforms: Key Considerations for Laos Employers*; International Labour Organization (ILO), Bureau for Employers' Activities (ACT/EMP): Geneve, Switzerland, 2019; p. 24.
38. Food and Agriculture Organization (FAO). *Special Report-2019 FAO/WFP Crop and Food Security Assessment Mission to the Lao People's Democratic Republic*; FAO: Rome, Italy, 2019; p. 20.
39. Ukkerman, R. *Bamboo Handicraft Products Fair trade/Marketing Study*; Final Report; SNV Netherlands Development Organisation: Hague, The Netherland, 2010; pp. 1–16.
40. Gender Development Association (GDA). Sangthong Bamboo Value Chain Producer Group. 2020. Available online: https://gdalaos.org/sangthong-bamboo-value-chain-producer-group/ (accessed on 18 December 2020).
41. Dwivedi, A.K.; Kumar, A.; Baredar, P.; Prakash, O. Bamboo as a complementary crop to address climate change and livelihoods—Insights from India. *Forest Policy Econ.* **2019**, *102*, 66–74. [CrossRef]
42. Lee, D.K.; Kwon, K.C.; Lee, Y.H. The role and contribution of sanlim-kyes during Seamaul Undong in the Korea in the 1970s. *Forest Sci Technol.* **2018**, *14*, 47–54. [CrossRef]
43. United Nations Development Programme (UNDP). *Saemaul Initiative Towards Inclusive and Sustainable New Communities: Implementation Guidance*; UNDP: New York, NY, USA, 2015; pp. 43, 97.
44. Endalamaw, T.B.; Lindner, A.; Pretzsch, J. Indicators and determinants of small-scale bamboo commercialization in Ethiopia. *Forests* **2013**, *4*, 710–729. [CrossRef]
45. Pérez, M.R.; Maogong, Z.; Belcher, B.; Chen, X.; Maoyi, F.; Jinzhong, X. The role of bamboo plantations in rural development: The case of Anji County, Zhejiang, China. *World Dev.* **1999**, *27*, 101–114. [CrossRef]
46. Ly, P.; Pillot, D.; Lamballe, P.; de Neergaard, A. Evaluation of bamboo as an alternative cropping strategy in the northern central upland of Vietnam: Above-ground carbon fixing capacity, accumulation of soil organic carbon, and socio-economic aspects. *Agric. Ecosyst. Environ.* **2012**, *149*, 80–90. [CrossRef]

Article

Global Trends in Research on Wild-Simulated Ginseng: Quo Vadis?

Seongmin Shin [1,2], Mi Sun Park [2,3,*], Hansol Lee [2], Seongeun Lee [4], Haeun Lee [2], Tae Hoon Kim [2] and Hyo Jin Kim [2,3]

1   Centre for International Forestry Research (CIFOR), Jalan CIFOR, Situ Gede, Bogor 16115, Indonesia; seongmin.shin@cgiar.org
2   Graduate School of International Agricultural Technology, Seoul National University, 1447 Pyeongchang-daero, Daehwa, Pyeongchang 25354, Gangwon, Korea; ihansol14@snu.ac.kr (H.L.); haeunl630@snu.ac.kr (H.L.); switch-thkim@snu.ac.kr (T.H.K.); erythritol@snu.ac.kr (H.J.K.)
3   Institutes of Green Bio Science and Technology, Seoul National University, 1447 Pyeongchang-daero, Daehwa, Pyeongchang 25354, Gangwon, Korea
4   Center for International Agricultural Partnership, Korea Rural Economic Institute, Naju-si 58217, Jeollanam-do, Korea; lse1229@krei.re.kr
*   Correspondence: mpark@snu.ac.kr; Tel.: +82-33-339-5858

Citation: Shin, S.; Park, M.S.; Lee, H.; Lee, S.; Lee, H.; Kim, T.H.; Kim, H.J. Global Trends in Research on Wild-Simulated Ginseng: Quo Vadis?. *Forests* **2021**, *12*, 664. https://doi.org/10.3390/f12060664

Academic Editor: Martin Lukac

Received: 26 April 2021
Accepted: 17 May 2021
Published: 24 May 2021

Publisher's Note: MDPI stays neutral with regard to jurisdictional claims in published maps and institutional affiliations.

Abstract: To the best of our knowledge, no study has systematically reviewed and analyzed the research trends of wild-simulated ginseng (WSG) used for food or medicinal purposes in many countries. WSG, a non-timber forest product, has been traditionally produced using agroforestry practices, and it has been consumed in various ways for a long time. WSG has a great demand in the market due to its medicinal effects, particularly in improving forest livelihoods and human health. Due to the significance of WSG, we conducted this research to explore the global research trends on WSG using systematic review methodology and keyword analysis. We used two international academic databases, the Web of Science and SCOPUS, to extract 115 peer-reviewed articles published from 1982 to 2020. The research subjects, target countries, and keywords were analyzed. Our results indicate four categories of WSG research subjects, namely growth conditions, components, effects on humans/animals, and the environment of WSG, and the case studies were mainly from the Republic of Korea, China, and the USA. Through topic modelling, research keywords were classified into five groups, namely medicinal effects, metabolite analysis, genetic diversity, cultivation conditions, and bioactive compounds. We observed that the research focus on WSG changed from the biological properties and cultivation conditions of WSG to the precise identification and characterization of bioactive metabolites of WSG. This change indicates an increased academic interest in the value-added utilization of WSG.

Keywords: wild-simulated ginseng; systematic review; keyword analysis; topic modelling; non-timber forest products (NTFPs); agroforestry

1. Introduction

Ginseng obtained from plant roots has been used as herbal medicine for thousands of years. It has been recorded as an important medicine in the oldest medicinal herb book, Sinnongbonchogyeong (B.C. 3300) [1]. Before ginseng was cultivated, wild ginseng was collected and consumed as an edible herb. Increasing demand for wild ginseng resulted in the overexploitation of wild ginseng to meet the demands of the industrial market. Agroforestry practices have been used to cultivate wild-simulated ginseng (WSG). WSG is a perennial semi-negative, semi-annotated pericardium belonging to the genus "Panax" of the family Araliaceae. WSG has been used for food or medicinal purposes in many countries, including the Republic of Korea, China, Japan, Russia, and the USA [2]. WSG is known by different names, including WSG, wild-cultivated ginseng, mountain-cultivated ginseng, and forest-cultivated ginseng, among others. In this study, we use the term

WSG because it is officially used by the governments of the Republic of Korea, USA, and Canada [3–5]. We define WSG as the ginseng grown under trees in mountain areas via artificial transplantation of seeds or seedlings.

The active components of WSG can be largely divided into saponins (known as ginsenosides) and non-saponins (polyacetylenes, phenolic compounds, acidic polysaccharides, peptides, alkoxides, and amino acid derivatives), depending on the characteristics of chemical structures. Other components of WSG include volatile oil, sugar, starch, pectin, and minerals [6]. WSG contains many functional components that exert excellent effects on human health, such as anti-obesity activity [7], activation of the sympathetic nervous system [8], anti-fatigue activity [9], memory loss improvement [10], spatial cognitive ability improvement [11], cancer prevention, anti-cancer activity [12], and liver functional resistance [13]. Since the health effects of WSG are more recognized than those of cultivated ginseng [14], WSG is mostly produced only as a health supplement [15]. WSG is rarely cultivated for processing as a general food or cosmetic product due to the high production cost of WSG and difficulties in cultivation. However, due to the increasing interest and demand for eco-friendly non-timber forest products (NTFPs), WSG has gained attention as a high-value product. Due to the increasing production and demand, especially in the Republic of Korea [16], WSG has become a major source of income to forest communities. Thus, it is necessary to build scientific databases and accumulate research data for WSG [17].

Due to the greater use of WSG, many studies have been conducted on its growth conditions and characteristics. Many studies have also been conducted on farming locations [18], harvest, growth [19], effectiveness [7–13], and identification of genetic traits [20] and genome sequences [21]. Research trend analysis reflects academic orientation based on the aggregate results of integrating and analyzing key topics or values investigated in previous studies, such as medicine [22], education [23], forestry [24], and cultivated ginseng [25,26]. Such research trend analysis is valuable for diagnosing the current status and suggesting a direction for subsequent research needed to compensate for the knowledge gaps from individual research [27]. Moreover, industrialization of WSG requires a scientific and systematic approach considering the environment, location conditions, cultivation technology, efficacy, and environmental impact. Nevertheless, to date, no study has systematically reviewed and analyzed the research trends of WSG. Thus, this study seeks to examine the trends in WSG research using systematic review and topic modelling which have been widely used to analyze the recent research trends in linguistic, political, medical and biomedical, geographical science, etc. [28]. The study aimed to achieve answers to the following research questions: (1) What are the dominant keywords in the WSG study? (2) What is the main topic of the WSG study? The study findings can contribute to designing future studies on WSG, including the selection of topics.

2. Wild-Simulated Ginseng

2.1. History of Wild-Simulated Ginseng in the World

The family *Panax*, which comprises thirteen species, is extensively spread over East Asia to North America. *Panax ginseng* C. A. Meyer has been grown in Asian countries for thousands of years. *P. ginseng* is the only *Panax* species with a long history of cultivation [29]. The *Shennong Bencao Jing* (Shennong's Herbal Classic), a traditional document focusing on medicinal herbs, was written during the 1st century AD [30]. Ginseng has been artificially planted since the 15th century [30]. Since then, wild ginseng, which has been reseeded in the mountains or forest areas, has been cultivated and consumed as a herbal remedy [30]. There are historical documents on WSG in China and the Republic of Korea. The first book was *Do-gyeong-bon-Cho* (A.D. 1061), which was recorded in China as a bibliographic research on wild-simulated farming archives on ginseng during the reign of King Injong of the Song dynasty. The document focuses on the specific shapes of WSG throughout its growth cycle for four, five, and ten years [1]. During the rule of King In-Jong of the Koryeo Dynasty (A.D. 1122), some documents were written on artificially transplanted and reseeded ginseng [31]. During the Sejong era, wild ginseng has been reported in "Jiriji" (A.D. 1419–1450), which includes different characteristics of wild and cultivated ginseng. During the Sukjong era (A.D. 1675–1720), technological advancements in farming began to be prevalent through books on ginseng cultivation methods, including its distribution and accumulation, and the pervasiveness of ginseng agriculture was declared. Jeongjo Annal of the Jeongjo period (A.D. 1777–1800) documents the plantation of simulated ginseng near villages, thus highlighting its active expansion in communities [1].

Compared to WSG, studies on American ginseng (*P. quinquefolius*) have been conducted more recently [32]. Wild American ginseng was first discovered and cultivated in the New England forests in the early 1700s [33]; however, the farmers moved to a new area without a stable settlement. The cultivation of ginseng facilitated settlement in the region [33,34]. Harvesting and trading ginseng promoted economic growth [33]. Once the trade value was surprisingly high, a report stated that at least 64 million roots were exported in 1841 [35]. During the mid-1800s, WSG cultivation decreased significantly due to the transition of land from forest to farmland [33,35]. Forest degradation was accelerated in the 20th century, and thus it is difficult to harvest mountain ginseng, which grew on steep slopes [35].

2.2. Physiology and Growth of Wild-Simulated Ginseng

Ginseng grows and reproduces well with specific conditions: acidic or weak acidic soil (pH 4–6); 10–30° slope; sufficient nitrogen (N), phosphorus (P), potassium (K), calcium (Ca), magnesium (Mg), and other organic contents [21]. As a perennial plant belonging to the genus *Panax*, WSG has similar physiological and growth characteristics to field-cultivated ginseng. However, WSG also has some unique characteristics. Both have main root, fine roots, and lateral roots. Main root is connected to the stem through rhizomes and the leaves are connected to the stem through leaf stalks [36]. However, they both have to be properly shaded for growth. The difference in the shading process is mentioned in the definition of WSG. Ginseng can be shaded using anthropogenic methods, whereas WSG must be shaded without any artificial measure. The biggest and most recognizable physiological characteristic is that WSG grows slower than field-cultivated ginseng, thus resulting in the smaller size of WSG. Furthermore, the photosynthesis rate is significantly lower in the leaves of WSG, mainly due to the cultivation environment, especially the lack of phosphorus in mountain soil [37]. The root volume is affected by the forest physiognomy of the cultivation area. In a mixed forest of soft and hardwood, the relationship between root volume and diameter at breast height (DBH) of surrounding trees shows a positive correlation [21]. The *Standard Manual for WSG Cultivation* from the Republic of Korea suggests that WSG should be cultivated in a mixed forest with trees having at least 15 cm breast height diameter and 10 m height for the overstory woods [38]. During proper growth conditions, WSG grows up to 92 mm (shoot) and 32 mm (root) in the second year and grows 187 mm (shoot) and 86 mm (root) in the fifth year [36]. In the Republic of Korea, 144 t of WSG (USD 36 million) was harvested in 2019 [39].

2.3. Effects of Wild-Simulated Ginseng

WSG has been regarded as a medicinal plant due to its significant functions [40]. The quality of WSG is considered superior to that of field-cultivated ginseng [37]. Furthermore, the biophysical activities of WSG are greater than that of ginseng [41]. For example, in skin cancer, ginseng has a chemopreventive effect, while WSG has a chemotherapeutic effect [41]. WSG shows diverse effects such as anti-oxidation [42], anti-inflammation [43], anti-tumor, and immune-enhancing activities [44]. Moreover, WSG has been used in the treatment of various health-related problems, including dry eye syndrome [45], obesity [46], high blood pressure [47], and cancer [48]. WSG has also been studied for human health and as a cosmetic due to its anti-wrinkle effect [42]. Additionally, many studies have reported the effects of WSG on the environment. Chowdhury and Bae [49] found some bacterial endophytes separated from WSG, which indicate significant inhibitory reactions against ginseng pathogens. Furthermore, some studies have reported an interconnection between mountain-cultivated ginseng and biodiversity [49–51]. Recent studies have shown that, due to the differences in anatomical structures and physiological conditions, different tissues of WSG retain diverse types of bacterial endophytes, and thus WSG transforms a wide variety of microbial communities [52]. Despite the benefits and effects of ginseng, which have been proven through various studies, few studies [37,53] have examined the effects of WSG. Thus, additional studies intensively focusing on the effects of WSG are needed.

2.4. Components of Wild-Simulated Ginseng

WSG comprises general components, such as water, crude fat, crude protein, crude ash, and nitrogen, and effective components such as crude saponin, including ginsenosides, polyphenol content, acid polysaccharides, and flavonoids. These WSG components have remarkable effects on blood pressure control, liver functions, anti-oxidation, anti-cancer, anti-diabetes, anti-inflammation, and anti-obesity effects [1,25,54,55]. Particularly, the polyphenol and flavonoid contents are related to anti-oxidation. The acidic polysaccharide content helps to boost immunity [54,55]. Ginsenoside, a type of glycoside combined to a glycone and aglycone, gets absorbed and transformed in the body to cause several benefits [54,56]. The components of WSG are closer to those of wild ginseng than to those of field-cultivated ginseng. The representative ginsenosides measured frequently are Rb1, Rb2, Rc, Rd, Re, Rf, Rg1, Rg2, Rg3, Rh2, F2, and CK. Rh2 and CK are not detected in cultivated ginseng, but they are detected only in wild ginseng and WSG [54,55,57]. The components of WSG vary with age, harvest time, and cultivation environment.

3. Topic Modelling

The explosive growth in structured/unstructured data and the development of methodologies for processing and analyzing such data have prompted interests in big data. Big data analysis is divided into data mining techniques with structured data, and text mining techniques with unstructured text data [58]. Text mining techniques determine the latent distribution of semantic patterns and topics from a text by classifying the unstructured text data into units that can be categorized in the computer's analytical process [59]. The difference between text mining and traditional analytical methods is that the data for text mining is larger in scale and utilizes various language processing techniques. The main issue in text mining is how to analyze text rather than the amount of text to be analyzed. Thus, there is a growing interest in text mining, which is specialized in extracting keywords from text on a large scale and conducting statistical analysis in automated environments, called topic modelling [59].

Topic modelling methods for analyzing topics in documents include latent Dirichlet allocation (LDA), probabilistic latent semantic analysis (pLSA), and latent semantic indexing (LSI) [28]. Among them, the LDA method is most commonly used in the analysis of academic research trends [28]. LDA is based on LSI, but it is modified by supplementing the shortcomings of pLSA, which has the advantage of facilitating the identification of relationships between words and concepts [28]. Furthermore, LDA-based topic modelling effectively classifies words with different meanings into context, although the algorithm is simple and the word used is the same [60]. Most importantly, we can attempt to classify research topics, such that we can examine the overall distribution and composition of the topics and the potential topic structures in the study; additionally, we can capture related topics [60]. The LDA method also has advantages over content analysis as LDA-based topic modelling consists of a single document of several topics, each of which consists of a combination of words belonging to that topic [28,60]. For this reason, a single word can be used for multiple topics and not just for one topic. Thus, a document is a mixture of topics, and the topic is a mixture of words [61]. By aggregating the topics, we can also analyze them based on the LDA algorithm to identify the research trends in a particular work. In topic modelling, the number of topics is determined by the researcher, who performs several pre-checks to determine the number of topics, and all words in the document are sequentially arranged in descending order of probability values.

4. Materials and Methods

4.1. Data Collection (Identification)

We used two major databases, SCOPUS and the Web of Science, to search bibliographic information dealing with WSG collected up to October 2020. The first search was done under the title, abstract, and keywords of the paper itself, which includes WSG synonyms (Appendix A). This research follows the preferred reporting items for systematic reviews and meta-analyses (PRISMA) flow diagram [62], which has four steps, namely identification, screening, eligibility, and inclusion (Figure 1).

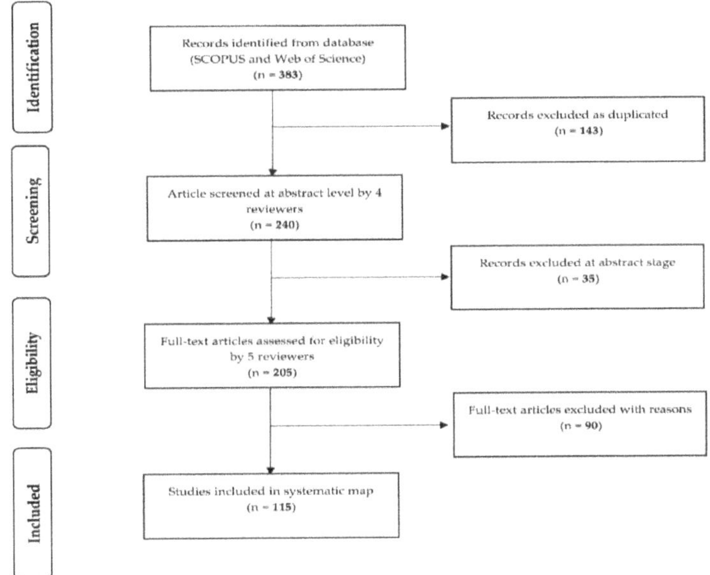

Figure 1. Preferred reporting items for systematic reviews and meta-analyses flow diagram.

4.2. Article Screening, Eligibility, and Inclusion

Three hundred and eighty-three articles collected from the identification stage were filtered through the next three stages, namely screening, eligibility, and inclusion. The literature search and selection were recorded independently by five coders, who compared the remaining records in each course [63]. Coders then screened the relevance of articles to WSG by reviewing the full text. Finally, a total of 115 articles of literature were used for the analysis (Figure 1).

4.3. Coding Strategy

A coding category system was developed to achieve the purpose of this study. After reviewing relevant literature, such as systematic reviews on cultivated ginseng [25,26,64], six categories were constructed, namely bibliographical data, study sites and fields, growth conditions, components, and effects on environment and animals (including humans), as stated previously in Section 2 (Table 1). This categorization was not designed to explore the quality/quantity but to examine general trends (e.g., WSG species, research fields, and publications years) and cause-and-effect relationships (e.g., WSG effect on human/animal or environments). By understanding the research trends in each category, we can determine the direction of WSG research. Five researchers independently coded, compared, and coordinated the results by discussing the mismatched data until reaching a consensus to increase the coding reliability.

Table 1. Coding category system.

Category	Sub-Category
Bibliographic data	Author, Title, Keywords, Abstracts, DOI, and Publication year.
Study site and fields	Country, Species, Farming area, and Study fields (Natural/Social).
Growth conditions	Site, Growth, Genetic diversity, Microstructure, Production, Photosynthesis, Age, Harvest, Pathogen/Disease, Season, and Fertilizer.
Components	Ginsenosides, Genome, Phenolic contents, Fatty acids, Proteins, Carbohydrates, Polyacetylene, Leaves, and Oligosaccharides.
Effects on environment	Soil, Carbon, Biodiversity, and Biocontrol.
Effects on animals (Humans)	Anti-oxidation, Anti-inflammation, Anti-cancer, Psychological (Stress), Obesity, Growth, Immune, Blood lipid profile, Alzheimer's Disease, Glycemic control, Semen, Fatigue, Tumor, Blood Pressure, Insulin Resistance, Liver function, Parkinson's disease, Diabetes, Dry eye syndrome, and Anti-apoptosis.

4.4. Topic Modelling

For topic modelling, this study utilized a list of keywords selected by researchers for topic modelling. Data preprocessing and cleaning are essential before starting LDA [60], as analysis results vary depending on the extracted keywords, or the author may choose unintended keywords, resulting in different results. To minimize the indexing effect of keywords extracted from the morpheme analysis process for data preprocessing, we refined the data by building dictionaries. Particularly, all keywords were reviewed to identify spacing, abbreviations, and word-form unification. For example, as the same words, "WSG" and "wild-cultivated ginseng", which are recognized as designated words with the same meaning, were in the same form to prevent them from being treated as different words when analyzing morphemes. Following this, we used term-frequency-inverse document frequency (TF-IDF) to avoid common errors that hinder the independent separation and grouping of keywords. By weighing words for information retrieval [64], the TF-IDF approach allows us to figure out the most frequent words in all documents, and those most likely to be indistinguishable to a specific topic group [65]. By using TF-IDF scores as a baseline, we extracted several words, including ginseng, wild ginseng, wild cultivated ginseng, *Panax ginseng*, ginsenoside, and antioxidant activities, to increase the independent topic distributions.

The number of topics in modelling is a major issue as it affects data interpretation [60]. The number of topics should be determined by focusing on the possibilities of interpretation of the analytical results and research purposes, rather than relying on probability values alone. During analysis, researchers run the LDA process and choose the number of topics, their iterations, α, β, and other values until topics are distinguishably divided [28]. In this regard, the researchers used a social network analysis software, NetMiner 4 [66]. For reasonable distributions of topic models, this study chose a set of parameters that affect the structure of word and topic distributions [60] with $\alpha = 0.01$, $\beta = 0.001$, and iteration = 10,000 times. After specifying five topics, all words in the papers were automatically calculated and arranged in each topic in descending order of their values, based on the main keyword of each topic with the highest probability. Further, linear regression analysis was conducted to identify the annual research trends in major topics.

5. Results

5.1. Article Distribution

The results indicate significant differences in the research on WSG, geographical imbalance (Figure 2), and specific period upsurge (Figure 3). First, the trends in publishing reveal three major countries, namely the Republic of Korea (56), China (25), and the USA (15), that have been actively conducting research on WSG. Approximately 73% of the total studies were published in the Republic of Korea and China. On the contrary, other countries such as Russia (3), Canada (2), and New Zealand (1) conducted relatively few studies on WSG. Figure 3 displays the fluctuation in the number of WSG studies since the first publication in 1982. After two reports in the 1980s, no article was published in the 1990s. The articles were published again between 2000 and 2001. The number of published articles increased from 2004. Most articles were published in the last 15 years. More than 50% of the studies were published after 2016, and most studies were published in 2019.

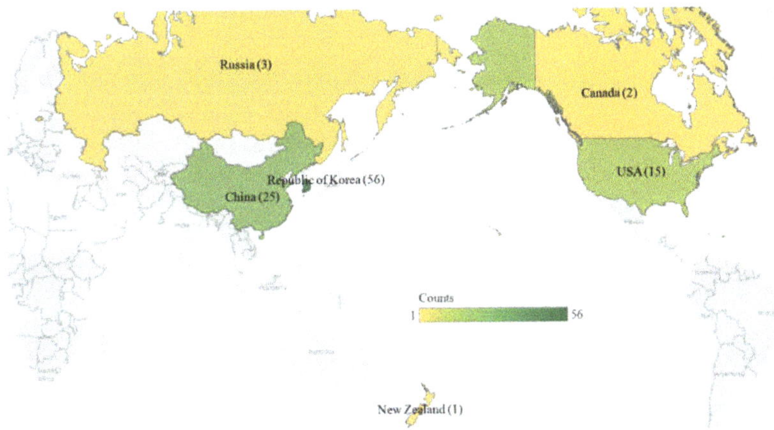

Figure 2. Geographic distribution of wild-simulated ginseng articles. The darker the green color, the higher the number of articles. This figure excludes five articles including more than one country cases and eight articles without specific country case (*N* = 115).

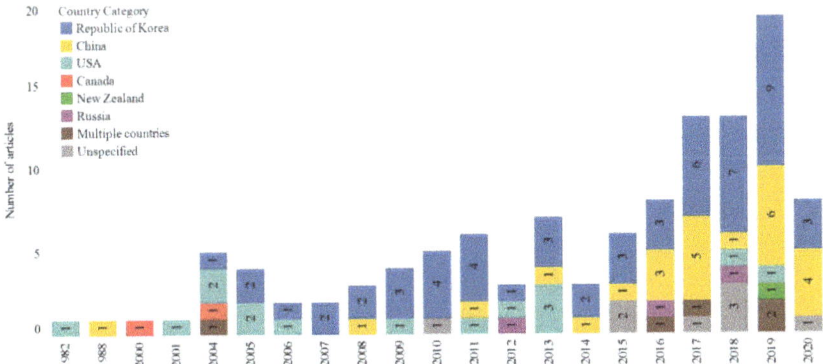

Figure 3. Distribution of wild-simulated ginseng articles by country and year (N = 115). The numbers indicate total articles published from January to December. However, the number of articles in 2020 includes articles from January 2020 to October 2020. The category of multiple countries means more than one country and the category of unspecified means no specific country.

Regarding WSG species, more than 65% of the selected articles investigated *P. ginseng* Meyer, followed by *P. quinquefolius* L. (18%) and *P. notoginseng* (1%) (Figure 4a). Only 5% of the publications belonged to the field of social science, whereas 95% of them belonged to the field of natural science (Figure 4b). The results indicate a lack of social science approach in WSG studies. Most WSG publications are based on the natural science approach and methodology for examining the components of WSG and their medicinal effects. Only a total of five studies applied social science aspects to WSG studies (Figure 4b). Burkhart et al. [67] conducted mixed-method studies on the perspective of stakeholders on federal government conservation efforts such as Convention on International Trade in Endangered Species of Wild Fauna and Flora (CITES), and they concluded that the top-down regulatory approach has limitations on conserving WSG in Pennsylvania [67]. McGraw et al. [35] reviewed scientific findings to understand how environmental changes and direct and indirect interactions with humans affect WSG in North America [35]. Schmidt et al. [68] applied geoadditive regression models to investigate the joint effects of environmental and social factors on ginseng harvest in the USA. The study reports that an overexploitation of WSG violates federal regulation, and a high correlation exists between harvesting and poverty [68]. Chamberlain et al. [69] examined the sociocultural, economic, and ecological elements of sustainable management of NTFPs, and they analyzed many medicinal species in the USA, including WSG. The review concluded that much more research and development is needed to ensure the long-term sustainability of NTFPs, their cultural values, and recognition of economic potential [69]. Jiang and Tian [70] investigated the saponin content in forests using high-performance liquid chromatography (HPLC), and they found that the saponin content of ginseng from Chinese forests is significantly different, suggesting policy support such as investment and optimization of marketing models to promote them [70].

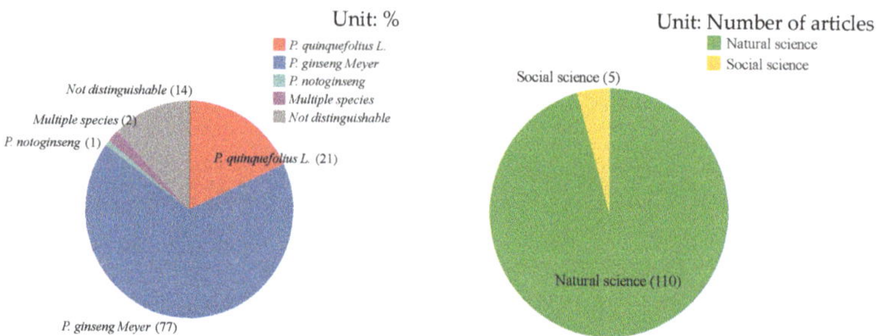

(**a**) Wild-simulated ginseng species   (**b**) Wild-simulated ginseng study fields

Figure 4. Distribution of wild-simulated species (**a**) and study fields (**b**) (*N* = 115).

5.2. Characteristics of Wild-Simulated Ginseng Research

The research analyzed WSG study areas by classifying them into four categories, namely growth conditions, components, effects on the environment, and effects on animals/humans. In the growth conditions category, site conditions (19) have been mostly studied, followed by growth (8), microstructure (7), genetic diversity (6), and photosynthesis (4) (Figure 5a). Among WSG components, most WSG articles covered ginsenoside (26), genome (7), and phenolic compounds (3) (Figure 5b). Regarding the effects of WSG, anti-oxidation (11) and anti-cancer (5) effects have been mostly examined. Most of the anti-oxidant and anti-cancer studies were related to ginsenoside. Lastly, biodiversity (5) was the hottest topic in the environmental effects of WSG (Figure 5c,d).

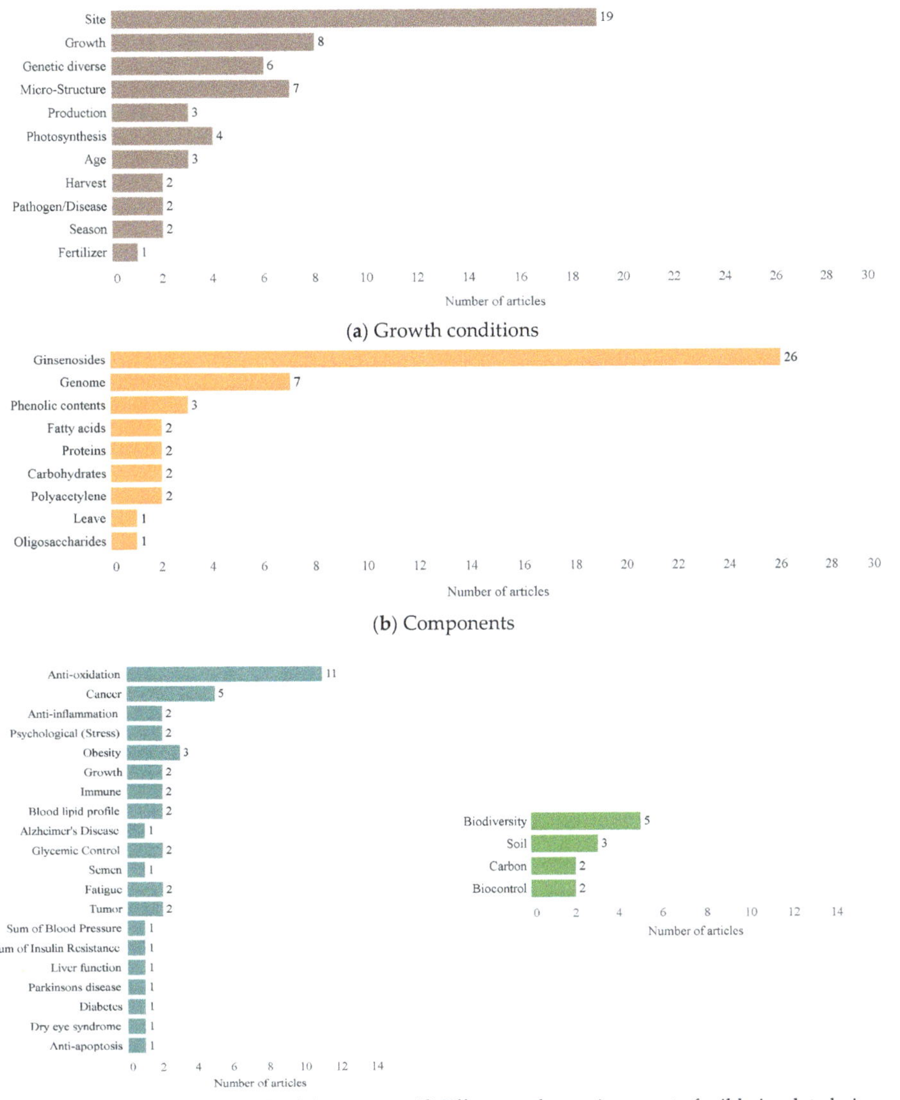

(**a**) Growth conditions

(**b**) Components

(**c**) Effects on humans (animals)   (**d**) Effects on the environment of wild-simulated ginseng

Figure 5. Characteristics of WSG research: (**a**) growth conditions, (**b**) components, (**c**) effects on humans (animals), and (**d**) effects on the environment of wild-simulated ginseng.

5.3. Keyword Frequency

Table 2 shows the frequency of the top 25 keywords out of a total of 481 keywords from the 115 selected articles. The top keywords are mostly ginseng types or species, such as Asian ginseng species, *P. ginseng* (49). As 11 anti-oxidation studies have been conducted (Figure 5c), the relevant keyword, antioxidant activity, was designated with the fifth-highest keyword frequency. However, since frequency analysis alone has limits to discovering the relationship between semantic structures and context of WSG research, it is necessary to look at the knowledge structure of the study through topic modelling analysis.

Table 2. Frequency of the top 25 keywords.

Rank	Keywords	Frequency
1	*Panax ginseng*	49
2	wild-simulated ginseng	30
3	ginsenoside	25
4	ginseng	16
5	antioxidant activity	10
6	wild ginseng	10
7	cultivated ginseng	9
8	metabolite	8
9	American ginseng	7
10	genetic diversity	7
11	cytotoxicity	5
12	HPLC	5
13	medicinal plant	5
14	plant conservation	5
15	UPLC-Q-TOF-MS	5
16	soil	4
17	Araliaceae	3
18	bacterial endophyte	3
19	*Burkholderia stabilis*	3
20	climate change	3
21	ginseng pathogens	3
22	harvest	3
23	identification	3
24	pharmacopuncture	3
25	saponin	3

5.4. Topic Analysis

As a result of topic modelling, five topics (Table 3) were derived from the large keyword network (Figure 6). The five topics included 336 keywords, excluding keywords with high TF-IDF scores as mentioned in the method Section 4.4 Topic Modelling. Table 3 displays the ranks of keywords according to the topic. Each topic was named depending on the characteristics of extracted keywords. Topic A was named "Medicinal effects", and it included keywords on the medical and pharmacological functions of WSG, such as medicinal plant, insulin, Chinese medicine, Tuina, and chronic fatigue syndrome. The studies related to Topic A indicate the wide use of WSG for various medicinal purposes, and they investigated methodologies to maintain or enhance the medicinal efficacy of WSG [71,72]. Furthermore, the remedial effects of WSG have been detected focusing on polyacetylene compounds, and many studies have compared WSG with other medicinal herbs [8,73]. Topic B was named "Metabolite analysis", including relevant keywords such as ultra-performance liquid chromatography–tandem mass spectrometry (UPLC-Q-TOF-MS), HPLC, bioreactor culture, discrimination model, quantitative analysis, orthogonal partial least squared-discriminant analysis (OPLS-DA), and vacuum freeze-drying. These keywords are related to the identification and quantification of WSG ingredients. HPLC, UPLC-Q-TOF-MS, and OPLS-DA are analytical techniques and methods to separate, identify, and quantify each component of WSG [74]. OPLS-DA, in particular, was used as a statistical technique to find the differences in components between different experimental groups by maximizing data visualization of metabolites in WSG [75]. Topic C was "Genetic diversity", which consisted of keywords such as genetic diversity, RAPD, 16S rDNA, UPGMA, genetic differentiation, Illumina MiSeq sequencing, and transcriptome. The terms cell, gene, and microunits belonged to Topic C. Various attempts have been made to identify genetic diversity across a wide variety of WSG species [76–78]. Particularly, keywords shown in Topic C (Table 3), such as RAPD, ISSR, and de novo RNA sequencing, are related to gene sequencing. Topic D was named "Cultivation conditions", which comprises keywords such as plant conservation, harvest, climate change, understory ginseng, photosynthesis, NTFPs, restoration ecology, endangered species, suitable ecological area, forest farming, winter warming, frost, and demography. Topic D focused on growth, cultivation, and cultivation environment. The demand for forest products is increasing worldwide [79], and the harvest pressure on ginseng is also increasing [68]. Along with the increase in the demand for WSG, several studies have investigated the environmental variables to predict the probability of ginseng growth using a species distribution model [68]. Furthermore, studies have been conducted on the ecology and conservation of WSG. Direct and indirect interacting threats, such as deer browsing, harvest, and climate change, and the long-term persistence of WSG have been examined based on the ecological and cultivation characteristics of WSG [35]. Lastly, Topic E was "Bioactive compounds", which included various bioactive compounds directly or indirectly related to WSG. Figure 7 indicates the structure of Topic E with three sub-clusters of bioactive compounds. The first sub-cluster contains keywords related to antimicrobial compounds such as ginseng pathogens, *Burkholderia stabilis*, bacterial endophyte, and biocontrol. *B. stabilis* is a bacterial endophyte isolated from WSG. *B. stabilis* EB159 is used as a biocontrol agent against ginseng pathogens [71]. The second sub-cluster includes the keywords related to brain health compounds such as prefrontal cortex, schizophrenia, and phencyclidine. The third sub-cluster includes the keywords related to antioxidant compounds such as DPPH radicals, kaempferol, and quercetin. Quercetin and kaempferol are the antioxidant compounds found in wild ginseng leaves [80].

Table 3. Five topic groups related to wild-simulated ginseng.

Topic A	Topic B	Topic C	Topic D	Topic E
Medicinal Effects	**Metabolite Analysis**	**Genetic Diversity**	**Cultivation Conditions**	**Bioactive Compounds**
medicinal plant	metabolite	cultivated ginseng	American ginseng	ginseng pathogens
cytotoxicity	UPLC-Q-TOF-MS	genetic diversity	plant conservation	*Burkholderia stabilis*
understory ginseng	HPLC	soil	Harvest	bacterial endophyte
fermentation	cultivated ginseng	cytotoxicity	climate change	quercetin
photosynthesis	saponin	leaf litter	medicinal plant	kaempferol
Korea	identification	RAPD	Araliaceae	metabolite
edible plants	phenolic compounds	pharmacopuncture	genetic diversity	bisphenol A
polyacetylene	bioreactor culture	transcriptome	trimethyltin	testicular toxicity
Allium tricoccum	adventitious roots	suitable site	IL-6	ethyl acetate extract
forest inventory	microorganism	calcium oxalate crystal accumulation	geoadditive model	biocontrol
Actaea racemosa	Araliaceae	soil microbial community	open access resource	DPPH radicals
local and traditional ecological knowledge	cultivation age	16s rDNA	forest herbaceous plant	ginseng leaves
panaxidol	discrimination model	inter-simple sequence repeat (ISSR)	timeseries	genome sequence
forestry economy	genomic	China	species distribution model	cell free supernatant
development countermeasures	subcritical water	genetic differentiation	life table response experiment	leaf
ovarian cancer cells	discrimination	terpenoids biosynthesis genes	non timber forest products	biological control
ginseng extracts	quantitative analysis	terpenoid phytohormones	restoration ecology	insulin
Bifidobacterium	shikonins	mulch	endangered species	panaxidol
growth	gamma irradiation	plant density	Jackknife test	complementary and alternative medicine
stomatal conductance	camptothecin	UPGMA	ecological suitable area	pharmacopuncture
age	biomass	population	deer browsing	medicinal plant

Table 3. *Cont.*

Topic A	Topic B	Topic C	Topic D	Topic E
Medicinal Effects	**Metabolite Analysis**	**Genetic Diversity**	**Cultivation Conditions**	**Bioactive Compounds**
insulin	OPLS-DA	Illumina MiSeq sequencing	extinction vortex	apoptosis
Chinese medicine	vacuum freeze drying	biomarker fungal	plant husbandry	polyacetylene
genome divergence	cell suspension	complementary and alternative medicine	forest farming	Korea
FISH	Chinese medicine	*de novo* RNA sequencing	woods grown American ginseng	photosynthesis
rRNA gene	peptide	microsatellites	cultivated American ginseng	productivity
Tuina	peptidomics	heterozygosity	winter warming	livestock
GISH	ginseng growth cycle	gene expression	frost	antibiotic
chronic fatigue syndrome	EC	Ultra-performance liquid chromatography-tandem mass spectrometry	demography	alternative
massage	agroforestry systems	Chinese medicine	testicular toxicity	hepatocellular carcinoma

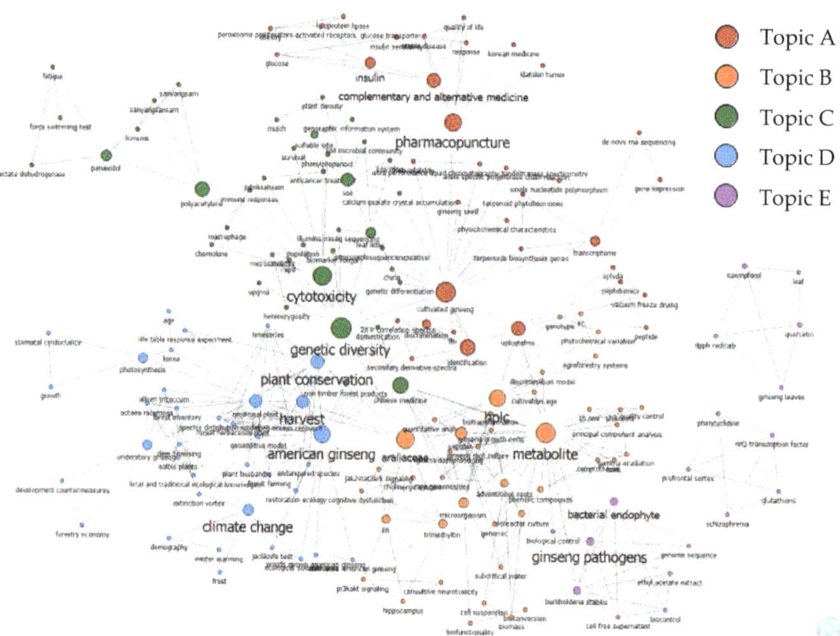

Figure 6. Clustering of topics related to wild-simulated ginseng. The five resulting clusters are colored by topic group. The size of the nodes is proportional to the sum of the links, called centralities. The bigger the size of the node is, the more influential the node is in the network. Topic A: Medicinal effects; Topic B: Metabolite analysis; Topic C: Genetic diversity; Topic D: Cultivation conditions; Topic E: Bioactive compounds.

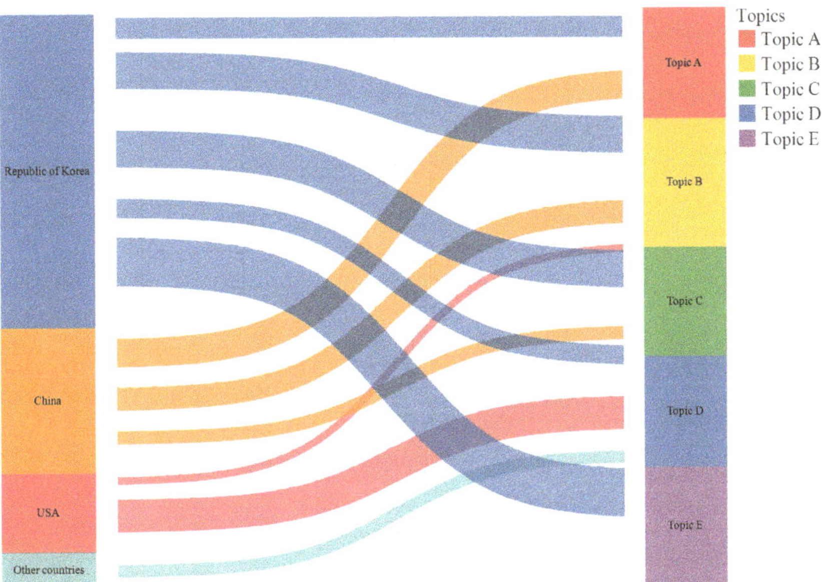

Figure 7. Sankey diagram on topics of wild-simulated ginseng research. The flow diagram shows the country names on the left and aggregation into the five topics on the right. The dimension of rectangles on the right is proportional to the fields' prevalence. Colors are the same as in Figure 2. Topic A: Medicinal effects; Topic B: Metabolite analysis; Topic C: Genetic diversity; Topic D: Cultivation conditions; Topic E: Bioactive compounds.

6. Discussion

6.1. Studies on Wild-Simulated Ginseng by Country

The overall aggregation and examination of topics by country were analyzed (Figure 7). Keywords from WSG articles published for approximately 40 years, including no publication in 11 years from 1989 to 1999, were evenly distributed into five topics, namely Topic A (19.05%), Topic B (22.02%), Topic C (19.64%), Topic D (19.05%), and Topic E (20.24%).

Studies on WSG were mainly conducted in three countries, namely the Republic of Korea, China, and the USA (Figure 2). WSG studies in the Republic of Korea were evenly distributed into all topics, but studies from China and the USA were conducted on some dominant topics. Chinese researchers have focused more on Topic A, "Medicinal effects", but less on Topic D, "Cultivation conditions". On the other hand, the USA shows an opposite pattern, with more studies on Topic D, "Cultivation conditions".

The Republic of Korea conducted the most number of studies on WSG based on a long history of WSG cultivation, since many studies have evaluated the therapeutic potency and cultivation technology of WSG since the first century [29]. Studies on Topic A (medicinal effects) in the Republic of Korea mainly focused on traditional medicinal remedies, including complementary and alternative medicine, such as pharmacopuncture. These studies evaluated the effects of WSG on patients with klatskin tumor [81], cancer [48], and obesity [7]. Topic B (Metabolite analysis) papers were related to the antioxidant effects of ginseng by analyzing the bioprocessing of saponins [82]. Few studies have reviewed microbial ginsenosides that can be used as special additives in the food and medicinal industry [83]. Several papers examined the genetic diversity (Topic C) of WSG populations. Specifically, the studies evaluated genetic differences [84], functional compounds [73], and genetic relationships among different types of cultivated ginseng [78]. Few studies have investigated the cultivation conditions (Topic D), and they indicate significant values of effective environments for ginseng growth, such as shade, soil, photosynthesis [85], mountain conditions [86], and forest types [87]. Lastly, most papers from the Republic of Korea included keywords on bioactive compounds (Topic E). The efficacy of WSG on the health of both humans and livestock has been highlighted through experiments. For instance, researchers found that WSG can be used as a potential biocontrol agent against some chemicals [71], and it possesses an ability to protect hormones for anti-apoptosis [28,88].

Next, the second-largest number of WSG studies was conducted in China (Figure 3). Publications on WSG in China were mainly related to "Medicinal effects" (Topic A) and "Metabolite analysis" (Topic B). The Chinese have been consuming ginseng since thousands of years as a traditional herbal medicine due to its medicinal effects [76,89]. Due its high medicinal value (Topic A), ginseng has been cultivated in large areas in China. Many studies have been conducted on its medicinal basis and application [90,91]. Based on advanced research on the noteworthy medicinal effects of ginseng, such as modulation of immune functions and metabolic processes, anti-stress activity, and improvement of the memory process, various preparations of ginseng have been approved through metabolite analysis (Topic B) for clinical application in China [91].

Lastly, the research on WSG in the USA is shown in Figure 7. A number of studies have been conducted on "Cultivation conditions", classified as Topic D. For nearly 300 years, American ginseng has been harvested in large quantities in eastern and central North America, and it has been exported to China [18,92]. In 1975, however, American ginseng was listed in the CITES Appendix II [35,93]. All species listed in Appendix II are considered susceptible to extinction without trade control, and in the case of American ginseng, there were concerns about existing high yields [35]. Since American ginseng was listed in Appendix II of CITES, early research on the growth and ecology of ginseng population was promoted, which can be explained as an opportunity to support many studies on WSG cultivation environment in the USA [32,94,95]. Moreover, in recent years, American ginseng has been recommended as a candidate for agroforestry crops that can receive premium prices through forest farming practices, especially the "wild-simulated" approach [18]. The production of WSG provides an economic benefit for forest landowners for generating short-term income. Several studies have identified a suitable environment for cultivating WSG [18,92].

6.2. Changing Topic Trends of Wild-Simulated Ginseng Studies

This study analyzed the changing trends in the share of topics over time by categorizing the results over five-year-periods (Figure 8). Before 2005 (Period 1), Topic E, "Bioactive compounds", was the least studied topic (11.76%), and other topics had a similar share. In Period 2, however, Topic A, "Medicinal effects", was prominent (30.95%) and Topic C, "Genetic diversity", also received more attention with an increase of more than 6%. However, studies on Topic B, "Metabolite analysis", decreased significantly (9.52%). In Period 3, the proportion of studies on "Cultivation conditions", Topic D, increased by 10% (31.71%). On the other hand, the proportion of studies dealing with Topic A, "Medicinal effects", decreased from 30.95% (Period 2) to 13.41% (Period 3). Recently, Topic E, "Bioactive compounds", became the most prominent research area in Period 4 (24.72%), but the proportion of studies on Topic D declined. A decrease in the ratio of topic groups does not indicate a decrease in the number of keywords. Both the number of articles and keywords on WSG increased over time (Figures 3 and 8). Figure 8 shows changes in the relatively dominant topics of keywords along with the ratio values.

We observed that in the beginning of WSG research, researchers were interested in the medicinal effects, metabolite analysis, genetic diversity, and cultivation conditions of WSG. However, significant differences were observed in Periods 2 and 3. In Period 2, most studies were conducted on "Medicinal effects", and in Period 3, most studies were conducted on "cultivation conditions" (more than 30%). In Period 4, the study focus changed to "metabolite analysis" and "bioactive compounds". The percentage and weight of the five topics changed in every period. Interestingly, the ratio of studies on bioactive compounds increased continuously while trends of other topics changed in an irregular pattern over time. The change in the ratio of dominant research topics from medicinal effects and cultivation conditions to metabolite analysis and bioactive compounds indicates a shift in research trends from biological and ecological to bioactive metabolite evaluation of WSG. The former includes biological growth, suitable cultivation, and medicinal components of WSG. It concentrates on the interaction between WSG and their cultivation environment. The latter includes identifying the functions of WSG components. Some studies have conducted a detailed analysis of WSG components, including the examination of single compounds of WSG such as kaempferol and quercetin. This approach is deeply related to the application of WSG in treating human diseases. It offers evidence of the contribution of WSG to enhance human health. The evidence is closely linked with value addition [82] in the utilization of WSG. As the global interests in ginseng consumption increase in the market of health foods [94,96], this trend highlights a hidden potential for the expansion of the industrial market of WSG with a focus on bioactive metabolite traits for human wellbeing.

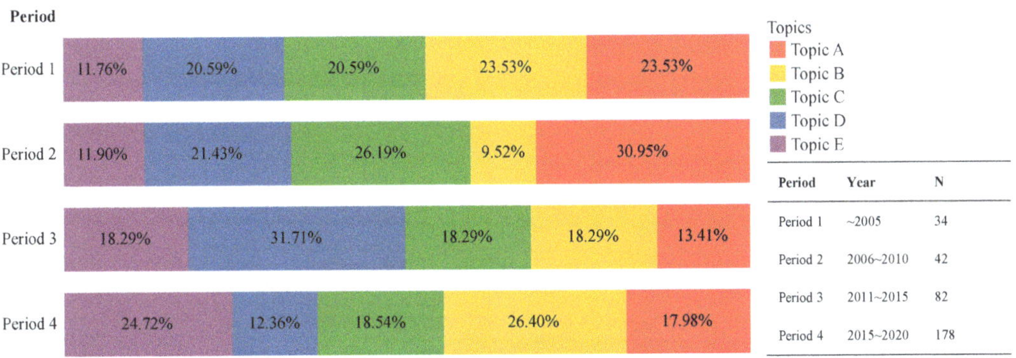

Figure 8. Change in topic share by period. Topic A: Medicinal effects; Topic B: Metabolite analysis; Topic C: Genetic diversity; Topic D: Cultivation conditions; Topic E: Bioactive compounds.

7. Conclusions

This study explored WSG research trends using systematic review and topic modelling. The systematic review approach allowed us to examine the existing literature thoroughly with an obvious, repeatable, and minimal bias process to search, identify, select, evaluate, and aggregate research evidence. Topic modelling allows us to analyze what topics have been conducted and how the topic has changed recently. We found that WSG studies mostly targeted ginsenoside (Figure 5b) and its effects on humans/animals (Figure 5c), the top keywords were related to anti-oxidation (Table 2), and research trends varied by country (Figure 7) and time period (Figure 8). Moreover, the research focus changed from the biological properties of WSG and its cultivation conditions to the precise identification and characterization of the bioactive metabolites of WSG. This change indicates an increased academic interest in the value-added utilization of WSG.

Considering the findings of this study, some suggestions are recommended for future studies on WSG. First, future studies need to consider diverse approaches. Most studies on WSG examined the effects of WSG in natural science (Figures 4b and 5c). For understanding and supporting the WSG industry and market, more studies should be conducted with a social science approach, including value chain analysis, household survey, consumer preference, and trade policy. Additionally, a meta-analysis is necessary to statistically analyze the collected data according to subjects, as our results provided specific subject categories (Figure 5c,d). Meta-analysis combines individual findings into a large study and then proves them statistically. Meta-analysis, starting from anti-oxidant, -cancer, and

-obesity studies, with a relatively sufficient sample size and more data can precisely analyze the trends. Theoretically, only two or three studies are sufficient to conduct a meta-analysis; however, understanding the statistical power of meta-analysis is most important [95].

Although this systematic approach identified the important keywords and research trends, this study had some limitations. First, our study showed a global trend using only English-written papers due to language barriers, although the Republic of Korea and China (Figure 2) are the two major countries. For a more comprehensive study, future research needs to include WSG research papers written in local languages. Moreover, morpheme analysis determines the quality of the analysis in topical modelling analysis. To overcome the disadvantages of unsupervised learning of LDA techniques and to enhance the appropriateness of topics, future studies should apply various transformed techniques and compare the results using new methodologies, such as supervised LDA, extended to conduct supervised learning, and hierarchical LDA, extended to allow multiple layers of the subject structure. Despite these limitations, the study offered dominant research keywords and topic categories of WSG studies. These results will contribute to understanding the trend of WSG research and designing future research on WSG.

Author Contributions: Conceptualization, S.S. and M.S.P.; methodology, S.S. and M.S.P.; software, S.S.; formal analysis, S.S., M.S.P., H.L. (Hansol Lee), H.L. (Haeun Lee), S.L., T.H.K., and H.J.K.; data curation, S.S.; writing—original draft preparation, S.S., H.L., H.L., S.L., and T.H.K.; writing—review and editing, M.S.P. and H.J.K.; visualization, S.S.; supervision, M.S.P.; project administration, M.S.P.; funding acquisition, M.S.P. All authors have read and agreed to the published version of the manuscript.

Funding: This work was supported by the Creative-Pioneering Researchers Program through Seoul National University (SNU).

Informed Consent Statement: Not applicable.

Conflicts of Interest: The authors declare no conflict of interest.

Appendix A

Table 1. Search Query.

Database	Search Query
SCOPUS	TITLE-ABS-KEY ("wild ginseng" OR "wild American ginseng" OR "wild panax ginseng" OR "wild Asian ginseng" OR "wild Asiatic ginseng" OR "wild Chinese ginseng" OR "wild Korean ginseng" OR "wild oriental ginseng" OR "forest ginseng" OR "forest American ginseng" OR "forest panax ginseng" OR "forest Asian ginseng" OR "forest Asiatic ginseng" OR "forest Chinese ginseng" OR "forest Korean ginseng" OR "forest oriental ginseng" OR "wild-simulated ginseng" OR "mountain cultivated ginseng" OR "forest cultivated ginseng" OR "wild-simulated ginseng" OR "wood cultivated ginseng")
Web of Science	TS ("wild ginseng" OR "wild American ginseng" OR "wild panax ginseng" OR "wild Asian ginseng" OR "wild Asiatic ginseng" OR "wild Chinese ginseng" OR "wild Korean ginseng" OR "wild oriental ginseng" OR "forest ginseng" OR "forest American ginseng" OR "forest panax ginseng" OR "forest Asian ginseng" OR "forest Asiatic ginseng" OR "forest Chinese ginseng" OR "forest Korean ginseng" OR "forest oriental ginseng" OR "wild-simulated ginseng" OR "mountain cultivated ginseng" OR "forest cultivated ginseng" OR "wild-simulated ginseng" OR "wood cultivated ginseng")

References

1. Yang, B.W.; Lee, J.B.; Lee, J.M.; Jo, M.S.; Byun, J.K.; Kim, H.C.; Ko, S.K. The comparison of seasonal ginsenoside composition contents in korean wild simulated ginseng (Panax ginseng) which were cultivated in different areas and various ages. *Nat. Prod. Sci.* **2019**, *25*, 1–10. [CrossRef]
2. Nah, S.Y. Ginseng; Recent advances and trends. *J. Ginseng Res.* **1997**, *21*, 1–12.
3. Korea Forest Service. Dictionary of Forest and Forestry. Available online: https://www.forest.go.kr/kfsweb/kfi/kfs/mwd/selectMtstWord&mn=NKFS_04_07_01&orgId= (accessed on 13 April 2021).

4. American ginseng: Non-Detriment Findings—Canada.ca. Available online: https://www.canada.ca/en/environment-climate-change/services/convention-international-trade-endangered-species/non-detriment-findings/american-ginseng.html#_02 (accessed on 13 April 2021).

5. United States Department of the Interior. General Advise for the Export of Wild and Wild-Simulated American Ginseng. 2018, p. 15. Available online: http://www.ahpa.org/Portals/0/PDFs/2018%20PDS/18_0926_FWS_2017_Wild_Ginseng_Sustainable_Harvest.pdf (accessed on 20 May 2021).

6. Nam, K.Y. Clinical applications and efficacy of Korean ginseng. *J. Ginseng Res.* **2002**, *26*, 111–131.

7. Mollah, M.L.; Kim, G.; Moon, H.; Chung, S.; Cheon, Y.; Kim, J.; Kim, K. Antiobesity effects of wild ginseng (Panax ginseng CA Meyer) mediated by PPAR-γ, GLUT4 and LPL in ob/ob mice. *Phyther. Res. An Int. J. Devoted to Pharmacol. Toxicol. Eval. Nat. Prod. Deriv.* **2009**, *23*, 220–225.

8. Taehan, Y.; Jungsuk, Y.; Hwiyong, L.; Beomyong, S.; Lakhyung, K.; Jungdu, R.; Jinchul, S.; Sungtaek, L. Comparing the effects of distilled Rehmannia glutinosa, Wild Ginseng and Astragali Radix pharmacopuncture with heart rate variability (HRV): A randomized, sham-controlled and double-blind clinical trial. *J. Acupunct. Meridian Stud.* **2009**, *2*, 239–247. [CrossRef]

9. Shin, I.S.; Kim, D.H.; Jang, E.Y.; Kim, H.Y.; Yoo, H.S. Anti-fatigue properties of cultivated wild ginseng distilled extract and tts active component panaxydol in rats. *J. Pharmacopunct.* **2019**, *22*, 68.

10. An, K.S.; Choi, Y.O.; Lee, S.M.; Ryu, H.Y.; Kang, S.J.; Yeon, Y.; Kim, Y.R.; Lee, J.G.; Kim, C.J.; Kang, B.J. Ginsenosides Rg5 and Rk1 enriched cultured wild ginseng root extract bioconversion of pediococcus pentosaceus HLJG0702: Effect on scopolamine-induced memory dysfunction in mice. *Nutrients* **2019**, *11*, 1120. [CrossRef] [PubMed]

11. Tu, T.T.; Sharma, N.; Shin, E.J.; Tran, H.Q.; Lee, Y.J.; Nah, S.Y.; Tran, H.P.; Jeong, J.H.; Ko, S.K.; Byun, J.K. Treatment with mountain-cultivated ginseng alleviates trimethyltin-induced cognitive impairments in mice via IL-6-dependent JAK2/STAT3/ERK signaling. *Planta Med.* **2017**, *83*, 1342–1350. [CrossRef]

12. Bae, K.; Kim, E.; Choi, J.J.; Kim, M.K.; Yoo, H.S. The effectiveness of anticancer traditional Korean medicine treatment on the survival in patients with lung, breast, gastric, colorectal, hepatic, uterine, or ovarian cancer: A prospective cohort study protocol. *Medicine* **2018**, *97*, 41. [CrossRef]

13. Hu, J.; Jiao, J.; Wang, Y.; Gao, M.; Lu, Z.; Yang, F.; Hu, C.; Song, Z.; Chen, Y.; Wang, Z. Effect of extract from ginseng rust rot on the inhibition of human hepatocellular carcinoma cells in vitro. *Micron* **2019**, *124*, 102710. [CrossRef] [PubMed]

14. Hong, J.Y.; Shin, S.R.; Bae, M.J.; Bae, J.S.; Lee, I.C.; Kwon, O.J.; Jung, J.W.; Kim, Y.H.; Kim, T.H. Pancreatic lipase inhibitors isolated from the leaves of cultivated mountain ginseng (Panax ginseng). *Korean J. Food Preserv.* **2010**, *17*, 727–732.

15. Park, C.K.; Kwak, Y.S.; Hwang, M.S.; Kim, S.C.; Do, J.H. Current status of ginseng (red ginseng) products in market of health functional food. *Korean Ginseng Res. Ind.* **2007**, *1*, 9–16.

16. Korea Forest Service. *Statistical Yearbook of Forestry*; FAO: Rome, Italy, 2010.

17. Gil, J.; Um, Y.; Byun, J.K.; Chung, J.W.; Lee, Y.; Chung, C.M. Genetic diversity analysis of wood-cultivated ginseng using simple sequence repeat Markers. *Korean J. Med. Crop Sci.* **2017**, *25*, 389–396. [CrossRef]

18. Burkhart, E.P. American ginseng (Panax quinquefolius L.) floristic associations in Pennsylvania: Guidance for identifying calcium-rich forest farming sites. *Agrofor. Syst.* **2013**, *87*, 1157–1172. [CrossRef]

19. McGraw, J.B.; Furedi, M.A.; Maiers, K.; Carroll, C.; Kauffman, G.; Lubbers, A.; Wolf, J.; Anderson, R.C.; Anderson, M.R.; Wilcox, B. Berry ripening and harvest season in wild American ginseng. *Northeast. Nat.* **2005**, *12*, 141–152. [CrossRef]

20. Lim, W.; Mudge, K.W.; Weston, L.A. Utilization of RAPD markers to assess genetic diversity of wild populations of North American ginseng (Panax quinquefolium). *Planta Med.* **2007**, *73*, 71–76. [CrossRef]

21. Kim, K.; Um, Y.; Jeong, D.H.; Kim, H.J.; Kim, M.J.; Jeon, K.S. The correlation between growth characteristics and location environment of wild-simulated ginseng (Panax ginseng CA Meyer). *Korean J. Plant Resour.* **2019**, *32*, 463–470.

22. Grimshaw, J.M.; Russell, I.T. Effect of clinical guidelines on medical practice: A systematic review of rigorous evaluations. *Lancet* **1993**, *342*, 1317–1322. [CrossRef]

23. Kim, B.S.; Lee, J.Y. Keyword Network Analysis on Research Trends on the Software Education. *Korean Assoc. Learn. Curric. Instr.* **2019**, *19*, 385–405.

24. Shin, S.; Soe, K.T.; Lee, H.; Kim, T.H.; Lee, S.; Park, M.S. A systematic map of agroforestry research focusing on ecosystem services in the Asia-Pacific Region. *Forests* **2020**, *11*, 368. [CrossRef]

25. Lee, N.H.; Son, C.G. Systematic review of randomized controlled trials evaluating the efficacy and safety of ginseng. *J. Acupunct. Meridian Stud.* **2011**, *4*, 85–97. [CrossRef]

26. Buettner, C.; Yeh, G.Y.; Phillips, R.S.; Mittleman, M.A.; Kaptchuk, T.J. Systematic review of the effects of ginseng on cardiovascular risk factors. *Ann. Pharmacother.* **2006**, *40*, 83–95. [CrossRef] [PubMed]

27. Kim, M.; Kim, C.; Kim, H. Research trends of connected education from kindergarten to elementary school from 2006 to 2011. *J. Korea Open Assoc. Early Child. Educ.* **2012**, *17*, 503–523.

28. Jelodar, H.; Wang, Y.; Yuan, C.; Feng, X.; Jiang, X.; Li, Y.; Zhao, L. Latent Dirichlet allocation (LDA) and topic modeling: Models, applications, a survey. *Multimed. Tools Appl.* **2019**, *78*, 15169–15211. [CrossRef]

29. Artyukova, E.V.; Kozyrenko, M.M.; Koren, O.G.; Muzarok, T.I.; Reunova, G.D.; Zhuravlev, Y.N. RAPD and allozyme analysis of genetic diversity in Panax ginseng CA Meyer and P. quinquefolius L. *Russ. J. Genet.* **2004**, *40*, 178–185. [CrossRef]

30. Park, H.J.; Kim, D.H.; Park, S.J.; Kim, J.M.; Ryu, J.H. Ginseng in traditional herbal prescriptions. *J. Ginseng Res.* **2012**, *36*, 225. [CrossRef]

31. Lee, F.C. *An Introduction to Korean Ginseng: The Elixir of Life*; Korea Ginseng & Tobacco Research Institute: Daejeon, Korea, 1983.
32. Bai, D.; Brandle, J.; Reeleder, R. Genetic diversity in North American ginseng (Panax quinquefolius L.) grown in Ontario detected by RAPD analysis. *Genome* **1997**, *40*, 111–115. [CrossRef]
33. Davis, J.; Persons, W.S. *Growing and Marketing Ginseng, Goldenseal and Other Woodland Medicinals*; New Society Publishers: Gabriola Island, BC, Canada, 2014; ISBN 1550925636.
34. Taylor, D.A. *Ginseng, the Divine Root*; Algonquin Books: Chapel Hill, NC, USA, 2006; ISBN 1565124014.
35. McGraw, J.B.; Lubbers, A.E.; Van der Voort, M.; Mooney, E.H.; Furedi, M.A.; Souther, S.; Turner, J.B.; Chandler, J. Ecology and conservation of ginseng (Panax quinquefolius) in a changing world. *Ann. N. Y. Acad. Sci.* **2013**, *1286*, 62–91. [CrossRef] [PubMed]
36. Yong, S.H.; Seo, Y.R.; Kim, H.G.; Park, D.J.; Seol, Y.; Choi, E.; Hong, J.H.; Choi, M.S. Growth characteristics and saponin content of mountain-cultivated ginseng (Panax ginseng C. A. Meyer) according to seed-sowing method suitable for cultivation under forest. *Forest Sci. Technol.* **2020**, *16*, 195–205. [CrossRef]
37. Choi, Y.E.; Kim, Y.S.; Yi, M.J.; Park, W.G.; Yi, J.S.; Chun, S.R.; Han, S.S.; Lee, S.J. Physiological and chemical characteristics of field-and mountain-cultivated ginseng roots. *J. Plant Biol.* **2007**, *50*, 198–205. [CrossRef]
38. National Institute of Forest Science. Standard Guideline for Wild Simulated Ginseng Cultivation; Seoul. 2018. Available online: https://nifos.forest.go.kr/kfsweb/cop/bbs/selectBoardArticle.do;jsessionid=RJBgvqF9aWNBxbHLrgoXmuaG3 JWpLSSInTrzeAea1TbE7c464ejdi2rPakZQQ4PL.frswas02_servlet_engine5?nttId=3130499&bbsId=BBSMSTR_1036&pageUnit= 10&pageIndex=17&searchtitle=title&searchcont=&searchkey=&searchwriter=&searchWrd=&ctgryLrcls=CTGRY150&ctgryMdcls= &ctgrySmcls=&ntcStartDt=&ntcEndDt=&mn=UKFR_03_03&orgId=kfri (accessed on 20 May 2021).
39. Korea Forest Service. Survey Report of Wild-Simulated Ginseng Production; Daejeon. 2020. Available online: https://www. forest.go.kr/kfsweb/kfi/kfs/cms/cmsView.do?mn=NKFS_02_01_07_03_01&cmsId=FC_000826. (accessed on 20 May 2021).
40. Um, Y.; Eo, H.J.; Kim, H.J.; Kim, K.; Jeon, K.S.; Jeong, J.B. Wild simulated ginseng activates mouse macrophage, RAW264. 7 cells through TRL2/4-dependent activation of MAPK, NF-κB and PI3K/AKT pathways. *J. Ethnopharmacol.* **2020**, *263*, 113218. [CrossRef]
41. Lee, M.H.; Choi, S.W.; Kim, E.J. Differential anti-carcinogenic effect of mountain cultivated ginseng and ginseng on mouse skin carcinogenesis. *J. Korean Soc. Food Sci. Nutr.* **2012**, *41*, 462–470. [CrossRef]
42. Kim, M.W.; Lee, E.H.; Kim, Y.J.; Park, T.S.; Cho, Y.J. Beauty food activities of wild-cultivated ginseng (Panax ginseng CA Meyer) ground part. *J. Appl. Biol. Chem.* **2018**, *61*, 33–38. [CrossRef]
43. Ok, S.; Kang, J.S.; Kim, K.M. Testicular antioxidant mechanism of cultivated wild ginseng extracts. *Mol. Cell. Toxicol.* **2016**, *12*, 149–158. [CrossRef]
44. Ahn, G.; Park, C.H.; Lee, E.O.; Lee, H.J.; Lee, J.H.; Kim, K.H.; Rhee, Y.H.; Jang, Y.S.; Kim, S.T.; Kim, S.H. Apoptotic effect of ethanol extracts of Bojungbangamtang and Acidic Polysaccharide of Korea Red Ginseng in a MCF7/adR multidrug-resistance breast cancer cells. *Yakhak Hoeji* **2006**, *50*, 272–277.
45. Pyo, J.S.; Kim, H.H.; Kim, K.M.; Kang, J.S. Amelioration of dry eye syndrome by oral administration of cultivated wild ginseng extract. *Bangladesh J. Pharmacol.* **2019**, *14*, 61–66. [CrossRef]
46. Mollah, M.L.; Cheon, Y.P.; In, J.G.; Yang, D.C.; Kim, Y.C.; Song, J.C.; Kim, K.S. Inhibitory effects of cultivated wild ginseng on the differentiation of 3T3-L1 pre-adipocytes. *J. Ginseng Res.* **2011**, *35*, 45–51. [CrossRef]
47. Moon, J.N.; Kim, J.K.; Lee, S.; Kwon, J.H. Antihypertensive effects of Korean wild simulated ginseng (Sanyangsam) extracts in spontaneously hypertensive rats. *Food Sci. Biotechnol.* **2019**, *28*, 1563–1569. [CrossRef] [PubMed]
48. Lee, J.H.; Kwon, K.R.; Cho, C.K.; Han, S.S.R.; Yoo, H.S. Advanced cancer cases treated with cultivated wild ginseng phamacopuncture. *J. Acupunct. Meridian Stud.* **2010**, *3*, 119–124. [CrossRef]
49. Chowdhury, M.D.E.K.; Jeon, J.; Rim, S.O.; Park, Y.H.; Lee, S.K.; Bae, H. Composition, diversity and bioactivity of culturable bacterial endophytes in mountain-cultivated ginseng in Korea. *Sci. Rep.* **2017**, *7*, 1–10. [CrossRef]
50. Sun, H.; Wang, Q.; Zhang, L.; Liu, N.; Liu, Z.; Lv, L.; Shao, C.; Guan, Y.; Ma, L.; Li, M. Distinct leaf litter drive the fungal communities in Panax ginseng-growing soil. *Ecol. Indic.* **2019**, *104*, 184–194. [CrossRef]
51. Sun, H.; Wang, Q.; Liu, N.; Li, L.; Zhang, C.; Liu, Z.; Zhang, Y. Effects of different leaf litters on the physicochemical properties and bacterial communities in Panax ginseng-growing soil. *Appl. Soil Ecol.* **2017**, *111*, 17–24. [CrossRef]
52. Chowdhury, M.D.E.K.; Jeon, J.; Rim, S.O.; Park, Y.H.; Lee, S.K.; Bae, H. Publisher Correction: Composition, diversity and bioactivity of culturable bacterial endophytes in mountain-cultivated ginseng in Korea. *Sci. Rep.* **2018**, *8*, 1. [CrossRef]
53. Kwon, K.R.; Park, W.P.; Kang, W.M.; Jeon, E.; Jang, J.H. Retracted: Identification and analysis of differentially expressed genes in mountain cultivated ginseng and mountain wild ginseng. *J. Acupunct. Meridian Stud.* **2011**, *4*, 123–128. [CrossRef]
54. Jeong, B.G.; Jung, G.R.; Kim, M.S.; Moon, H.G.; Park, S.J.; Chun, J. Ginsenoside contents and antioxidant activities of cultivated mountain ginseng (Panax ginseng CA Meyer) with different ages. *Korean J. Food Preserv* **2019**, *26*, 90–100. [CrossRef]
55. Lee, G.; Choi, G.S.; Lee, J.Y.; Yun, S.J.; Kim, W.; Lee, H.; Baik, M.Y.; Hwang, J.K. Proximate Analysis and Antioxidant Activity of Cultivated Wild Panax ginseng. *Food Eng. Prog.* **2017**, *21*, 208–214. [CrossRef]
56. Han, Y.J.; Kwon, K.R.; Cha, B.C.; Kwon, O. Component analysis of cultivated ginseng, cultivated wild ginseng, and natural wild ginseng by structural parts using HPLC method. *J. Korean Inst. Herb. Acupunct.* **2007**, *10*, 37–53.
57. Lee, D.Y.; Cho, J.G.; Lee, M.K.; Lee, J.W.; Lee, Y.H.; Yang, D.C.; Baek, N.I. Discrimination of Panax ginseng roots cultivated in different areas in Korea using HPLC-ELSD and principal component analysis. *J. Ginseng Res.* **2011**, *35*, 31–38. [CrossRef]

58. Kroeze, J.H.; Matthee, M.C.; Bothma, T.J.D. Differentiating between data-mining and text-mining terminology. *SA J. Inf. Manag.* **2004**, *6*, 4. [CrossRef]
59. Tan, A.-H. Text mining: The state of the art and the challenges. In Proceedings of the PAKDD 1999 Workshop on Knowledge Disocovery from Advanced Databases, Beijing, China, 26–28 April 1999; Citeseer: University Park, PA, USA, 1999; Volume 8, pp. 65–70. Available online: https://citeseerx.ist.psu.edu/viewdoc/download?doi=10.1.1.132.6973&rep=rep1&type=pdf (accessed on 20 May 2021).
60. Maier, D.; Waldherr, A.; Miltner, P.; Wiedemann, G.; Niekler, A.; Keinert, A.; Pfetsch, B.; Heyer, G.; Reber, U.; Häussler, T. Applying LDA topic modeling in communication research: Toward a valid and reliable methodology. *Commun. Methods Meas.* **2018**, *12*, 93–118. [CrossRef]
61. Lee, S.-S. A study on the application of topic modeling for the book report text. *J. Korean Libr. Inf. Sci. Soc.* **2016**, *47*, 1–18.
62. Moher, D.; Liberati, A.; Tetzlaff, J.; Altman, D.G. Preferred reporting items for systematic reviews and meta-analyses: The PRISMA statement. *Int. J. Surg.* **2010**, *8*, 336–341. [CrossRef] [PubMed]
63. Littell, J.H.; Corcoran, J.; Pillai, V. *Systematic Reviews and Meta-Analysis*; Oxford University Press: Oxford, UK, 2008; ISBN 0195326547.
64. Salton, G.; Harman, D. *Information Retrieval*; John Wiley and Sons Ltd: Hoboken, NJ, USA, 2003; pp. 858–863. ISBN 0470864125.
65. Ramos, J. Using tf-idf to determine word relevance in document queries. *Proceedings of the First Instructional Conference on Machine Learning*; Citeseer, 2003; pp. 29–48. Available online: http://citeseerx.ist.psu.edu/viewdoc/download?doi=10.1.1.121.1424&rep=rep1&type=pdf (accessed on 20 May 2021).
66. Cyram. Cyram Netminer 4.1. Seoul Cyram 2013. Available online: http://www.cyram.com/library/library01.jsp (accessed on 20 May 2021).
67. Burkhart, E.P.; Jacobson, M.G.; Finley, J. A case study of stakeholder perspective and experience with wild American ginseng (Panax quinquefolius) conservation efforts in Pennsylvania, USA: Limitations to a CITES driven, top-down regulatory approach. *Biodivers. Conserv.* **2012**, *21*, 3657–3679. [CrossRef]
68. Schmidt, J.P.; Cruse-Sanders, J.; Chamberlain, J.L.; Ferreira, S.; Young, J.A. Explaining harvests of wild-harvested herbaceous plants: American ginseng as a case study. *Biol. Conserv.* **2019**, *231*, 139–149. [CrossRef]
69. Chamberlain, J.; Small, C.; Baumflek, M. Sustainable forest management for nontimber products. *Sustainability* **2019**, *11*, 2670. [CrossRef]
70. Jiang, X.; Tian, Z. Study on Ginsenoside Nutrition Content of Forest Ginseng and Factors of Forestry Economic Grow. *Arch. Latinoam. Nutricion* **2020**, *70*, 127–134.
71. Kim, H.; Rim, S.O.; Bae, H. Antimicrobial potential of metabolites extracted from ginseng bacterial endophyte Burkholderia stabilis against ginseng pathogens. *Biol. Control.* **2019**, *128*, 24–30. [CrossRef]
72. Beon, M.S.; Park, J.H.; Kang, H.M.; Cho, S.J.; Kim, H. Geographic information system-based identification of suitable cultivation sites for wood-cultivated ginseng. *J. Ginseng Res.* **2013**, *37*, 491. [CrossRef]
73. Shin, I.S.; Jo, E.; Jang, I.S.; Yoo, H.S. Quantitative analyses of the functional constituents in SanYangSam and SanYangSanSam. *J. Pharmacopunct.* **2017**, *20*, 274.
74. Wang, J.; Leung, C.; Ho, H.; Chai, S.; Yau, L.; Zhao, Z.; Jiang, Z. Quantitative comparison of ginsenosides and polyacetylenes in wild and cultivated American ginseng. *Chem. Biodivers.* **2010**, *7*, 975–983. [CrossRef]
75. Xu, X.; Xu, S.; Zhang, Y.; Zhang, H.; Liu, M.; Liu, H.; Gao, Y.; Xue, X.; Xiong, H.; Lin, R. Chemical comparison of two drying methods of mountain cultivated ginseng by UPLC-QTOF-MS/MS and multivariate statistical analysis. *Molecules* **2017**, *22*, 717. [CrossRef] [PubMed]
76. Li, S.; Li, J.; Yang, X.L.; Cheng, Z.; Zhang, W.J. Genetic diversity and differentiation of cultivated ginseng (Panax ginseng CA Meyer) populations in North-east China revealed by inter-simple sequence repeat (ISSR) markers. *Genet. Resour. Crop. Evol.* **2011**, *58*, 815–824. [CrossRef]
77. Cruse-Sanders, J.M.; Hamrick, J.L. Genetic diversity in harvested and protected populations of wild American ginseng, Panax quinquefolius L. (Araliaceae). *Am. J. Bot.* **2004**, *91*, 540–548. [CrossRef] [PubMed]
78. Hwang, H.Z.; Yeon, M.J.; Seong, J.Y.; Park, H.Y.; Kim, S.O.; Jeon, H.W.; Kim, K.C.; Choi, K.J.; Lee, Y.J.; Suh, D.S. Genetic variations in Korean wild ginsengs and their genetic relationships. *Korean J. Genet.* **2006**, *28*, 163–169.
79. Schippmann, U.; Leaman, D.J.; Cunningham, A.B. Impact of Cultivation and Gathering of Medicinal Plants on Biodiversity: Global Trends and Issues. In *Biodiversity and the Ecosystem Approach in Agriculture, Forestry and Fisheries*; FAO: Rome, Italy, 2002.
80. Jung, C.H.; Seog, H.M.; Choi, I.W.; Cho, H.Y. Antioxidant activities of cultivated and wild Korean ginseng leaves. *Food Chem.* **2005**, *92*, 535–540. [CrossRef]
81. Han, J.B.; Im, C.R.; Lee, J.W.; Kim, S.S.; Seong, S. A case of inoperable klatskin tumor showing response to wild ginseng pharmacopuncture. *Orient. Pharm. Exp. Med.* **2013**, *13*, 235–237. [CrossRef]
82. Murthy, H.N.; Georgiev, M.I.; Kim, Y.S.; Jeong, C.S.; Kim, S.J.; Park, S.Y.; Paek, K.Y. Ginsenosides: Prospective for sustainable biotechnological production. *Appl. Microbiol. Biotechnol.* **2014**, *98*, 6243–6254. [CrossRef]
83. Eom, S.J.; Kim, K.T.; Paik, H.D. Microbial bioconversion of ginsenosides in Panax ginseng and their improved bioactivities. *Food Rev. Int.* **2018**, *34*, 698–712. [CrossRef]
84. Jang, W.; Jang, Y.; Kim, N.H.; Waminal, N.E.; Kim, Y.C.; Lee, J.W.; Yang, T.J. Genetic diversity among cultivated and wild Panax ginseng populations revealed by high-resolution microsatellite markers. *J. Ginseng Res.* **2020**, *44*, 637–643. [CrossRef]

85. Suh, H.; Seo, S.M.; Woo, S.Y.; Lee, D.S. Forest cultivated ginseng in Korea: All cure medicinal plants. *J. Med. Plants Res.* **2011**, *5*, 5331–5336.
86. Suh, H.; Seo, S.M.; Lee, D.-S.; Woo, S.Y. Net photosynthesis of Panax ginseng on the different age in mountain area, Korea. *J. Med. Plants Res.* **2011**, *5*, 5933–5935.
87. Woo, S.Y.; Lee, D.S.; Kim, P.G. Growth and eco-physiological characteristics of Panax ginseng grown under three different forest types. *J. Plant. Biol.* **2004**, *47*, 230–235. [CrossRef]
88. Ok, S.; Kang, J.S.; Kim, K.M. Cultivated wild ginseng extracts upregulate the anti-apoptosis systems in cells and mice induced by bisphenol A. *Mol. Cell. Toxicol.* **2017**, *13*, 73–82. [CrossRef]
89. Fu, Y.; Yin, Z.; Wu, L.; Yin, C. Fermentation of ginseng extracts by Penicillium simplicissimum GS33 and anti-ovarian cancer activity of fermented products. *World J. Microbiol. Biotechnol.* **2014**, *30*, 1019–1025. [CrossRef]
90. Pan, H.Y.; Qu, Y.; Zhang, J.K.; Kang, T.G.; Dou, D.Q. Antioxidant activity of ginseng cultivated under mountainous forest with different growing years. *J. Ginseng Res.* **2013**, *37*, 355. [CrossRef]
91. Chang-Xiao, L.; Pei-Gen, X. Recent advances on ginseng research in China. *J. Ethnopharmacol.* **1992**, *36*, 27–38. [CrossRef]
92. Van der Voort, M.E.; McGraw, J.B. Effects of harvester behavior on population growth rate affects sustainability of ginseng trade. *Biol. Conserv.* **2006**, *130*, 505–516. [CrossRef]
93. Frey, G.E.; Chamberlain, J.L.; Prestemon, J.P. The potential for a backward-bending supply curve of non-timber forest products: An empirical case study of wild American ginseng production. *For. Policy Econ.* **2018**, *97*, 97–109. [CrossRef]
94. Baeg, I.H.; So, S.H. The world ginseng market and the ginseng. *J. Ginseng Res.* **2013**, *37*, 1–6. [CrossRef] [PubMed]
95. Chung, H.S.; Lee, Y.; Kyung Rhee, Y.; Lee, S. Consumer acceptance of ginseng food products. *J. Food Sci.* **2011**, *76*, S516–S522. [CrossRef] [PubMed]
96. Pigott, T. *Advances in Meta-Analysis*; Springer Science & Business Media: Berlin/Heidelberg, Germany, 2012; ISBN 1461422787.

 *forests*

Article

Transformation of Agro-Forest Management Policy under the Dynamic Circumstances of a Two-Decade Regional Autonomy in Indonesia

Dodik Ridho Nurrochmat [1,*], Ristianto Pribadi [2], Hermanto Siregar [3], Agus Justianto [4] and Mi Sun Park [5]

1   Department of Forest Management, Faculty of Forestry and Environment, IPB University Bogor, Kampus IPB Dramaga, Bogor 16680, Indonesia

2   Directorate General of Forestry Planning and Environmental Governance, Ministry of Environment and Forestry of the Republic of Indonesia, Gedung Manggala Wanabakti Jl. Gatot Subroto, Jakarta 12190, Indonesia; ristianto.p@gmail.com

3   Department of Economics, Faculty of Economics & Management, IPB University Bogor, Kampus IPB Dramaga, Bogor 16680, Indonesia; hsiregar@apps.ipb.ac.id

4   Forestry and Environment Research, Development and Innovation Agency, Ministry of Environment and Forestry of the Republic of Indonesia, Gedung Manggala Wanabakti Jl. Gatot Subroto, Jakarta 12190, Indonesia; ajustianto@gmail.com

5   Graduate School of International Agricultural Technology, Institutes of Green Bio Science & Technology, Seoul National University, 1447, Pyeongchangdaero, Daehwa, Pyeongchang, Gangwon 25354, Korea; mpark@snu.ac.kr

*   Correspondence: dnurrochmat@apps.ipb.ac.id; Tel.: +62-813-1484-5101

Citation: Nurrochmat, D.R.; Pribadi, R.; Siregar, H.; Justianto, A.; Park, M.S. Transformation of Agro-Forest Management Policy under the Dynamic Circumstances of a Two-Decade Regional Autonomy in Indonesia. *Forests* **2021**, *12*, 419. https://doi.org/10.3390/f12040419

Academic Editor: Martin Lukac

Received: 10 March 2021
Accepted: 25 March 2021
Published: 31 March 2021

Publisher's Note: MDPI stays neutral with regard to jurisdictional claims in published maps and institutional affiliations.

Abstract: Agro-forest management policy is one of the most trending issues in Indonesia under the dynamics circumstances of regional autonomy. Regional autonomy has been recognized in the formal governance system of the Republic Indonesia through Regional Governance Law 5/1972 and Village Governance Law 5/1979. A strong political reform following deep economic crisis in 1998 has forced Indonesian President Suharto to step down, and the new government has to accommodate political reform agendas, included a broader regional autonomy, which has been implemented under a Regional Governance Law 22/1999, then replaced by Law 32/2004 and Law 23/2014. The existing Regional Governance Law has shifted almost all authorities in forest management from the regency to the province, and associated with the new established Law 11/2020 on job creation, a single license of multi-purpose forest utilization was introduced, including agroforestry, that will potentially reduce deforestation and improve the community welfare. This study evaluates key elements of local development goals, risks and barriers, as well as basic capitals for agro-forest management in Tebo Regency, Jambi Province, using an Interpretive Structural Modelling approach. Overall, this study concludes that weak coordination, low quality of human capital, inappropriate communication with stakeholders, and lack of financial resources are the greatest challenges to the implementation of agro-forest management, particularly agroforestry, as a part of social forestry schemes.

Keywords: agro-forest management; interpretive structural modelling; regional autonomy; social forestry

1. Introduction

Since the enactment of Regional Autonomy Law 22/1999, the decentralization of forest management has been one of the mainstream policies in Indonesia. This study evaluates policy transformation from an era of large regional autonomy in 1999, where almost all government authorities—except defence and security affairs, fiscal and monetary affairs, foreign affairs, judicature, religion, and other policies in strategic areas such as technologies, conservation, and national standardization—are decentralized to the regency, to a new decentralized regional system that shifted almost all government authorities from the regency to province and central government, but at the same time the central government giving more opportunities to the local communities to manage forests through social forestry license. This means, although most of the forest management authorities have been shifted from the regency to the central and provincial government, more and more numbers of local communities are getting direct benefits from forest management under the new decentralization system.

Agro-forest management means a synergy of agriculture and forest management in a landscape unit [1]. This has a broader meaning rather than agroforestry, i.e., a pattern of landuse for a combination of the cultivation of agriculture and forest commodities in a piece of land. Agro-forest management is very important to give more livelihood alternatives, particularly in a larger village population. It is argued that larger areas of forests allocated for community forests are associated with the increased of community welfare [2,3].

Regional autonomy is one of the most important topics on the agenda of political reform following the collapse of Soeharto's New Order regime in May 1998 [4]. After regional autonomy, some administrative, fiscal, and political authorities were transferred from the central government level in Jakarta to the local level, i.e., provincial, regency (kabupaten) or municipal (kota), and village level. (The hierarchy of the government system in Indonesia from the top to the bottom consists of central government, province, regency or municipal, and village. Both regency and municipal are an autonomous government under the province. Municipal is autonomous city government led by a mayor, while regency is led by a regency head, called "Bupati". Usually, a regency is laid in the rural region, covering a much larger area rather than municipal. In this study, the term "state" refers to "negara" or the government of Indonesia. So, "state forest" means forest belong to the government. The term "regional government" involves all levels of local government, i.e., province, regency or municipal, and village (other than central government)). With regard to decentralization in the agro-forest management, the issuance of a large number of small-scale logging and forest conversion licenses by the Regency Head and the imposition of taxes and regulatory restrictions on timber concessions by the regency governments at the early stage of regional autonomy two decades ago marked the onset of the decentralization era [4,5].

In 2004, Regional Autonomy Law 22/1999 was replaced by Law 32/2004 that removed the authority of the regency head to give small-scale logging license. Since 2004, all logging licenses were not given by the regency or province, but central government, through the Ministry of Forestry (now Ministry of Environment and Forestry). The 2004 regional governance law was leaving some authorities in forest management, particularly protection forests and forest management unit (FMU), to the regency. Ten years after the implementation of Law 32/2004, this law has been replaced by Law 23/2014. Referring to the existing regional governance law, the regency governments have lost most of their authority in forest management, only Taman Hutan Raya or forest garden is under the authority of regency. Some issues regarding the distribution of authority and budget between central and local governments, the lack of local institutions' capacity to implement the decentralization agenda, issues with sustainability in forest use and the livelihoods of the nearby local communities, and the importance of clarity in governance and legal structure of decentralized forest management are widely covered by previous studies. However, the results from analyses of forest management decentralization outcomes vary between previous studies [4,6–10]. Therefore, further empirical studies that take into account local contexts, including governance and resource availability, are required.

This study aims to identify and understand the transformation of agro-forest management policies as a consequence of local development goals, risks and barriers, as well as basic capitals for agro-forest management after two decades of regional autonomy in Tebo Regency, Jambi Province in Indonesia. To achieve its objective, this study therefore addresses the following research questions: (1) What are the key elements of the local development goals and how do these link to agro-forest management policies? (2) What are the main risks and barriers to optimal implementation of the decentralized forest management in the Tebo Regency? (3) What basic capitals does the Tebo Regency have to enhance agro-forest management under social forestry schemes, consisting of community forest (Hutan Kemasyarakatan/HKm), community forest plantation (Hutan Tanaman Industri/HTR), village forest (Hutan Desa/HD), collaborative forest management (Pengelolaan Hutan Pola Kemitraan), customary forest (Hutan Adat), and smallholder private forest (Hutan Rakyat/HR). Within the Indonesian context, basic capitals incorporate not only natural resources but also include various strengths and supporting factors such as education, infrastructure, rule of law, and many others.

2. Conceptual Framework: Forest Governance and Agro-Forest Management Policy

Forest governance is an important element in agro-forest management [3,11,12]. The concept refers to "how the forests and the people participating in management and utilization of the forestry resources are governed" [12–17]. A set of indicators with respect to governance key dimensions can be used as an assessment tool at the local level, i.e., social, economy, ecology, legal and institutional, and accessibility and technology [18]. Amidst the ongoing process of regional autonomy, which reflects a shift in the governance structure, one policy question being imposed is how this would impact agro-forest management. Some previous studies show that decentralization has resulted in a significant difference in forest governance amongst regencies, and that this change in forest governance is associated with the deforestation rates [4,5,19]. Nowadays, two decades after the onset of regional autonomy, the topic of decentralized forest management remains an empirical and important policy matter. According to the existing Regional Governance Law 23/2014, all types of forestry licenses given to corporations or to the community are the authority of the central government. The central government, i.e., Ministry of Environment and Forestry, provides forestry business licenses according to the proposal from the company and social forestry license for farmer group, after verifying and consulting with the regency head and governor through province forest service.

There are variations between regions in terms of human resource capacity, economic growth, public participation, and regulatory and institutional robustness. The Law 11/2020 on job creation has replaced a partial license of forest commodity such as utilization of timber, non-timber, ecotourism, and environmental services (e.g., carbon sequestration) with a single license for multiple products for the forestry business. Thus, the license holder is able to utilize all forest products following the business plan, approved by the Ministry of Environment and Forestry. This law enabled to implement several schemes of agro-forest management. The grey boxes indicate all types of forest and activities related directly with agro-forest management (Figure 1).

The current law on job creation has simplified procedures for getting licenses of any types of forest utilization, including social forestry schemes. Social forestry involves two areas of activity, i.e., community forestry, the community-based activities conducted on forests, and community development, the activities to empower community who live around the forests through, for instance, a forest empowerment program. This program is carried out by the forest authority, together with forest farmer groups, called Lembaga Masyarakat Desa Hutan (LMDH). The farmer group may get a license of forest utilization from the central government, called a recognition of the forest and environment partnership (KULINKK). The government also gives license for "on forest" activities under social forestry schemes, i.e., HKm, HTR, HD, or collaborative forest management, and recognizes customary forests as well as smallholder private forests.

The authority to manage the production forest and protection forest is now held by the provincial government, while the central government holds the authority to manage the conservation forest. Since the enactment of regional autonomy, the forestry sector contribution to Tebo Regency's Gross Domestic Regional Product (GDRP) continued to decline from 10.12% in 2002 to only 5.10% in 2013 [20]. Furthermore, the forest cover in Tebo Regency has also declined over the past two decades (Figure 2).

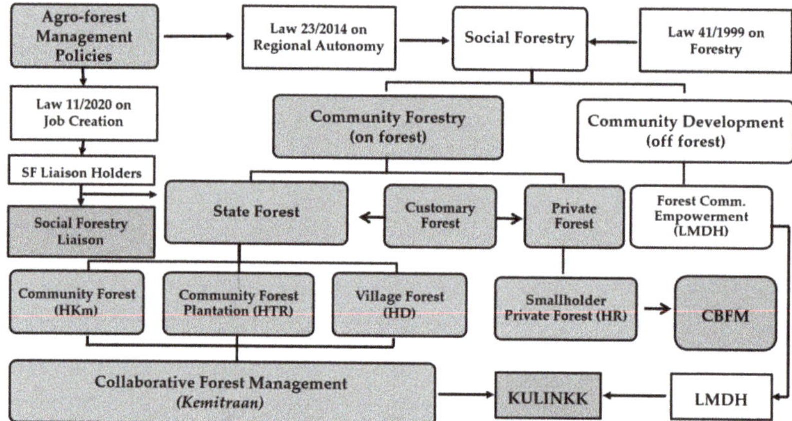

Figure 1. Agro-forest management policies under the current regulations.

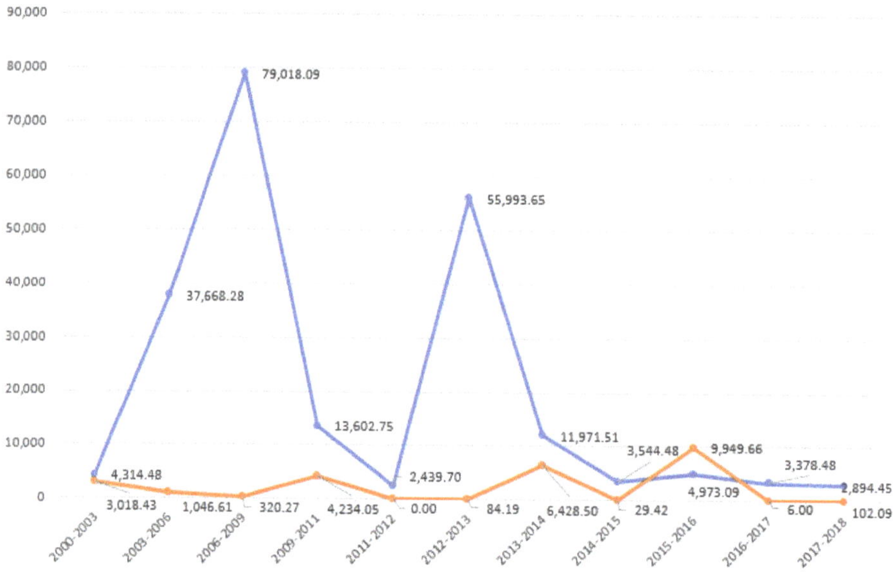

Figure 2. Deforestation and reforestation in Tebo Regency, 2000–2018. Source: Directorate General of Forestry Planning and Environmental Governance, data processed.

The current regulation on a single license of multi-purpose forest utilization, including agro-forest management, is a very important change to not only administrative procedure but also forest business practice in order to reduce deforestation by increasing financial direct benefits of forests and improving the community welfare.

3. Methods

The study was carried out in Tebo Regency, Jambi Province, Indonesia. It handled two decades of the implementation of regional autonomy in Indonesia, and its implication to the agro-forest management, by focusing on the case of the respected regency. The regency has an area of 646,100 hectares [21], and 43.33% or 280,100 ha is covered by forests [22].

This study applies the Interpretive Structural Modelling (ISM) technique. The ISM technique, first introduced by Warfield [23], is usually used to evaluate interrelated elements linked to complex issues, for example as experienced during strategic policy planning [23–26]. In this study, data are collected through a two-stage method. First, it reviews relevant forestry development planning documents. The purpose of this stage is to identify the chain of agro-forest management policies with the national, provincial, and regency levels from the selected planning documents. This qualitative document analysis has been widely used in previous studies [27], using the ISM technique [28,29].

As the first step, we developed a structural self-interaction matrix (SSIM) to represent the contextual relationships between elements listed in Table 1. Four symbols—namely V, A, X, and O—were used to represent the relationship between two variables, i.e., variables i and j. The relationship is denoted by V if variable i influences variable j; A if variable j influences variable i; X if both variables influence each other; and O if there is no relationship between the two variables.

Table 1. List of local development goals.

Category	No	Conditions of Agro-Forest Management
	A1	Increased quantity and quality of land transportation infrastructure
	A2	Increased quantity and quality of public facilities
	A3	Increased quality of irrigation infrastructure
	A4	Increased quality of education
	A5	Increased quality of health
	A6	The realization of the balance growth between population and social development
Local development goals	A7	Decreased gap in community income
	A8	Decreased unemployment rate
	A9	Increased rate of economic growth in leading sectors
	A10	Growth in the tourism sector
	A11	Increased food security based on local food sources and science & technology
	A12	Increased quality of environment
	A13	Increased stability of public order, legal and political awareness
	A14	Increased social and community protection
	B1	Forest management unit is not able to run its functions and authorities appropriately
	B2	High potential of forest and land fires
	B3	High rate of land encroachment, especially for agriculture and settlements
	B4	Limited number of human resources for managing the forest areas
	B5	Limited capacity of human resources for managing the forest areas
	B6	The poverty level of the community surround forest areas is still high
	B7	Low legal awareness of the community
	B8	The potential of forest resources has not been utilized optimally
	B9	Weak working relations between district, provincial and central governments
	B10	The community-based forestry industry has not yet developed
	B11	Low interest in private investment in the forestry sector
	B12	There is no cross-sector based regional development integration yet
	B13	Lack of communication among stakeholders in the context of the synergy of forestry
	B14	Limited funding for forest management
	B15	Low public awareness and understanding of the importance of forests
	C1	Quality of Education and human resources
	C2	Quality of leadership in government agencies
	C3	Access and availability of development funding
	C4	Working infrastructure and technology
	C5	Community social capital
Basic capitals for agro-forest management	C6	Availability of development land
	C7	Governance system
	C8	Networking and partnership
	C9	Multi stakeholders involvement in development process
	C10	Law Enforcement and Assurance
	C11	Stakeholders commitment

A futher step was to develop the Reachability Matrix (RM). The initial RM is obtained by changing the SSIM symbols (i.e., V, A, X, and O) into a binary matrix following standard rules that exist within the ISM literature. (According to Liu et al. [29], the following rules apply: (i) if the (i,j) entry in the SSIM is V, the entry (i,j) in the reachability matrix is set to one; while entry (j,i) is set to zero; (ii) if the (i,j) entry in the SSIM is A, the entry (i,j) in the reachability matrix is set to zero; while entry (j,i) is set to one; (iii) if the (i,j) entry in the SSIM is X, the both (i,j) and (j,i) entries in the reachability matrix are set to one; and (iv) if the (i,j) entry in the SSIM is O, the both (i,j) and (j,i) entries in the reachability matrix are set to zero.) The final RM can be derived after the application of the transivity rules. (The transitivity property states that if attribute X is related to attributed Y, and attribute Y is related to attribute Z, then attribute X is necessarily related to attribute Z.) A further analysis was then performed by partitioning the RM. Based on the final RM, then reachability sets and antecedent sets for a particular attribute can be obtained. (A simple definition noted by Fathi et al. [28] is that the reachability set for a particular element includes the element itself and all of the other elements that it determines; while the antecedent set comprises the element itself and all of the other elements that determine it.) Intersection occurs when the same element numbers are found in both reachability and antecedent sets. We then performed an iteration process (the first iteration process determines the level of a variable that has the same value or an intersection number with the value or a number in the reachability set, which is then denoted as Level 1. Level 1 is then ignored in the second iteration process which produces Level 2. Subsequently, Levels 1 and 2 are ignored in the third iteration process, and so on) to obtain the hierarchic levels of elements. Based on results from the earlier stages to produce the SSIM, RM, and level's partitioning, three digraphs were derived showing the hierarchy of levels of the elements of interest with the bottom level showing the key elements.

Based on the review, we obtained three categories of agro-forest management: (1) local development goals, (2) risks and barriers of agro-forest management, and (3) basic capitals for agro-forest management (Figure 3).

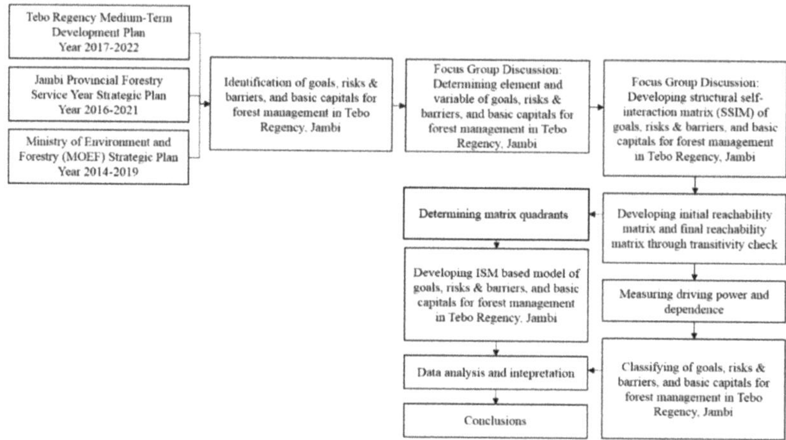

Figure 3. Flow chart of the development goals, risks and barriers, and basic capital for agro-forest management in Tebo Regency, Jambi.

In the second stage, three categories for agro-forest management are discussed in focus group discussions (FGDs) in the Tebo Regency. The FGDs were attended by six key participants (experts) representing institutions related to agro-forest management, namely Jambi Provincial Forestry Service, Tebo Regency Development Planning Agency, Jambi Natural Resources Conservation Unit, East Tebo forest management unit, Jambi University, and a Jambi forestry observer. The selection of those six FGD participants was based on the results from a preliminary visit to Tebo Regency, particularly having discussions with numbers of people from various stakeholders for their recommendations on expert respondents. While this selection basis narrowed down the number of experts relevant to the research theme, information derived from the FGD presents reliable and comprehensive insights into discussion points [28]. The experts ensured that there was no key element missing from the three categories, and they scrutinized the contextual relationships between the listed elements of interests.

4. Results

4.1. Qualitative Planning Document Analysis and Results from FGDs

Following a search of relevant planning documents and consultations with relevant experts, we selected three main planning documents related to forestry development in the Tebo Regency, namely (1) the MoEF Strategic Plan [30]; (2) the Jambi Provincial Forestry Service Year Strategic Plan 2016–2021 [22]; and (3) the Tebo Regency Mid-Term Development Plan (RPJMD) 2017–2022 [21].

The Ministry of Environment and Forestry (MoEF) Strategic Plan Year 2014–2019 is regulated under the MoEF Decree 39/2015. It serves as the legal basis for developing the policy and strategy of environment and forestry sector development applicable to all working units. The Strategic Plan provides a list of programs and activities, barriers, basic capitals, goals, and a financial framework for achieving the MoEF's vision of preserving environmental conditions to improve people's livelihoods and sustainable resource management and, in parallel, enhance natural resources' contribution to the national economy.

Table 1 shows the list of development goals, risks and barriers, and basic capitals for agro-forest management as discussed by FGD participants.

4.2. Identifying and Classifying Key Elements

The ISM technique was used to determine the key elements of the local development goals, risks and barriers, and basic capitals for agro-forest management in Tebo Regency, Jambi. Figure 4 shows the increased quality of environment (A12); increased stability of public order, legal, and political awareness (A13); and increased social and community protection (A14) are the key elements of the Tebo Regency development goals. The well-functioning of the last two aspects (A13 and A14) and progress towards achieving the first aspect (A12) should also contribute to a successful implementation of agro-forest management in Tebo Regency. The increased quality of irrigation infrastructure (A3) is an element that does not have influences on the other elements. Based on discussion during the FGD, irrigation infrastructure as one of the priority development targets is not expected by the public. The agriculture sector in Tebo Regency is dominated by rubber plantations and oil palm plantations, which do not depend on irrigation infrastructure.

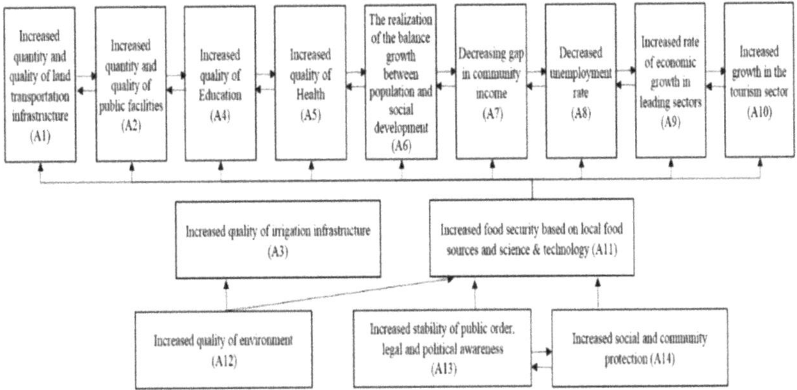

Figure 4. Diagraph of local development goals.

Figure 5 shows five key elements: limited capacity of human resources for managing the forest areas (B5); weak working relations between regency, provincial, and central governments (B9); lack of communication among stakeholders in the context of the synergy of forestry development (B13); limited funding for agro-forest management (B14); and the low public awareness and understanding of the importance of forests (B15).

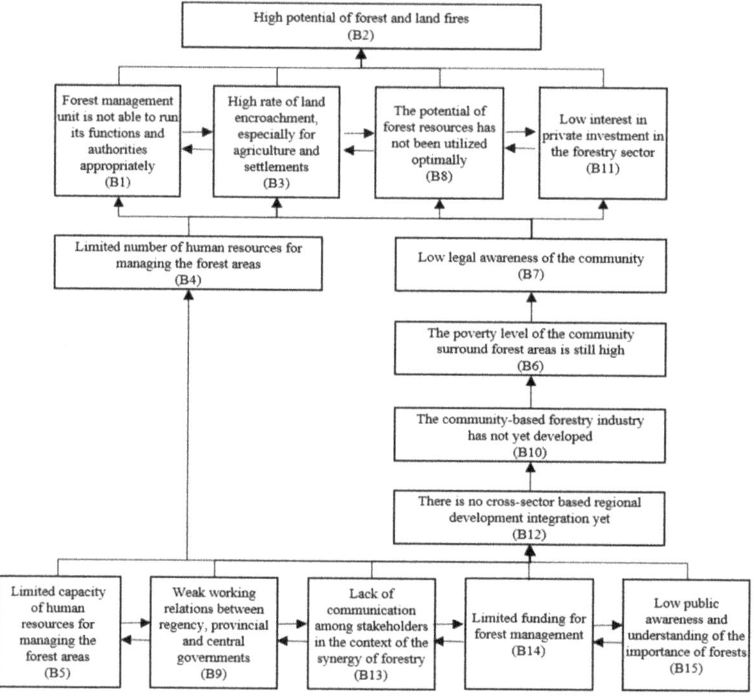

Figure 5. Diagraph of risks and barriers to forest management.

Furthermore, Figure 6 presents seven key elements of basic capitals for agro-forest management. These include the quality of education and human resources (C1); quality of leadership in government agencies (C2); community social capital (C5); governance system (C7); networking and partnership (C8); multi stakeholder's involvement in the development process (C9); and stakeholder commitments (C11). The results further confirm an earlier observation of the importance of the broad dimensions of social and human capital in the agro-forest management from quality education to coordination and good governance.

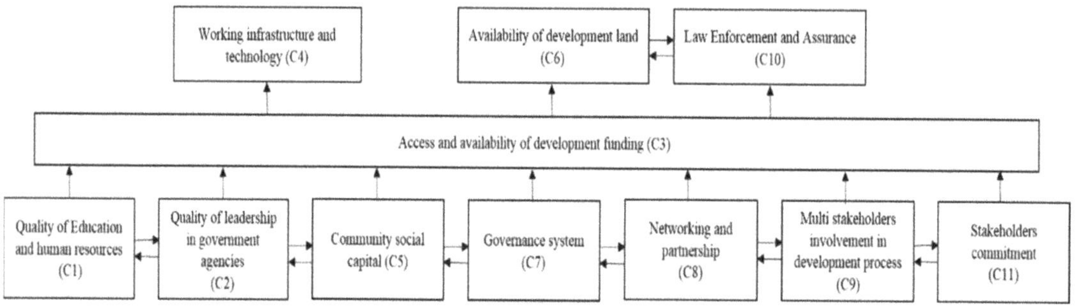

Figure 6. Diagraph of development basic capital.

As the last stage of this analysis, the Matrice d'Impacts croises technique multiplication appliqúe an classment (MICMAC) analysis was applied to identify and assess the driving and dependence powers of each element of interest related to local development goals, risks, and barriers to the agro-forest management and basic capitals for agro-forest management in Tebo Regency, Jambi. Three two-dimensional diagrams were developed using the dependence value and the driving power to determine Y axis and X axis coordinates, respectively. Results from the MICMAC analysis categorized elements into four clusters or quadrants, namely (1) autonomous, (2) dependent, (3) linkage, and (4) independent. The classification results are presented in Figures 7–9.

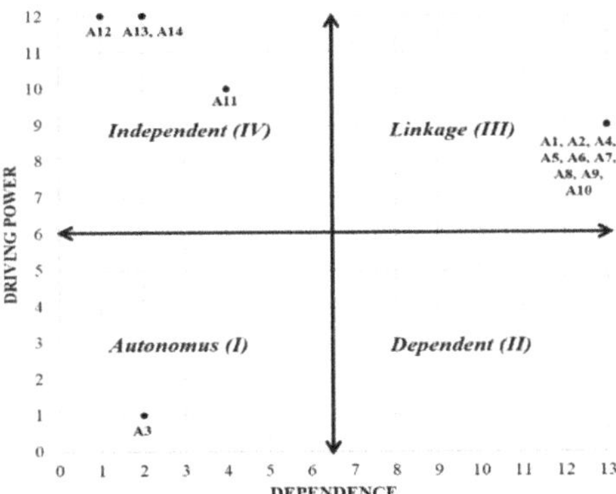

Figure 7. Classification of development goals.

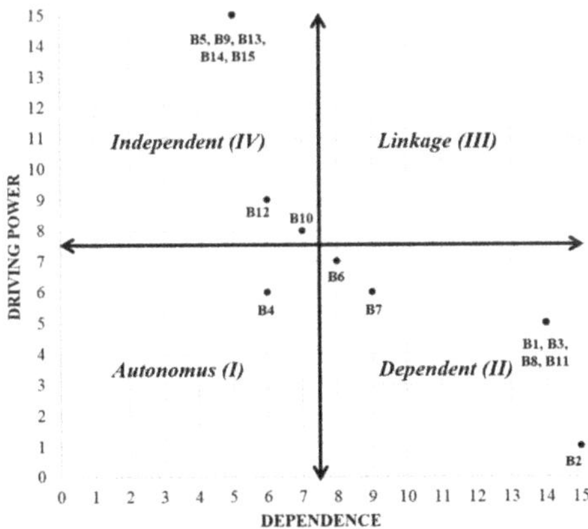

Figure 8. Classification of risks and barriers to agro-forest management.

Figure 7 shows the increased quality of irrigation infrastructure (A3) is in the autonomous category. This implies that this element has weak driving power and weak dependence power, and it is disconnected from the local development goals of Tebo Regency, Jambi. This finding is interesting given that more than 50% of the Tebo Regency's gross domestic regional income comes from the agriculture sector. There is quite a large number of elements in the linkage category, i.e., strong driving and dependence powers. The category reflects unstable relationships, which mean that any action on these elements will impact others and have a feedback effect on themselves.

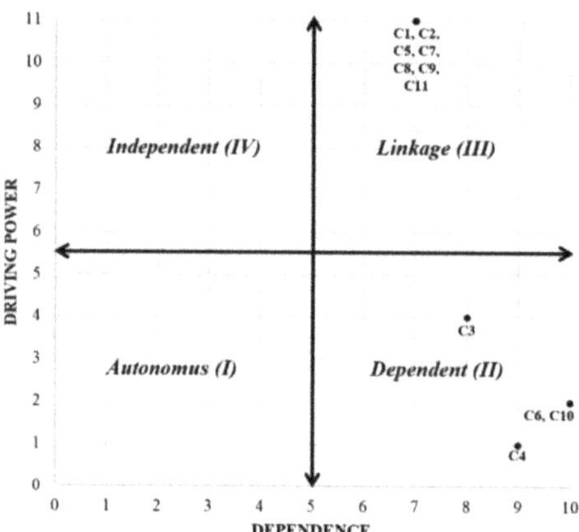

Figure 9. Classification of basic capitals.

Meanwhile, Figure 8 suggests that the limited number of human resources for managing the forest areas (B4) has no relevance to the risks and barriers to agro-forest management. The results indicate less concern over the quantity aspect of human capital, but instead a concern about the quality of human resources involved in the programs. Figure 8 also shows a large number of elements in the independent category, i.e., strong driving power but weak dependence power. With regard to basic capitals, Figure 9 shows seven elements in the linkage category with strong driving as well as dependence powers. Except the element of working infrastructure and technology (C4), it is important to note that all of these elements are associated with human and social capital, including education, leadership, social capital, governance, networking, stakeholder involvement, and law enforcement. The relationships between these elements are interdependent. Hence, an action affecting one element may impact the relevance of other elements to the focus dimension.

5. Discussion

5.1. Addressing Local Development Goals

Whilst there is no agro-forest management-specific document currently in place, there are other documents that look at the environmental aspects of the local development in the Tebo Regency. For instance, one of Tebo Regency's development missions for 2017–2022 is to encourage regional economic growth and individual incomes based on agribusiness and agro-industry development by paying attention to environmental sustainability. The mission is further devided into five programs, two of which are the Program for the Protection and Conservation of Natural Resources; and the Program for Improving the Quality and Access to Information on Natural Resources and the Environment. One of these programs' targets is to develop three 'Adiwiyata schools' in the period of 2017–2022 as an effort to improve social and environmentally friendly culture at local schools (the implementation of which is regulated by the Minister of Environment Regulation Number 5 Year 2013). Forest conservation should be included in the supporting educative materials, which can be further disseminated to other local schools. This can be achieved through a collaboration between the Tebo Regency Government, the Jambi Provincial Forestry Service, and the Office of the Technical Implementation Unit under the coordination of the Ministry of Environment and Forestry.

The "missions" of the Regional Mid-term Development Plan (RPJMD) of Tebo Regency 2017–2022 also encourage the creation of peace and order in community life. The mission is further translated into three development strategies, namely (1) increasing community participation in maintaining security and order; (2) increasing security control and environmental order; and (3) increasing disaster prevention and mitigation efforts. While these do not specifically refer to the forestry sector, the inclusion of environmental and disaster prevention elements should have some linkages with efforts in the forestry sector. Furthermore, earlier studies also highlight that low public and government legal awareness is one of the reasons for poor law obidience [31,32]. To this end, raising legal awareness should continue to be a priority. Improving social and community protection is also very important and is one of the strategic targets in the RPJMD of Tebo Regency 2017–2022. This plan defines a strategy to increase public access to basic social services such as health, education, and employment. Resources allocated by the Tebo Regency government should be integrated with the ongoing forestry development programs to the extent possible, such as Social Forestry that can support various community activities that support agro-forest management and conservation efforts.

5.2. Addressing Risks and Barriers to the Agro-Forest Management

Referring to Nurtjahjawilasa [33], a task force can be set up to perform the assessment. This task force can be formed by the Jambi Provincial Forestry Service, which has an authority based on Law 23/2014 on Regional Governance. The task force should be multidisciplinary and consist of representatives from various relevant institutions for addressing both forestry-specific human resource needs and those required to do cross-sectoral, cross-institution, and stakeholder engagement programs.

Addressing sub-optimal coordination in forest land management between relevant institutions has also been pointed out by previous studies [33–36]. The complexity of agro-forest management and the changing institutional structure requires an agile governance and appropriate work relation arrangements between involved agencies including matters pertaining to accountability, transparency, clarity in work allocation, and knowledge of relevant regulatory frameworks. To this end, there has been an ongoing effort to improve the implementation of agro-forest management in the decentralization era.

Enhancing stakeholder participation is also very important to achieve a good governance [37]. To optimize participation by the stakeholders, the role of effective communication is paramount. The changing technologies, shifting demographics, and regulatory frameworks require an alteration in the way information should be communicated to direct stakeholders and the general public. Learning from the marketing strategies used in the private sector, a communication plan should be an integral part of agro-forest management, the core of which should be around providing regular, useful, and accurate information to audiences.

In addition to an assessment for human resources, improved coordination, and stakeholder participation, sources of funding are also pivotally important for successful agro-forest management. The Food and Agriculture Organization lists challenges facing agro-forest management financing [38]. Many challenges stem from the absence of, or minimal monetary incentives from, agro-forest management programs. Another challenge is the complexity of agro-forest management costs and risks compared to other land uses. Thus, agro-forest management should apply a multiple utilization strategy, such as ecotourism and agroforestry, in addition to timber management.

Our last recommendation on addressing risks and barriers to agro-forest management is to improve public awareness and understanding of the importance of forests. Learning from the United States (US), noting promotional activities undertaken by the US government [39], an innovative and user-friendly approach is suggested. For instance, the United States of Forest Service (USFS) introduced a forest fire prevention mascot named 'Smokey Bear' in 1944, which has been recorded as the longest running US government campaign in history. The popularity of the mascot with "only you can prevent forest fires" motto is comparable to that of Mickey Mouse and Santa Claus. According to the USFS, Smokey Bear has a strong bearing on the decline in forest fires in the US, from 22 million acres per year to under eight million acres per year in 2004. However, it is also important to note that although many fires are bad, fire is a natural part of ecosystems, and some ecosystems are fire-dependent. Thus, prescribed burning may be necessary in certain managed forests.

In addition to a general campaign on agro-forest management, consideration should also be given to improve awareness of select topics such as the role of women in agro-forest management. Women play roles in gathering and utilizing food sources originating from forests such as leaves, fruits, roots, tubers, seeds, nuts, mushrooms, sap, and animal products (such as eggs and honey), while men are invovled in hunting and fishing [40]. Women have the potential to bridge the "gap" between environmental and development issues and pass this knowledge on to their children [41].

5.3. Enhancing Basic Capitals

The first and foremost basic capital to implement any management and development program, including agro-forest management, is education and human resources. In the context of decentralization, the human resource capacity is very important in determining regional development priorities. The lack of human resource capacity is a weakness of developing countries that becomes an obstacle to implement decentralized forest management [42]. Strategies to address issues with quality human resources should include the development of a clear career pathway in the forestry sector, consolidation between the sector's needs and education, and training programs including knowledge specific to the implementation of decentralized agro-forest management.

Second, a leadership role is critical for successful agro-forest management. Leadership is seen as one of the success factors for community participation or collective actions in addition to the allocation of resources such as costs, time and energy, and expected benefits from the collection action [43]. Effective leadership can encourage collective action through inspiring the community, urging the implementation of institutional norms, resolving conflicts, building networks with development partners, and ensuring the achievement of benefits expected by the community. To achieve these outcomes, strategies such as leadership management exercises, encouraging participation, and capacity development of early-career officials should be considered.

In addition to education and a leadership role, social capital is another important basic capital for agro-forest management. Social capital according to Putnam is defined as a social institution that involves networks, norms, and social trust that encourage social collaboration (coordination and cooperation) for the common good [44], while Bourdieu [45] defines social capital as a collection of resources owned by each member in a group that is used together. As summarized by Ishihara and Pascual [46], since the 1990s the concept of social capital has received growing attention in the literature on the management of common pooled resources and collective actions, especially in relation to the use and development of sustainable natural resources. Social capital contributes to collective actions to manage complex environments through knowledge creation and diffusion. It can also improve the community's ability to overcome social and environmental problems including challenges facing agro-forest management. The Tebo Regency Government should therefore assess social capital currently existing in the community before developing a program to increase social capital, and its contribution to the agro-forest management.

Fourth, we highlight the need to improve the governance system. A key element of this is enhancing the rule of law. For instance, Yasmi et al. [47] review conditions under which decentralization can achieve its objectives, which include addressing issues related to property rights such as de jure and de facto claims of forests. In addition to the involvement of governments and communities, a solution to such issues may benefit from the involvement of a mediator, making reference to scientific findings and results from research, and planning and monitoring by multi stakeholders [48] to gain an optimum benefit [49].

Part of good forest governance is enhanced partnerships and an efficient forest business license. The establishment of Law 11/2020 on job creation, has changed a complex and partial forest business license to a single license of multi-purpose forest utilization. In addition, the complexity of agro-forest management that requires scientific support and significant human, financial, and technological inputs as well as governments' resource constraints provide the rationale for partnerships between different levels of the government and stakeholders including the private sector, NGOs, and international institutions. A stakeholder engagement plan is advisable to provide practical guidance towards achieving a participatory agro-forest management. The plan may include activities such as the regular holding of multi-stakeholder forums, and the use of electronic platforms such as a regency government-managed website, to invite partnership opportunities. These would only succeed, however, if involved stakeholders have clarity about the intended outcomes of and benefits from such partnerships. This study formulated a number of policy recommendations as summarized in Table 2.

Table 2. Recommendations to address issues pertaining local development goals, risks and barriers to forest management, and provision of basic capitals.

Local Development Goals (LD)	Risks and Barriers to Forest Management (RB)		Basic Capitals (BC)	
1. Improve the environment quality 2. Increase the stability of public order, political and legal awareness 3. Improve social and community protection	1. Assess the need and quality of human resources for managing the forest areas 2. Strengthen coordination between relevant institutions and groups 3. Enhance communication between stakeholders 4. Improve forest management financing 5. Increase public awareness and understanding of the importance of forests		1. Education and human resources 2. Effective leadership role 3. Improve social capital 4. Governance including rule of law and multi-stakeholder engagement 5. Partnerships	

In addition to a periodic comprehensive forest inventory (Inventarisasi Hutan Menyeluruh Berkala/IHMB), this study also recommends conducting a periodic comprehensive social inventory (Inventarisasi Sosial Menyeluruh Berkala/ISMB) to define the best institutional option of agro-forest management according to state capacity and social capital [5,50,51], in order to pursue local development goals, to address risk and barriers to agro-forest management, and to strengthen basic capitals (Figure 10).

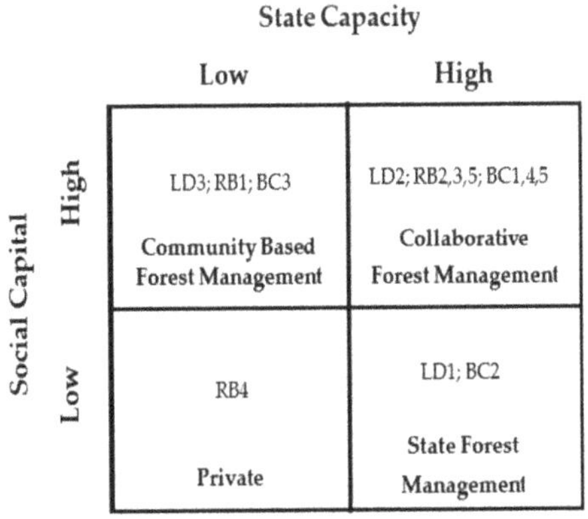

Figure 10. Recommended institution of the agro-forest management according to state capacity and social capital.

According to the matrix in Figure 10, this study recommends a community-based forest management approach as the most appropriate institutional choice for agro-forest management due to the situation of high social capital and low state capacity. In this situation, LD3—Improve social and community protection, RB1—Assess the need and quality of human resources for managing the forest areas, and BC3—Improve social capital are recommended to address issues pertaining to local development goals, risks, and barriers to forest management and provision of basic capitals. When both social capital and state capacity are high, then a collaborative forest management will be recommended for agro-forest management, by implementing LD2—Increase the stability of public order, political and legal awareness, RB2—Strengthen coordination between relevant institutions and groups, RB3—Enhance communication between stakeholders, RB5—Increase public awareness and understanding of the importance of forests, BC1—Education and human resources, BC4—Governance including rule of law and multi-stakeholder engagement, and BC5—Partnership.

State forest management is recommended for agro-forest management in the situation of low social capital and high state capacity, by implementing LD1—Improve the environment quality and BC2—Effective leadership role. Finally, private agro-forest management is recommended in the situation where both social capital and state capacity are low, to avoid open access and the tragedy of the commons [52], amongst other by implementing RB4—Improve forest management financing.

6. Concluding Remarks

The Indonesian democratic decentralization process is a work in progress. After two decades of the implementation of regional autonomy in Indonesia, state, provincial, regency, and also village governments continue to face challenges to find an appropriate distribution of authority and responsibility among themselves, and a way to interact and partner with other stakeholders. The dynamic socio-economic environments alongside pressures on the regulatory frameworks and governance systems to be agile, responsive, and inclusive have put even more pressures for government agencies at the regency level to pursue their own development goals and be in a position to implement agro-forest management appropriately.

We found that out of the 14 development goals identified in the planning document review and FGDs, 3 are viewed as key elements for Tebo Regency, namely the increased environmental quality; increased stability of public order, legal, and political awareness; and increased social and community protection. These three elements indicate a balance between social and security expectations from the community and the environmental needs. If achieved, these development goals will provide a strong base for further progress in the region's agro-forest management efforts.

Furthermore, we identified that five out of the 15 risks and barriers to agro-forest management, identified in the planning document review and FGDs are perceived to have a major influence on agro-forest management. These include the limited capacity of human resources for managing the forest area; weak inter-agency work relationships; lack of communication among stakeholders; limited funding for agro-forest management; and low awareness and public understanding of the importance of forests. It is evident that most of these are centred at the needs to address human dimension of the agro-forest management including human capital, coordination, and communication.

Our analysis also found that 7 out of the 11 basic capitals identified in an earlier stage are perceived as key elements. These cover the quality of education and human resources; quality of leadership in government agencies; community social capital; governance system; networking and partnership; multi stakeholders' involvement in the development process; and stakeholder commitments. The interdependency between these elements is obvious given that many of these are part of what constitutes good governance.

Based on results from the analysis, several recommendations were put forward. Whilst most, if not all, recommendations that we propose here are not novel, the persistent issues with regard to the lack of coordination and communication, quality human capital, stakeholder participation, as well as financing issues call for concrete actions from governments and other stakeholders to focus on practical ways to address these long-standing issues. Overall, this study highlights the importance of reviewing planning documents and understanding how stakeholders perceive the planning goals, challenges, and basic capitals, hence positioning themselves to contribute to the success of the implementation of agro-forest management.

Author Contributions: Conceptualization D.R.N., H.S., R.P., M.S.P.; methodology D.R.N., H.S., R.P., M.S.P.; software R.P., H.S.; validation D.R.N., R.P., H.S.; formal analysis D.R.N., R.P., H.S., A.J., M.S.P.; investigation R.P.; resources D.R.N., R.P., A.J., M.S.P.; data curation R.P.; writing—original draft preparation R.P., D.R.N.; writing—review and editing D.R.N., M.S.P.; visualization D.R.N., R.P.; supervision D.R.N., H.S., A.J.; project administration R.P., D.R.N.; funding acquisition R.P., D.R.N. All authors have read and agreed to the published version of the manuscript.

Funding: This research was funded by the Ministry of Forestry of the Republic of Indonesia, Nr. SK. 6038/Menhut-II/Peg/2014 and partly supported by WCR Program Kemristek BRIN 2020–2021 and ABS Fund of CRC 990-EFForTS (DFG).

Institutional Review Board Statement: Not applicable.

Informed Consent Statement: Not Applicable.

Data Availability Statement: The data that support the findings of this study are available from the corresponding author upon request.

Acknowledgments: The authors would like to thank the Ministry of Environment and Forestry of the Republic of Indonesia for financing this study and the partial funding of WCR Program Kemristek BRIN 2020–2021 and ABS Fund of CRC 990-EFForTS (DFG). We pay tribute to our respected advisor, the late Endang Suhendang, who shared with us not only scientific knowledge but also wisdom. Our gratitude also goes to Risti Permani of Deakin University, Australia, for her contributions in English editing.

Conflicts of Interest: The authors declare no conflict of interest. The funders had no role in the design of the study; in the collection, analyses, or interpretation of data; in the writing of the manuscript, or in the decision to publish the results.

References

1. Marchi, M.; Ferrara, C.; Biasi, R.; Salvia, R.; Salvati, L. Forest Management and Soil Degradation in Mediterranean Environments: Towards a Strategy for Sustainable Land Use in Vineyard and Olive Cropland. *Sustainability* **2018**, *10*, 2565. [CrossRef]
2. Dewi, S.; Belcher, B.; Puntodewo, A. Village economic opportunity, forest dependence, and rural livelihoods in East Kalimantan, Indonesia. *World Dev.* **2005**, *33*, 1419–1434. [CrossRef]
3. Erbaugh, J.T.; Nurrochmat, D.R.; Purnomo, H. Regulation, formalization, and smallholder timber production in northern Central Java, Indonesia. *Agroforest. Syst.* **2017**, *91*, 867–880. [CrossRef]
4. Pribadi, R.; Nurrochmat, D.R.; Suhendang, E.; Siregar, H. Enhancing the role of the district government in decentralized forest management. *J. Manaj. Hutan Trop.* **2020**, *26*, 114–122. [CrossRef]
5. Nurrochmat, D.R. *The Impacts of Regional Autonomy on Policitcal Dynamics, Socio-Economics and Forest Degradation Case of Jambi—Indonesi*; Cuvillier Verlag: Goettingen, Germany, 2005.
6. Barr, C.M.; Resosudarmo, I.A.P.; Dermawan, A.; McCarthy, J.; Moeliono, M.; Setiono, B. (Eds.) *Decentralization of Forest Administration in Indonesia: Implications for Forest Sustainability, Economic Development, and Community Livelihoods*; CIFOR: Bogor, Indonesia, 2006.
7. Larson, A.M. Decentralization and forest management in Latin America: Towards a working model. *Public Adm. Dev.* **2003**, *23*, 211–226. [CrossRef]
8. Xu, J.; Ribot, J.C. Decentralisation and accountability in forest management: A case from Yunnan, Southwest China. *Eur. J. Dev. Res.* **2004**, *16*, 153–173. [CrossRef]
9. Turyahabwe, N.; Geldenhuys, C.J.; Watts, S.; Obua, J. Local organisations and decentralised forest management in Uganda: Roles, challenges and policy implications. *Int. For. Rev.* **2007**, *9*, 581–596. [CrossRef]
10. Mbwambo, L.; Eid, T.; Malimbwi, R.E.; Zahabu, E.; Kajembe, G.C.; Luoga, E. Impact of decentralised forest management on forest resource conditions in Tanzania. *For. Trees Livelihoods* **2012**, *21*, 97–113. [CrossRef]
11. Sahide, M.A.K.; Supratman, S.; Maryudi, A.; Kim, Y.S.; Giessen, L. Decentralisation policy as recentralisation strategy: Forest management units and community forestry in Indonesia. *Int. For. Rev.* **2016**, *18*, 78–95. [CrossRef]
12. Maryudi, A.; Nurrochmat, D.R.; Giessen, L. Research trend: Forest policy and governance—Future analyses in multiple social science disciplines. *For. Policy Econ.* **2018**, *91*, 1–4. [CrossRef]
13. Erbaugh, J.T.; Nurrochmat, D.R. Paradigm shift and business as usual through policy layering: Forest-related policy change in Indonesia (1999–2016). *Land Use Policy* **2019**, *86*, 136–146. [CrossRef]

14. Mohanty, B.; Sahu, G. An Empirical Study on Elements of Forest Governance: A Study of JFM Implementation Models in Odisha. *Procedia Soc. Behav. Sci.* **2012**, *37*, 314–323. [CrossRef]
15. Nurrochmat, D.R.; Nugroho, I.A.; Hardjanto; Purwadianto, A.; Maryudi, A.; Erbaugh, J.T. Shifting contestation into cooperation: Strategy to incorporate different interest of actors in medicinal plants in Meru Betiri National Park, Indonesia. *For. Policy Econ.* **2017**, *83*, 162–168. [CrossRef]
16. Harbi, J.; Erbaugh, J.T.; Sidiq, M.; Haasler, B.; Nurrochmat, D.R. Making a bridge between livelihoods and forest conservation: Lessons from Non-Timber Forest Products' utilization in South Sumatera, Indonesia. *For. Policy Econ.* **2018**, *94*, 1–10. [CrossRef]
17. Astuti, E.W.; Hidayat, A.; Nurrochmat, D.R. Community forest scheme: Measuring impact in livelihood. case study Lombok Tengah Regency, West Nusa Tenggara Province. *J. Manaj. Hutan Trop.* **2020**, *26*, 52–58. [CrossRef]
18. Sukwika, T.; Darusman, D.; Kusmana, C.; Nurrochmat, D.R. Evaluating the level of sustainablity of privately managed forest in Bogor, Indonesia. *Biodiversitas* **2016**, *17*, 241–248. [CrossRef]
19. Suwarno, A.; Hein, L.; Sumarga, E. Governance, decentralisation and deforestation: The case of central Kalimantan Province, Indonesia. *Q. J. Int. Agric.* **2015**, *54*, 77–100.
20. Badan Pusat Statistik Kabupaten Tebo. *Kabupaten Tebo Dalam Angka 2018*; BPS: Tebo, Mozambique, 2018.
21. Pemerintah Kabupaten Tebo. *Rencana Pembangunan Jangka Menengah Daerah Kabupaten Tebo Tahun 2017–2022*; Badan Perencanaan Pembangunan Daerah Kabupaten Tebo: Tebo, Mozambique, 2017.
22. Dinas Kehutanan Provinsi Jambi. *Rencana Strategis Dinas Kehutanan Provinsi Jambi 2016–2021*; Dinas Kehutanan Provinsi Jambi: Jambi, Indonesia, 2015.
23. Warfield, J.N. Developing Interconnected Matrices in Structural Modeling. *IEEE Trans. Syst. Man Cybern.* **1974**, *SMC-4*, 81–87. [CrossRef]
24. Janes, F.R. Interpretive structural modelling (ISM): A methodology for structuring complex issues. *Trans. Inst. Meas. Control* **1988**, *10*, 145–154. [CrossRef]
25. Marimin. *Pengambilan Keputusan Kreteria Majemuk: Teknik dan Aplikas*; Gramedia Widiasarana: Jakarta, Indonesia, 2004.
26. Attri, R.; Dev, N.; Sharma, V. Interpretive structural modelling (ISM) approach: An overview. *Res. J. Manag. Sci.* **2013**, *2319*, 1171.
27. Katila, P. Forestry development priorities in Finnish national forest programs. *Int. For. Rev.* **2017**, *19*, 125–138.
28. Fathi, M.; Ghobakhloo, M.; Syberfeldt, A. An Interpretive Structural Modeling of teamwork training in higher education. *Educ. Sci.* **2019**, *9*, 16. [CrossRef]
29. Liu, P.; Li, Q.; Bian, J.; Song, L.; Xiahou, X. Using interpretive structural modeling to identify critical success factors for safety management in subway construction: A china study. *Int. J. Environ. Res. Public Health* **2018**, *15*, 1359. [CrossRef] [PubMed]
30. Kementerian Lingkungan Hidup dan Kehutanan. *Rencana Strategis Kementerian Lingkungan Hidup dan Kehutanan Tahun 2015–2019*; KLHK: Jakarta, Indonesia, 2015.
31. Arnold, L.L. Deforestation in decentralized Indonesia: What's law got to do with it? *Law Environ. Dev. J.* **2008**, *4*, 77–101.
32. Usman, A.H. Kesadaran hukum masyarakat dan pemerintah sebagai faktor tegaknya negara hukum di Indonesia. *J. Wawasan Huk.* **2014**, *30*. [CrossRef]
33. Nurtjahjawilasa. *Kelembagaan dan Kebijakan Pengembangan Sumber Daya Manusia Kehutanan: Studi Kasus di Bidang Perizinan Kehutanan*; Institut Pertanian Bogor: Bogor, Indonesia, 2016.
34. Ekawati, S. Tata hubungan kerja antar institusi kehutanan dalam pengelolaan hutan lindung di era otonomi daerah. *J. Anal. Kebijak. Kehutan.* **2010**, *7*, 211–225. [CrossRef]
35. Syahadat, E.; Suryandari, E.Y. Pola tata hubungan kerja dalam pembangunan hutan kemasyarakatan. *J. Anal. Kebijak. Kehutan.* **2016**, *13*, 127–145. [CrossRef]
36. Nurrochmat, D.R.; Boer, R.; Ardiansyah, M.; Immanuel, G.; Purwawangsa, H. Policy forum: Reconciling palm oil targets and reduced deforestation: Landswap and agrarian reform in Indonesia. *For. Policy Econ.* **2020**, *119*, 102291. [CrossRef]
37. United Nations Development Program. *Governance for Sustainable Human Development*; UNDP: New York, NY, USA, 1997.
38. Food and Agriculture Organization. Financing Sustainable Forest Management. In *Forestry Policy Brief*; FAO: Rome, Italy, 2008.
39. The Advertising Council. *Public Service Advertising that Changed a Nation*; The Advertising Council: New York, NY, USA, 2004.
40. Shackleton, S.; Paumgarten, F.; Kassa, H.; Husselman, M.; Zida, M. Opportunities for enhancing women's economic empowerment in the value chains of three African non-timber forest products (NTFPs). *Int. For. Rev.* **2011**, *13*, 136–151.
41. Aditya, S.K. Role of women in environmental conservation. *Int. J. Political Sci. Dev.* **2016**, *4*, 140–145.
42. Morell, M. FAO Experience in Decentralization in the Forest Sector. In *Interlaken Workshop*; FAO: Rome, Italy, 2004.
43. Sinha, H.; Suar, D. Leadership and People's Participation in Community Forestry. *Int. J. Rural Manag.* **2005**, *1*, 125–143. [CrossRef]
44. Putnam, R.D. The prosperous community social capital and public life. *Am. Prospect* **1993**, *13*, 35–42.
45. Bourdieu, P. The Forms of Capital. In *Handbook of Theory and Research for the Sociology of Education*; Richardson, J.G., Ed.; Greenwood Press: New York, NY, USA, 1986.
46. Ishihara, H.; Pascual, U. Social capital in community level environmental governance: A critique. *Ecol. Econ.* **2009**, *68*, 1549–1562. [CrossRef]
47. Yasmi, Y.; Guernier, J.; Colfer, C.J.P. Positive and negative aspects of forestry conflict: Lessons from a decentralized forest management in Indonesia. *Int. For. Rev.* **2009**, *11*, 98–110. [CrossRef]
48. Badan Penelitian Pengembangan dan Inovasi Kehutanan. *Desain Perencanaan Pengelolaan Multiguna Hutan Menghadapi Era Industri 4.0. Policy Brief*; FORDIA: Bogor, Indonesia, 2018.

49. Nurfatriani, F.; Darusman, D.; Nurrochmat, D.R.; Yustika, A.E.; Muttaqin, M.Z. Redesigning Indonesian forest fiscal policy to support forest conservation. *For. Policy Econ.* **2015**, *61*, 39–50. [CrossRef]
50. Birner, R. *Analytical Methods in the Social Sciences*; Institute of Rural Development, Georg-August University: Goettingen, Germany, 2001.
51. Nurrochmat, D.R.; Darusman, D.; Ekayani, M. *Kebijakan Pembangunan Kehutanan dan Lingkungan: Teori dan Implementasi*; IPB Press: Bogor, Indonesia, 2016.
52. Hardin, G. The tragedy of the commons. *Science* **1968**, *162*, 1243–1248.

Article

Adoption of Agroforestry in Northwest Viet Nam: What Roles Do Social and Cultural Norms Play?

Mai Phuong Nguyen [1,*], Tim Pagella [2], Delia C. Catacutan [3], Tan Quang Nguyen [1] and Fergus Sinclair [2,4]

[1] World Agroforestry, ICRAF Viet Nam, Ha Noi 100000, Vietnam; N.QuangTan@cgiar.org
[2] School of Natural Sciences, Bangor University, Bangor, Gwynedd LL57 2DG, UK; t.pagella@bangor.ac.uk (T.P.); F.Sinclair@cgiar.org (F.S.)
[3] World Agroforestry, ICRAF Southeast Asia, P.O. Box 161, Bogor 16115, Indonesia; D.C.Catacutan@cgiar.org
[4] World Agroforestry, Headquarter Office, P.O. Box 30677, Nairobi 00100, Kenya
[*] Correspondence: n.maiphuong@cgiar.org

Citation: Nguyen, M.P.; Pagella, T.; Catacutan, D.C.; Nguyen, T.Q.; Sinclair, F. Adoption of Agroforestry in Northwest Viet Nam: What Roles Do Social and Cultural Norms Play? *Forests* **2021**, *12*, 493. https://doi.org/10.3390/f12040493

Academic Editors: Mi Sun Park and Himlal Baral

Received: 10 March 2021
Accepted: 12 April 2021
Published: 16 April 2021

Publisher's Note: MDPI stays neutral with regard to jurisdictional claims in published maps and institutional affiliations.

Abstract: This article presents research about the influences of social and cultural norms on the adoption of agroforestry in the northwest mountainous region of Viet Nam. The farming systems practiced by various ethnic groups in the northwest mainly occur on sloping land, which extends over 70% of the land area in the region. Decades of intensive monoculture of annual crops has resulted in severe soil erosion, contributing to soil degradation and decline in crop yields. Integrating agroforestry practices on sloping land has the potential to halt and reverse soil degradation and improve local livelihoods, but its adoption is conditioned by the diverse social and cultural norms of different ethnic groups. This research applies knowledge-based system methods in order to understand local opportunities, preferences, and constraints influencing the adoption of agroforestry practices, using a purposive, gender-balanced sample of sixty farmers from six villages across three provinces in the northwest region comprising people from Kinh, Thai and H'mong ethnic groups. Our results show that although farmers from all groups are aware of the economic and ecological benefits of trees for soil conservation in general, they have different perceptions about the benefits of particular agroforestry practices. Behavioural norms controlling agroforestry adoption vary amongst ethnic groups, and farmers' individual social and cultural preferences influence their aspirations and adoption decisions. We conclude that developing appropriate agricultural interventions in a culturally rich environment such as northwest Viet Nam requires understanding of the context-specific needs and interests of socially and culturally disaggregated populations. Policies supporting agroforestry are more likely to contribute to more sustainable livelihoods and ecological benefits where they are tailored to the requirements of different ethnic groups.

Keywords: agroforestry; adoption; perception; behavioural controls; ethnic minorities; Viet Nam

1. Introduction

Agroforestry is where trees interact with agriculture and farmers manage interactions with the aim of achieving net economic and ecological gains [1,2]. "Agroforestry", therefore, is a broad term used to describe a wide range of practices at a range of scales [3]. At its simplest, agroforestry is a set of practices that combine woody components and agricultural crops and/or livestock, but this often involves quite complex interactions between people and trees, requiring systems analysis to understand why people adopt particular practices [4]. Based on spatial and temporal arrangements, agroforestry practices are defined as either simultaneous or sequential and in relation to the juxtaposition of tree and crop components. In the simultaneous practices, all components are intercropped at the same time, while in sequential practices, components occupy land at different times [5]. Sequential practices have varying degrees of overlap between the crop and tree components. Concomitant systems overlap in the beginning (such as in taungya systems), and superimposed systems have overlaps between components only at certain times (such as the temporary grazing of orchards) [6].

Under the pressure of population increase and global demand for food without further damaging the environmental resources, smallholder farming needs to change the way it is conducted [7]. Agroforestry provides a broad set of regulating ecosystem service benefits at both plot and landscape levels. Agroforestry provides benefits to soil health by increasing soil biodiversity, increasing the supply of nutrients, and reducing soil loss compared to monoculture agricultural systems [8,9]. Contour-based cropping systems reduced soil loss by 30–60% in the first year, and up to 72–98% by the third year in Thai Lan [10]. Adding trees into agroecosystems is critical for addressing climate change by slowing deforestation [11,12]. A good example is shaded agroforestry systems, with a number of studies highlighting their value to adapt to climate change [13,14].

Agroforestry can contribute to improving food security through the provision of fodder, fruit, and the use of fertiliser trees [15], or planting trees on farms in multi-strata and intercropping practices [16]. Agroforestry plays an important role in improving food quantity and nutrient provision by diversifying food products [17,18].

In Viet Nam, agroforestry has been present for a long time with many different forms across the country, but there has not been widespread adoption. Extant practices fall into several categories including traditional models of (i) forest–garden–fishpond–livestock or garden–fishpond–livestock without forests in the lowland; and (ii) home gardens with fruit trees, perennial tree-based alley cropping systems at field level on higher land [19]. Given the need for effective soil stabilisation in farming systems in the northwest, the integration of trees using agroforestry has significant potential. However, realising this potential requires moving beyond understanding the biophysical pre-requisites for agroforestry expansion to incorporate better knowledge of the key social factors that may influence adoption [20].

Intensive monoculture of food crops on steep sloping land in northwest Vietnam (which accounts for about 70% of the area of the northwest region) has resulted in significant environmental problems. This includes soil erosion, decline in soil quality, and loss of biodiversity [21,22]. Consequently, many of these smallholder farming systems now face economic uncertainty because increasing costs for fertiliser and seedlings may force farmers to reconsider their cultivation practices to find more sustainable options.

About 3.4 million people live in the northwest provinces of Viet Nam, in culturally diverse communities made up of nearly 30 ethnic groups [23]. In this region, there is a strong link between ethnicity and topography [24], with different ethnic groups associated with different elevations. The main ethnic group occupying lowland areas (below around 600 m) is the Kinh. Although Kinh is the most common ethnic group in Viet Nam, making up 88% of the national population, they are only the second largest group in the northwest, accounting for 26% of the region's population in 2015 [23]. The Thai group is the third most common ethnic minority of the country (accounting for about 2% of the total population), but the largest group in the northwest (28% of the population in 2015) [23]. Thai people in this region generally live in the middle altitudinal zone (around 600–800 m). The Hmong are the third most common ethnic minority in the northwest (14% of the population in 2015) [23], and generally live at higher altitudes (approximately above 800 m).

Agroforestry adoption is not just a "copy and paste" process; it is highly dependent on the biophysical, socio-economic context of the households [25]. Farmers will primarily adopt tree species based on their own needs [26], therefore understanding farmers' knowledge, interests and challenges is essential in order to provide appropriate support that meets their actual needs and capacities. The research reported here aimed to acquire information about local opportunities, preferences and constraints influencing the adoption of agroforestry practices and how they vary amongst ethnic groups and thus can be used to tailor agroforestry options to suit local contexts [27].

2. Materials and Methods

2.1. Theoretical Framework

Participatory approaches developed in the 1990s were viewed as a paradigm shift in research and development, providing tools to capture the views of local people and moving away from top-down prescriptions [28]. Earlier participatory approaches were, however, often applied without critically thinking about the issues of social barriers to farmers' decision-making. Several theories have been applied to understand farmers' behaviours. For example, the "Diffusion of innovation" theory [29] looks into how and why a technology is adopted and spread. Value–belief–norm theory [30] provides an approach to analyse social supports for the environmental movement. The theory of planned behaviour [31] is a conceptual framework that has often been used to explore the social dimensions of technology acceptance and adoption [32–35] and forms the basis for the methods in use in this paper. The key aspects of this theory are briefly discussed here.

2.1.1. Social Norms

Social norms are the expectations from a community on individuals to perform a behaviour in a specific situation. Social norms interact with social relationships to influence a farmer's adoption decision through, for example, their neighbours' adoption patterns, social expectation, and pressure or social status that may result from engaging in an activity. Current recommendations suggest that analysis of these norms should be incorporated in best practice adoption studies [36]. Adjusting agricultural intervention to socio-cultural factors can improve how farmers value innovations and their motivation to adopt them [37]. This has led to a major paradigm shift to support local innovation where farmers adapt practices to suit their local circumstances rather than the widespread promotion of prefabricated technology packages [27].

2.1.2. Farmer Capacity to Adopt

A farmer's capacity to adopt agroforestry is defined by both biophysical and socio-economic conditions. Biophysical conditions are derived from understanding of the ecological conditions of the farm, including characteristics such as soil type, slope [38], farm size [39], or the geographical context of the plots [40]. On the other hand, socio-economic factors include: market incentives, household preferences [38], economic benefits of land tenure and available time [41], the amount of social capital and human capital (including knowledge), and the influence of local and national policies [42].

Applying the theory of planned behaviour [31], willingness to adopt is influenced by personal beliefs and attitudes, social norms, and perceived behavioural controls. Farmers' attitudes encompass how personal beliefs act on behaviour, which can be negative, positive, or neutral. Farmers' attitude towards adoption is strongly correlated with farmers' perceived behavioural control (how they perceive the level of difficulty in adopting) and their self-belief in the capacity to adopt and then maintain the practice [43]. For example, farmers' attitudes towards agroforestry were strongly related to the level of access to information and extension support in Bangladesh [44].

2.1.3. Farmer Perceptions

Farmers' adoption decisions regarding agroforestry practices are influenced by the degree of difficulty associated with acquiring accurate information about the benefits of such innovations [32,42], alongside perceptions of risks and barriers associated with tree planting [45] or knowledge of agroforestry techniques. The selection of tree species for agroforestry adoption highly depends on farmers' attitude or knowledge concerning the impact of trees on food, soil, water and crops from their experiences and observation [46].

2.1.4. Perceived Behavioural Controls

Perceived behavioural control is the personal perception of the difficulty involved in performing a specific behaviour [31]. High financial cost and lack of knowledge on tree management techniques are commonly perceived challenges of planting trees on farms [47]. Limited skills and techniques were highlighted in studies in Pakistan [48], while the unavailability of capital and quality seed were identified in Rwanda [49,50].

Social and cultural values maintain a strong link with ethnicity because different ethnic groups have their own religions, beliefs, values, and resources which influence their attitudes, social norms, and behavioural controls related to agricultural innovation [51]. If communities are viewed as homogenous groups, the result may be that only the voices of a small number of powerful people are included, and thus the design of interventions may not be appropriate for those who are supposed to be empowered [52].

In this study, we explore how social factors influence pathways to adoption for agroforestry practices designed to stabilise soils and support local livelihoods for different ethnic groups in northwest Viet Nam. This includes: (i) identifying and understanding the social factors including personal attitude, subjective norms and behavioural controls towards farmers' willingness to adopt agroforestry (and whether this varies with ethnicity); (ii) recording preferences for different forms of agroforestry systems; and (iii) identifying potential constraints to the adoption of agroforestry from three main ethnic groups, the Kinh, Thai, and H'mong.

This study contributes to the existing global literature in understanding cultural and ethnicity aspects of local ecological knowledge on agroforestry and other land uses [53–56]. Scaling up agroforestry adoption need to be adapted to fine-scale variation in ecological and social contexts, including local needs [57]. Adoption is more likely to happen when farmers have knowledge, labour, and secure land tenure [58,59]. Social factors including farmers' preferences, attitude, cultural or social constraints, and local knowledge strongly influence farmers' adoption decisions [60]. These factors can be categorised as farmers' capacity to adopt and farmers' willingness to adopt [61].

Farmers' capacity and willingness are two key elements for agroforestry adoption (Figure 1), based on the theory of planned behaviour [31] and the practical evaluation of agroforestry adoption [61]. This study hypothesises that agroforestry adoption will occur when farmers have both the capacity and willingness to adopt agroforestry.

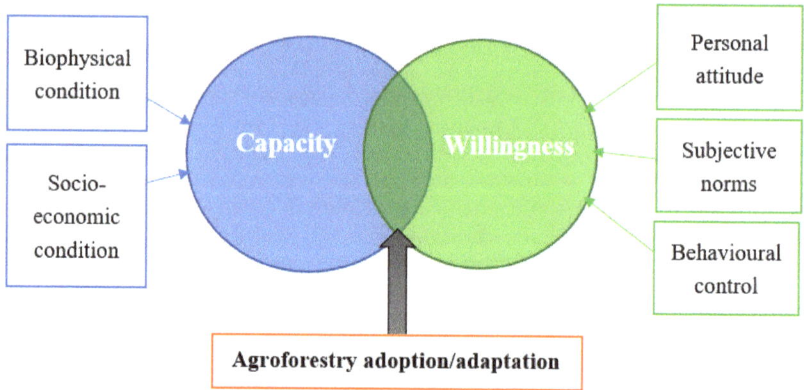

Figure 1. Farmer's capacity and willingness for pathways towards agroforestry adoption (adapted from Ajzen [31] and Mills et al. [61]).

2.2. Data Acquisition Methods

Data used in this study were collected from two surveys in 2016 and in 2017 in the northwest region of Viet Nam. Households covered in the surveys were in the sites of the Agroforestry for Livelihoods of Smallholder Famers in Northwest Viet Nam (AFLi) project, which is representative of the ecological landscape and social–cultural dynamic of the northwest region.

The surveys are described briefly below.

2.2.1. Survey 1 (2016): Adoption Capacity Survey

The first survey explored farmers' agroforestry adoption preferences. This involved a purposive sampling of current adopters and non-adopters of agroforestry. The first group was current adopters, consisting of two sub-groups: (1) farmers who worked with the AFLi project, called "project adopters" ($n = 166$); and (2) farmers who adopted agroforestry but were not involved in the AFLi project, called "spontaneous adopters" ($n = 7$). The second group consisted of farmers who had not adopted agroforestry on their farms, classified as non-adopters ($n = 56$). Questions in the first survey were designed to understand the capacity of farmers to adopt agroforestry, and their perceptions of the degree to which their biophysical context (elevation, cultivation, traditional practice) and social–economic conditions (finance, labour, knowledge) affected their capacity to adopt agroforestry systems (for non-adopters) or expand these systems (for adopters).

2.2.2. Survey 2 (2017): In-Depth Household Survey

The second survey was an in-depth household survey which focused on acquiring farmers' local ecological knowledge regarding tree planting following a knowledge-based systems approach [62,63]. Indigenous knowledge is culturally specific [64]; therefore, this survey looked at how local knowledge was shaped by attitudes and perceptions, as well as behavioural controls towards agroforestry adoption and hence farmers' preferences with regard to potential agroforestry options and the degree to which these were influenced by their ethnicity. This survey used key informant interviews of six commune representatives, combined with farmer focus group discussions and semi-structured interviews. One focus group discussion was conducted in each village, with the same farmers participating in the semi-structured interviews.

Key informant interviews were designed to understand the overall context of the communes, including the distribution of ethnic groups across elevations, socio-economic contexts of three ethnic groups, supporting policies of tree planting, and agroforestry development.

Farmer focus group discussions aimed to understand village history, culture, tradition, cultivation practices, agroforestry opportunities, and constraints using village map sketching, historic mapping, faming calendars, and strength–opportunity–weakness–threats analysis (SWOT).

Finally, semi-structured interviews with fifty-eight farmers (50% female) were used to explore their knowledge about agroforestry management and obtain a deeper understanding about agroforestry management, the values of trees, social norms related to tree planting, farmers' attitude on the benefits of agroforestry, and preferences relating to agroforestry adoption. The interviews were conducted together with farm visits. Farmers were again purposively selected for these interviews. All interviewees were agroforestry adopters who were currently not involved in the AFLi project (to avoid influence from project training). We consulted six village leaders to choose farmers implementing agroforestry practices by themselves. The total number of farmers interviewed was 58, with 10 farmers invited per village representing the minimum number of agroforestry adopters per village across the study sites. This was gender-balanced; thus, equal numbers of men and women were interviewed. Two villages were selected for each ethnic group (Kinh, Thai and H'mong), amounting to six villages in total (Figure 2).

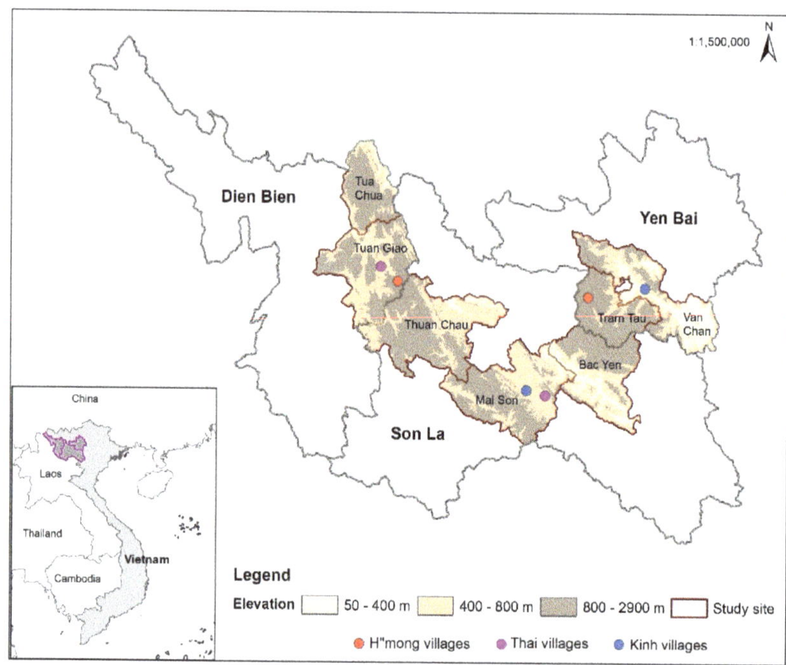

Figure 2. Map of the study villages within the Agroforestry for Livelihoods of Smallholder Famers in Northwest Viet Nam (AFLi) project site in northwest Viet Nam.

2.3. Study Site

The study was conducted in the AFLi project sites in Yen Bai, Son La and Dien Bien of the northwest region of Viet Nam. The northwest of Viet Nam (21–23° N and 103–105° E) is one of eight eco-regions in Viet Nam. It is the most mountainous, remote, and poorest area of Viet Nam. Forest cover accounts for about half of the area, although the region is prone to severe deforestation, land degradation, and soil erosion. Extreme weather such as landslides, droughts, frost, and hailstorms significantly affect agricultural production and the economic development of the region. Key agricultural products of three provinces include annual crops such as rice (*Oryza sativa* L.), soybean (*Glycine max* (L.) Merr.), sugarcane (*Saccharum barberi* Jeswiet), amomum (*Amomum villosum* Lour.), perennial crops such as Shan tea (*Camellia sinensis* (L.) Kuntze), Arabica coffee (*Coffea arabica* L.), macadamia (*Macadamia* spp.). Popular fruit trees include H'mong apple— son tra *(Docynia indica* (Colebr. ex Wall.) Decne), plum (*Prunus salicina* Lindl.), peach (*Prunus persica* (L.) Batsch), mango (*Mangifera indica* L.) orange (*Citrus sinensis* (L.) Osbeck), longan (*Dimocarpus longan* Lour.). Common timber trees in three provinces consist of acacia (*Acacia mangium* Willd.), manglietia (*Manglietia conifera* (Dandy) V.S.Kumar), melia (*Melia azedarach* L.), teak (*Tectona grandis* L.f.), cinnamon (*Cinnamomum verum* J.Presl), pine (*Pinus* spp.), leucaena (*Leucaena leucocephala* (Lam.) de Wit), cassia (*Cassia siamea* Lam), vernicia (*Vernicia montana* Lour.). Wild peanut (*Arachis pintoi* Krapov. & W.C.Greg), peanut (*Arachis hypogaea* L.) or soybean (*Glycine max* (L.) Merr.) are often cultivated to improve soil fertility [65–67].

Six villages representing the biophysical conditions and ethnicity within project sites were selected, with two villages per ethnic group. These six villages are described below and shown in Figure 2.

- Kinh ethnic group: Van Thi 3 village in Van Chan district, Yen Bai and Tan Que village in Co Noi commune, Mai Son district, Son La.

- Thai ethnic group: Na Ban village in Mai Son district, Son La and Giang village in Tuan Giao district, Dien Bien.
- H'mong ethnic group: Hua Xa A village in Tuan Giao district, Dien Bien and Sang Pao village in Tram Tau district, Yen Bai.

2.4. Analytical Method

In this paper, qualitative data analysis was used as the key analytical method to provide insight about farmers' perceptions about agroforestry adoption and how this affects their decisions, through looking at the reasons and supporting farmers' explanations of their adoption decisions [68].

Qualitative data analysis has been widely used in the literature regarding planned behaviour [31], which is a cognitive process involving subjective norms, perceived behavioural control, and attitudes, as well as in the acquisition and analysis of local agroecological knowledge [62,63].

Textual analysis of interview transcripts was the main tool used in the paper to analyse and describe the actors involved, their subjective norms, perceived behavioural control, and attitude. In addition, illustrations are provided through the use of matrix, chart, and figure formats for both descriptive and explanatory purposes.

3. Results

3.1. Farmer Capacity to Adopt Agroforestry

3.1.1. Farmers' Perceptions of Biophysical Conditions Required for Agroforestry Expansion

Biophysical attributes include factors defining the suitability of an agroforestry practice such as elevation, rainfall, temperature, slope, and soil type. Elevation is an important biophysical condition linked to ethnic groups in northwest Viet Nam, with the H'mong group inhabiting areas over 800 m above sea level (asl), the Thai at 400–800 masl, and the Kinh below 400 masl. Farming practices and choice of tree-crops differ by elevation and ethnic group (Table 1).

Table 1. Farming characteristics of three ethnic groups.

Ethnic Group	Range of Elevation (masl)	Traditional Cultivation Techniques	Suitable and Preferred Tree Crops
Kinh	0–600	Intensive cultivation and intercropping	Tea Longan, mango, plum, pomelo
Thai	400–800	Partly shifting cultivation	Cffee, macadamia Plum, mango, longan manglietia, melia
H'mong	>800	Shifting cultivation	Shan tea, coffee son tra Pine

3.1.2. Socio-Economic Conditions

Socio-economic attributes include financial capacity, size of landholding, distance of farm to market, education, and knowledge, awareness, and proficiency with techniques (Table 2). The Kinh are lowland migrants that settled in the lower part of the region under the post-war migration policy of the Vietnamese government. The Kinh have advantages in terms of language, enabling easy access to education, technology, market information, and other productive resources. The Thai live just above the Kinh communities. They are also able to speak the Vietnamese language and have reasonable access to information and technology. In contrast, the H'mong live in high-elevation areas that are often far from the main roads and local markets. The H'mong have sparse access to external information and government support because of their social distance from other communities in Viet Nam, which is exacerbated by language barriers. Today, many H'mong farmers are still practicing shifting cultivation with short fallow periods, while some, possessing less land, have already shifted to sedentary farming. Poverty is manifest in the economic status and educational level of households, which influence the way farmers learn new techniques. Land holding influences the availability of land to be used under agroforestry. The Kinh people have long been using agroforestry as a traditional practice, while the H'mong people rarely did so in the past.

Table 2. Typical social characteristics of three main ethnic groups in northwest Viet Nam.

Ethnic Group	Poverty	Distance to Market	Distance to the Field *	Education	Average Land Holding
Kinh	Low	0–2 km	<1 km	High school, university	<1 ha
Thai	Medium	0–3 km	1–3 km	High school	1–3 ha
H'mong	High	5–10 km	3–6 km	Primary school	2–5 ha

* distance to field represents the distance from the homestead to their fields.

3.1.3. Farmer Willingness to Adopt Agroforestry

(1) Farmers' Attitude to Benefits from Agroforestry Adoption

Although farmers from all groups were aware of the benefits of using trees in soil conservation, they had different perceptions on the benefits of specific agroforestry practices, which was likely to influence the types of agroforestry adopted (see Figure 3). All groups stated that agroforestry practices had some provisioning function relating to income generation but had differing needs in relation to regulating functions. The H'mong group were interested in increased land, labour and fertiliser use efficiency, while the Thai highlighted soil erosion reduction, and the Kinh were motivated by the idea of soil fertility improvement. This study suggests that farmers' specific social circumstances influence their aspirations and constraints related to adopting agroforestry interventions. When it comes to tree planning, H'mong farmers have myriad considerations even with increased demand for fuelwood and timber for house construction—first is to save time walking from and to their farms and homes; second is their limited family labour; and third is the high-cost of fertilisers. Meanwhile, Thai farmers think that timber and firewood, food for humans and animals (buffalo and cow), and soil erosion mitigation are important benefits that can be derived from agroforestry. Most Thai households have confined animals, so fodder grasses in agroforestry practices are useful to augment their cut-and-carry animal feeding strategies. However, reluctance to adopt agroforestry also grew from concerns that trees grow slowly, affecting cash flow. Finally, apart from income, Kinh farmers think that agroforestry is an efficient strategy to address having a limited size of land holding. Having more access to technical information from local extension workers, the Kinh are more willing to intercrop various tree species, especially in home-gardens. In Tan Que district, Son La province, many Kinh farmers are already intercropping peanuts and beans with perennials to improve soil fertility and optimise land productivity.

(2) Subjective Norms Influencing Agroforestry Adoption

Subjective norms constructed by the beliefs of ethnic groups about others' expectations toward forest protection, tree-growing and agroforestry were revealed in the in-depth interviews and focus group discussions. Farmers hold the same subjective norms, i.e., expectations of their behaviours regarding forest protection. Government reforestation programs supported the establishment of new forest plantations, and natural forest regeneration and protection [69]. In these programs, production forest lands were allocated for farmers to grow timber trees such as *Acacia* spp., *Melia azedarach*, *Manglietia conifera*, *Pinus* spp., and son tra (*Docynia indica*). In turn, households and communities were expected to protect the forest. In Thai villages, however, cultural norms exist wherein community forests are considered "ghost forests" or burial grounds that families come to visit every year. The Thai also believe they have to protect the old wild trees in their village because they are holy and revered for use in traditional ceremonies. When the Kinh migrated to the uplands, they were expected to retain forest patches on hill tops while annual crops, fruit trees or cash crops could be planted in the mid-lower portions of the hill, resulting in a forest–garden–fishpond–livestock system. This comes from a farming design called "vuon-ao-chuong"-VAC (garden-fishpond-livestock), which literally is the combination of a vegetable garden, fishpond, and livestock recommended by government extension programs. Kinh farmers believe they are expected to adopt the modified VAC practice to help reduce soil erosion and prevent flash floods. All Kinh farmers concurred that they learnt fruit tree management techniques and bought seedlings from their hometown to grow in their new upland environment; 26% of Kinh correspondents indicated that intercropping was a traditional technique to address limitations of land holding size. Farmers from all ethnic groups also mentioned that they had been trained and encouraged by local authorities and extension workers to gradually shift from maize/rice cultivation to perennial tree-cropping on slopes. Farmers often choose the most degraded plot with low crop yield to grow trees intercropped in the first few years before later removing annual crops when the tree canopy closes. Today, there are a number of programs for growing avocado, macadamia, son tra (*Docynia indica*), Shan tea (*Camellia sinesis*), and various citrus species that are expected to shift the norm away from annual crop cultivation to tree-based cultivation on sloping lands.

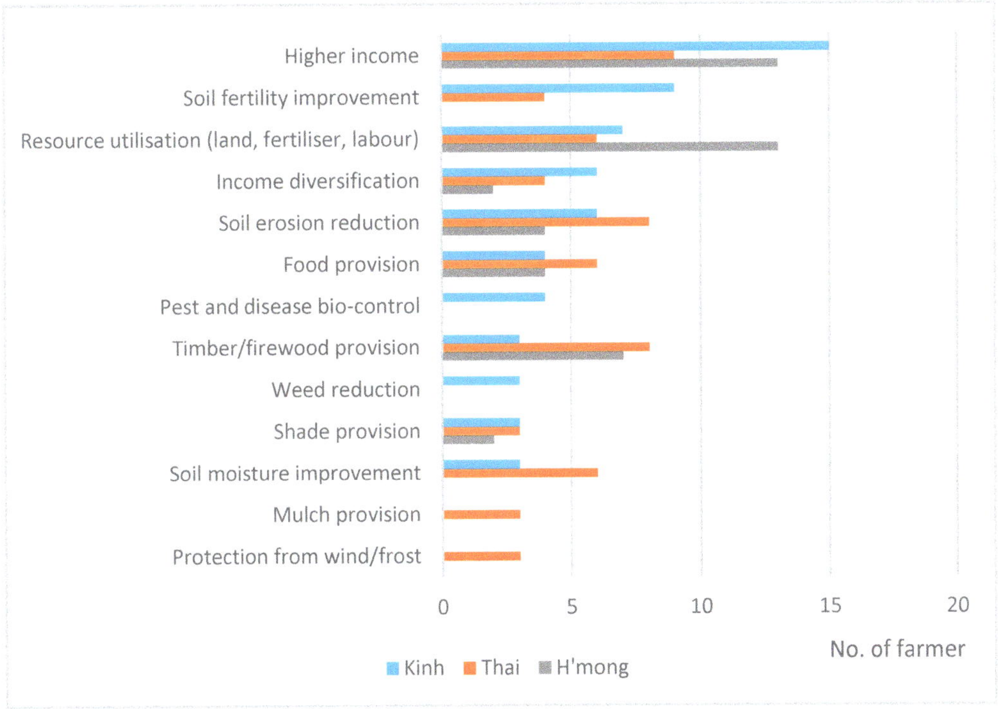

Figure 3. Absolute number of farmers from different ethnic groups mentioning different benefits that can be derived from agroforestry practices.

(3) Behavioural Controls Affecting Agroforestry Adoption

Behavioural controls or challenges to adopting agroforestry varied amongst the three ethnic groups; however, lack of land, labour, and financing are common problems. Most farmers also lacked quality planting materials and market links. Land as a limitation, however, manifests in different ways—the Kinh have smaller farmlands and therefore have nothing to spare for agroforestry. In contrast to the Kinh, the Thai and H'mong have larger landholdings, but prefer to establish agroforestry on plots near their homestead and on fertile soils. In addition, H'mong farmers had concerns about the technical management of agroforestry because of their low access to technical information exacerbated by their language barrier with Vietnamese-speaking extension workers. The study also found gender-specific behavioural controls. Most of the concerns about marketing fruit amongst the Thai were from female respondents, while only H'mong men had concerns about market. Both male and female farmers from all ethnic groups were concerned about the technical aspects of agroforestry and expressed a strong desire to learn management techniques.

3.2. Typologies of Agroforestry Adopters and Non-Adopters

Farmer typologies were derived based on criteria identified for the capacity and willingness to adopt agroforestry practices. (Table 3).

Table 3. Farmer characterisations based on capacity and willingness to adopt/scale out agroforestry.

Capacity to Adopt/Scale out Agroforestry	Willingness to Adopt/Scale out Agroforestry	
	Unwilling	Willing
Positive capacity	Type 3: Have available land, labour, finance, techniques Do not like to adopt agroforestry (for non-adopter farmers) or only want to maintain current agroforestry adoption (for adopter farmers)	Type 1: Have available land, labour, finance, techniques Have positive attitude towards agroforestry, willing to adopt (for non-adopter farmers) or expand agroforestry (for adopter farmers)
Negative capacity	Type 4: Lack of available land, labour, finance, techniques Do not like to adopt agroforestry (for non-adopter farmers) or only want to maintain current agroforestry adoption (for adopter farmers)	Type 2: Lack of available land, labour, finance, techniques Have positive attitude towards agroforestry, willing to adopt (for non-adopter farmers) or expand agroforestry (for adopter farmers)

We examined the capacity and willingness of non-adopters, wherein more than half (53%), and mostly Thai and Kinh were Type 4 farmers who were unwilling to adopt or expand agroforestry due to limited capacity in land, labour, financing and technical management (Figure 4), and also because of a lack of evidence on agroforestry benefits. Type 3, comprising 14% of non-adopters, mostly belonged to the H'mong ethnic group who expressed a lack of know-how on agroforestry.

Thirty percent of non-adopters had resources and were willing to adopt agroforestry on their farms. Most of these were H'mong, because they saw the benefits of agroforestry from project adopters. One of the techniques they wanted to adopt was planting fruit trees together with grass so that they had more food for cattle in the winter. Cows and buffalos play an important role in H'mong livelihoods because they help farmers in land preparation and the transport of materials between homes and farms. Two H'mong farmers did not have enough land but they were interested in agroforestry and willing to adopt it, whereas none of the Kinh or Thai farmers who lacked land wanted to adopt agroforestry.

The majority of the non-adopters (53%) did not want to adopt agroforestry because of a perceived lack of capacity. Most of these were Thai and Kinh farmers. Only a few H'mong farmers were in this group, primarily because they lacked cultivatable land (which, in this context, meant land that was fertile and less than 10 km from their home). For the Kinh and Thai farmers, they stated that they did not have spare land for agroforestry.

Of those that were unwilling but had capacity (Type 3: 14%), a large proportion were H'mong; the reasons given were a lack of knowledge about agroforestry. For the Kinh and Thai, they had not yet seen the benefits, therefore they were not willing to adopt agroforestry at present.

In Type 4, the number of non-adopters was much higher compared to adopters, clearly because they did not receive direct support from the AFLi project. This is consistent with the impact of low access to information leading to low rates of adoption [70], although in other contexts farmers owning more resources were found to adopt agroforestry more readily [41]. H'mong farmers expressed willingness to adopt agroforestry despite their limited capacity in comparison to the Kinh and Thai, consistent with positive perception being an important factor in agroforestry adoption [71]. The research also shows that ethnicity linked to socio-economic contexts highly influenced agroforestry adoption. For example, the Thai displayed the highest potential for adopting or expanding agroforestry, which can be linked to the wider range of tree-species suitable in mid-level altitudes and socio-economic factors such as the access to market, seedlings, and information. The choice of agroforestry practices also differed with access to market by the same ethnic group—the H'mong in Dien Bien province were more interested in fruit trees than the Hmong in Yen Bai because of the former's greater access to the highway and local markets. Furthermore, despite having the same farming tradition, the Thai in Son La have less concern about marketing agroforestry products in contrast to those in Dien Bien, whose primary concern is market distance and transportation.

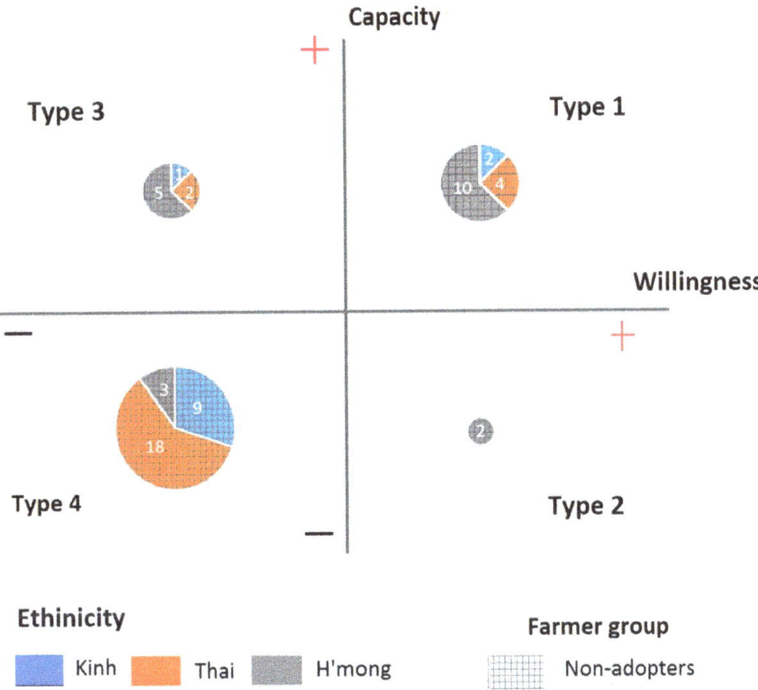

Figure 4. Non-adopter farmers' capacity and willingness to adopt agroforestry (*n* = 56).

3.3. Preferred Agroforestry Options for Different Ethnic Groups

Based on the understanding of capacity, willingness and motivation relating to agroforestry adoption, farmers were able to identify suitable tree-based options which fitted their local social–ecological context and that were economically viable. Fruit tree intercropping systems were the common interest of all groups because fruits have high selling prices. The other systems were identified based partly on individual farmer motivations for agroforestry. For example, H'mong farmers tended to prefer an annual crop component in their practices to provide food. In contrast, Thai farmers liked to have grass for livestock, and Kinh farmers wanted to improve soil by growing nitrogen-fixing species such as peanut or soybean (Table 4).

Understanding local preferences and motivation to adopt agroforestry helps project teams to modify trials and match them with the local interest. For example, the grass component was removed from practices for H'mong farmers, while peanut/pineapple was added into the practices offered to Kinh farmers. Grass was maintained for the Thai group.

Table 4. Preferred agroforestry options identified by three ethnic groups and local contexts.

Ethnicity		H'mong	Thai	Kinh	
Common System		Mixed Fruits (Peach, Plum, Mango)–Lime/Maize			
By Elevation		High (Above 800 m)	Medium (500–800 m)	Medium (500–800 m)	Low (Below 500 m)
By Location	Dien Bien	Son tra/Rice/Maize/Cassava Coffee/Leucaena /Mixed fruits Coffee/Maize	Cassia /Vernicia /Grass Coffee—Cassia /Leucaena /Longan	N/A	
	Son La	N/A	Fruit trees/Cana/Maize/ Soybean/Cucumber/Pumpkin Macadamia/Coffee/Fruit trees/Grass/ amomum	N/A	Mixed fruit trees/Wild peanut/Peanut/Soybean Pomelo/Guava
	Yen Bai	Shan tea/Rice/Maize/Cassava Son tra/Rice/Maize/Cassava	N/A	Melia /Vernicia /Tea Manglietia /Melia /Vernicia Plum/Pineapple/ Soybean/Peanut Tea/Maize	N/A

(N/A: not applicable).

3.4. Discussion

3.4.1. Ethnicity and Agroforestry Adoption

Results of this research clearly demonstrate that ethnicity associated with specific contexts highly influences agroforestry adoption as well as the relevance of potential designs. The results suggest that the Thai ethnic group had the highest potential for adopting agroforestry and for expanding their current practices in the near future. This is a combination of the advantages of living at medium elevations, which were suitable for more tree species coupled with moderate access to market, seedlings, and information. The Kinh were more technologically advanced and had good access to markets, but their land size was restricted. The H'mong were more isolated and had the most difficult agricultural conditions, although they had the largest land holding sizes.

The number of non-adopters in Group 4 (i.e., those with no capacity and unwillingness to adopt) was much smaller than the number of the adopter group (Figures 4 and 5). Adopter farmers had received training by the project teams about agroforestry benefits. This is consistent with low access to information leading to low rates of adoption [70], although in other contexts farmers owning more resources have been found to adopt agroforestry more readily [41]. In this research, a higher percentage of H'mong farmers were willing to adopt agroforestry compared to Kinh and Thai, although they had more limited resources. However, only the H'mong farmers who had spontaneously adopted wanted to expand their agroforestry practices (Figures 4 and 5). Famers' positive perceptions towards agroforestry is important for adoption [71]; therefore, this brings up a high potential for scaling out agroforestry adoption for the H'mong, who are particularly vulnerable to poverty, climate change, and degraded landscapes.

Farmers' perspectives on benefits from agroforestry generally align with scientific findings about positive impacts of agroforestry such as soil loss reduction [10], improving soil fertility [8,71], providing food and fodder [18,71], and reducing the impact of extreme weather [72], but it was difficult for farmers to recognise some benefits that they could not observe, such as climate resilience or carbon sequestration. People's attitudes to future agroforestry options were also heavily context- and ethnic-specific. For example, H'mong farmers in two different provinces, Dien Bien and Yen Bai, had different preferred options. This can be explained by variations in their access to the highway (i.e., a context variable rather than an ethnic variable). The options that farmers wanted to have in Table 4 were consistent with their motivations to adopt agroforestry in Figure 3, and fits with their context. H'mong famers preferred cash crops or fruit trees with short-term crops for food and high income in a sufficiently short time period (3–4 years). Thai farmers wanted to have fodder grass for their livestock, and Kinh farmers wanted to have legume plants in their systems. However, it is clear that market factors play an overwhelmingly important role in the choices of farmers and their likelihood of change over the short term [73]. Except for the Thai in Dien Bien and the H'mong in Yen Bai whose fruit trees do not grow well, all three ethnic groups in other areas liked to grow fruit trees with the species dependent on market availability.

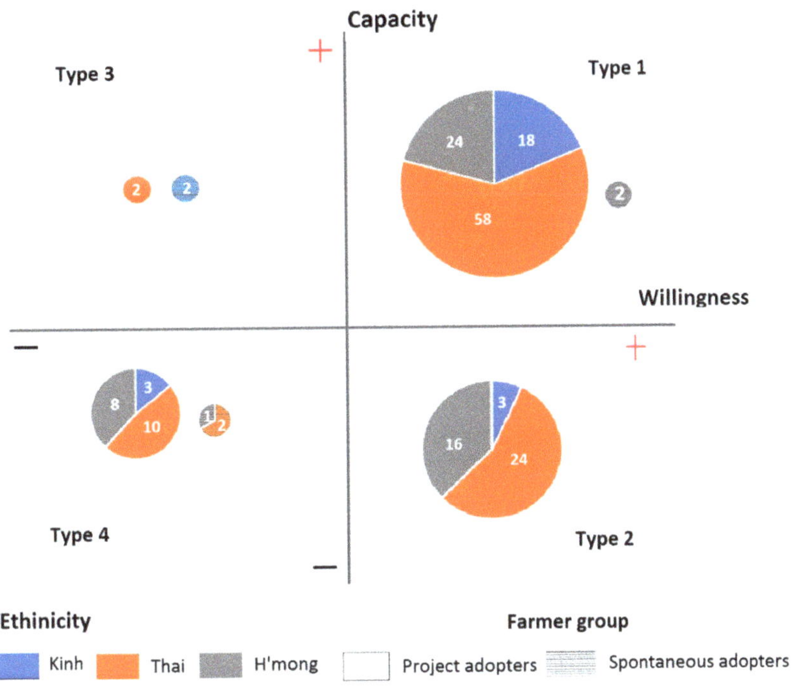

Figure 5. Farmer-adopters' capacity and willingness to scale out agroforestry (project-supported adopters: n = 166; spontaneous adopters: n = 7).

Farmers adopt tree species and modify how they adopt them based on their own needs [74]. Various adaptations were observed from the survey, such as planting one or double grass strips between tree rows, and growing trees at different spacings and densities. This was quite similar to the adoption patterns in Nepal [71], where farmers in the lowlands did not opt to grow fodder trees with paddy rice, but upland farmers integrated fodder trees with rainfed crops to feed livestock, which then provide manure to fertilise crops. Unlike the farmers planting trees to improve soil fertility in Malawi [75] or for fuel and timber in Ethiopia [76], most farmers in northwest Vietnam wanted high-value fruit trees suitable for their biophysical condition.

The complexity and diversity of agroforestry adoption makes it difficult to monitor and evaluate the benefits of agroforestry accrued to adopters and non-adopters. The long-term nature of trees makes this even more challenging, because farmers intercrop tree and crops for the first few years but may focus entirely on the trees after a few years once they have become productive. Attitudes to adoption may change when the context changes (such as project support and market availability); our observations represent farmers attitudes at the time we conducted our survey, although this will include their accumulated experience.

3.4.2. How Farmers Transition from Being Non-Adopters to Adopters

Previously, adopters have been classified as "real adopters", "testers" or "pseudo adopters" depending on the permanency of their use of particular practices [8] conditioned by how farmers may have benefited from projects socially or materially beyond what they derive from the agroforestry practices that they may be using. Therefore, the adopters might implement the technologies just because of the incentives from a project rather than the intrinsic value of the practice. In this study, it is unclear how permanently the farmers in the project adopter group will use agroforestry practices, although 61% expressed a capacity and willingness to adopt agroforestry. On the other hand, non-adopter farmers who want to adopt agroforestry might be more committed because they are willing to do it on their own without any support. This requires further research to track adoption and adaptation over long periods of time, encompassing the entire length of agroforestry rotations.

The social condition that enables the scaling out of diversified farming systems requires changes in people's aspirations and actions [77]. In order to scale out agroforestry adoption, it is necessary to shift farmers from other types into Type 1 (have capacity and willingness to adopt) from both adopters and non-adopters. Moving from Type 2 to Type 1 requires them to improve their capacity factors, such as labour, finance, or land, which is quite difficult and dependent on external support. The most feasible option is to increase their access to social capital in order to hire more labour or rent more land. Moving from Type 3 to Type 1 requires a change in farmers' attitudes and perceptions, addressing behavioural controls constraining agroforestry adoption. Farmers can change by themselves if they see the success from project agroforestry trials, or learn from neighbours, friends, and social media. This is sustainable because farmers combine the new information and techniques with their indigenous knowledge to generate locally adapted practice, consistent with farmers adopting agroforestry in Malawi [78]. Successful agroforestry trials should be promoted widely through different channels which farmers can easily access. The proportion of Type 4 farmers who are adopters is much smaller than the non-adopters, meaning that more farmers wanted to continue adopting after some time working with the project.

3.4.3. Policy Implications for Scaling out Agroforestry

Policy plays an important role in the widespread adoption of agroforestry [79]. This study suggests that for the effective adoption of agroforestry, government policy must respond to the needs at farmer level and fit into their specific contexts, which is consistent with recommendations from various authors [80]. Findings from this study show that addressing perceived behaviour controls enables the conditions for farmers to adopt agroforestry. Strong correlation often exists between farmers' intention to adopt or maintain agroforestry and their behavioural controls [43], but behavioural controls vary amongst ethnic groups and are related to their contexts, which differ from one community to another [81]. Therefore, policies from governmental or development projects should be tailored to different ethnicities in different locations and modified in light of their local knowledge and practices. This research also revealed that men and women often appeared to have different concerns towards agroforestry adoption. For example, Thai women were concerned about the market for fruits, while H'mong men wanted to learn more about agroforestry techniques. Effective policies will, therefore, need to be based on a clear understanding of the needs of men and women. Further research on gender would help shed light on design principles for scaling out agroforestry. Furthermore, understanding different motivation for agroforestry adoption and the preferred options of different farmers could help policy makers and development projects design best-fit agroforestry practices for specific contexts. If an option does not fit with their existing practices, farmers are unlikely to be willing to adopt it [80].

4. Conclusions

This research has revealed contrasting social contexts for agroforestry adoption amongst three ethnic groups in northwest Viet Nam, the Kinh, Thai, and H'mong. Non-adopter and adopter farmers were categorised into four groups with different levels of capacity and willingness to adopt agroforestry. Farmers' behavioural controls constraining the adoption of agroforestry also varied among the three ethnic groups and were influenced by their location, accessibility to market, and different cultivation traditions. Most farmers lacked access to high-quality tree seedlings and connections to markets. Kinh and Thai farmers in lowland areas were concerned about climate change and the high cost of managing agroforestry and preferred high-value fruit trees. H'mong people, in contrast, were more concerned about the efficiency of using fertiliser and labour and accessing required financial resources to support the purchase of seedlings and fertilisers. This understanding contributes to selecting which agroforestry interventions are most likely to be suitable for different ethnic groups, and what supporting policies are required to enable adoption.

Author Contributions: Conceptualisation, D.C.C., T.P., M.P.N., T.Q.N. and F.S.; Methodology, D.C.C., M.P.N. and T.P.; Software, M.P.N.; Validation, T.P. and M.P.N.; Formal analysis: M.P.N.; Investigation: M.P.N.; Resources: D.C.C. and F.S.; Data curation, M.P.N.; Writing—original draft preparation, M.P.N. and T.P.; Writing—review and editing, T.P., D.C.C., T.Q.N. and F.S.; Visualization: M.P.N.; Supervision: T.P., D.C.C. and F.S.; Project Administration: M.P.N.; Funding acquisition, D.C.C. and F.S. All authors have read and agreed to the published version of the manuscript.

Funding: The study was supported by the AFLi project funded by ACIAR (Australian Centre for International Agricultural Research) and FTA (CGIAR Research Program on Forest, Trees and Agroforestry).

Institutional Review Board Statement: The study was conducted according to the guidelines of the Declaration of Helsinki and the research was approved by Bangor University Ethical Review Committee and follows World Agroforestry's Policy Guidelines Series on Research Ethics (2014). In particular all participation in the research was voluntary, and all information, including any confidential personal data, was anonymised and stored in appropriately secure facilities, following current World Agroforestry protocols.

Informed Consent Statement: Informed consent was obtained from all subjects involved in the study.

Data Availability Statement: The data presented in this study are available on request from the corresponding author. The data are not publicly available due to privacy.

Acknowledgments: We thank the survey respondents for sharing their information and our ICRAF colleagues Vu Thi Hanh, Pham Huu Thuong, Nguyen Van Thach, Do Van Hung, the master's student Marika Simelson, and the intern Nguyen Tuan Anh for their contributions in farmer interviews, farmer rankings, and focus group discussions.

Conflicts of Interest: The author declares no conflict of interest.

References

1. Lundgren, B.O.; Raintree, J.B. Sustained Agroforestry. In *Agricultural Research for Development: Potentials and Challenges in Asia*; Nestel, B., Ed.; International Service for National Agricultural Research (ISNAR): The Hague, The Neithelands, 1983.
2. Nair, P.K.R. Definition and concepts of agroforestry. In *An Introduction to Agroforestry*; Springer Science & Business Media: Berlin/Heidelberg, Germany, 1993.
3. De Foresta, H.; Temu, A.; Boulanger, D.; Feuilly, H.; Gauthier, M. *Towards the Assessment of Trees Outside Forests: A Thematic Report Prepared in the Framework of the Global Forest Resources Assessment*; Food and Agriculture Organization of the United Nations: Rome, Italy, 2013.
4. Sinclair, F.L. A general classification of agroforestry practice. *Agrofor. Syst.* **1999**, *46*, 161–180. [CrossRef]
5. Sánchez, P.A.; Woomer, P.L.; Palm, C.A. Agroforestry approaches or rehabilitating degraded lands after tropical deforestation. In *Proceedings of the JIRCAS International Symposium Series (Japan)*; Ibrahim, M., Beer, J., Eds.; CATIE: Turrialba, Costa Rica, 1994.
6. Somarriba, E.; Kass, D.; Ibrahim, M. Definition and classification of agroforestry system. In *Agroforestry Prototypes for Belize*; CATIE/GTZ: Turrialba, Costa Rica, 1998.
7. Pretty, J.; Bharucha, Z.P. Sustainable intensification in agricultural systems. *Ann. Bot.* **2014**, *114*, 1571–1596. [CrossRef] [PubMed]

8. Nerlich, K.; Graeff-Hönninger, S.; Claupein, W. Agroforestry in Europe: A review of the disappearance of traditional systems and development of modern agroforestry practices, with emphasis on experiences in Germany. *Agrofor. Syst.* **2013**, *87*, 475–492. [CrossRef]

9. Torralba, M.; Fagerholm, N.; Burgess, P.J.; Moreno, G.; Plieninger, T. Do European agroforestry systems enhance biodiversity and ecosystem services? A meta-analysis. *Agric. Ecosyst. Environ.* **2016**, 150–161. [CrossRef]

10. Hilger, T.; Keil, A.; Lippe, M.; Panomtaranichagul, M.; Saint-Macary, C.; Zeller, M.; Pansak, W.; Vu Dinh, T.; Cadisch, G. Soil Conservation on Sloping Land: Technical Options and Adoption Constraints. In *Sustainable Land Use and Rural Development in Southeast Asia: Innovations and Policies for Mountainous Areas*; Fröhlich, H.L., Schreinemachers, P., Stahr, K., Clemens, G., Eds.; Springer: Dordrecht, The Netherlands, 2012; pp. 229–279.

11. Swinton, S.M.; Lupi, F.; Robertson, G.P.; Hamilton, S.K. Ecosystem services and agriculture: Cultivating agricultural ecosystems for diverse benefits. *Ecol. Econ.* **2007**, *64*, 245–252. [CrossRef]

12. Bucheli, V.J.P.; Bokelmann, W. Agroforestry systems for biodiversity and ecosystem services: The case of the sibundoy valley in the colombian province of putumayo. *Int. J. Biodivers. Sci. Ecosyst. Serv. Manag.* **2017**, *13*, 380–397. [CrossRef]

13. Philpott, S.M.; Lin, B.B.; Jha, S.; Brines, S.J. A multi-scale assessment of hurricane impacts on agricultural landscapes based on land use and topographic features. *Agric. Ecosyst. Environ.* **2008**, *128*, 12–20. [CrossRef]

14. Lin, B.B. Resilience in Agriculture through Crop Diversification: Adaptive Management for Environmental Change. *Bioscience* **2011**, *61*, 183–193. [CrossRef]

15. Kiptot, E.; Franzel, S.; Degrande, A. Gender, agroforestry and food security in Africa. *Curr. Opin. Environ. Sustain.* **2014**, *6*, 104–109. [CrossRef]

16. Mbow, C.; Van Noordwijk, M.; Luedeling, E.; Neufeldt, H.; Minang, P.A.; Kowero, G. Agroforestry solutions to address food security and climate change challenges in Africa. *Curr. Opin. Environ. Sustain.* **2014**, *6*, 61–67. [CrossRef]

17. Place, F.; Roothaert, R.; Maina, L.; Franzel, S.; Sinja, J.; Wanjiku, J. Leguminous trees help to raise milk yields. *Appropr. Technol.* **2010**, *37*, 47–49.

18. Maliki, R.; Cornet, D.; Floquet, A.; Sinsin, B. Agronomic and economic performance of yambased systems with shrubby and herbaceous legumes adapted by smallholders. *Outlook Agric.* **2012**, *41*, 171–178. [CrossRef]

19. Nguyen, T.H.; Catacutan, D.C. History of agroforestry research and development in Vietnam: A review of literature. In *Proceedings of the Realizing the Potential of Agroforestry in Vietnam*; Catacutan, D.C., Bui, N., Bo, N.V., Hop, B.T.H., Eds.; World Agroforestry Centre: Nairobi, Kenya, 2012; pp. 15–16.

20. Irshad, M.; Khan, A.; Inoue, M.; Ashraf, M.; Sher, H. Identifying factors affecting agroforestry system in Swat, Pakistan. *African J. Agric. Res.* **2011**, *6*, 2586–2593. [CrossRef]

21. Wezel, A.; Steinmüller, N.; Friederichsen, J.R. Slope position effects on soil fertility and crop productivity and implications for soil conservation in upland northwest Vietnam. *Agric. Ecosyst. Environ.* **2002**, *91*, 113–126. [CrossRef]

22. Schweizer, S.A.; Fischer, H.; Häring, V.; Stahr, K. Soil structure breakdown following land use change from forest to maize in Northwest Vietnam. *Soil Tillage Res.* **2017**, *166*, 10–17. [CrossRef]

23. Tung, P.D.; Cuong, N.V.; Thinh, N.C.; Nhung, N.T.; Van, T.T.K. *Report Ethnic Minorities and Sustainable Development Goals: Who Will be Left Behind? Results from Analyses of the Survey on the Socio-Economic Situation of 53 Ethnic Minorities in 2015*; Report; UNDP: Hanoi, Vietnam, 2016; pp. 83–88. Available online: https://www.google.com.hk/url?sa=t&rct=j&q=&esrc=s&source=web&cd=&ved=2ahUKEwifv8PY84HwAhWaeN4KHWFUBWkQFjABegQIBRAD&url=https%3A%2F%2Fwww.vn.undp.org%2Fcontent%2Fdam%2Fvietnam%2Fdocs%2FPublications%2FFinal%2520report%2520on%2520the%2520Overview%2520of%2520socio-economic%2520status%2520of%252053%2520ethnic%2520minorities%2520E.pdf&usg=AOvVaw0Nsj5LO1MLWfIOZBsj0Qgm (accessed on 16 April 2021).

24. Michaud, J.; Turner, S.; Roche, Y. Mapping ethnic diversity in highland Northern Vietnam. *GeoJournal* **2002**, *57*, 305–323. [CrossRef]

25. Kiptot, E.; Hebinck, P.; Franzel, S.; Richards, P. Adopters, testers or pseudo-adopters? Dynamics of the use of improved tree fallows by farmers in western Kenya. *Agric. Syst.* **2007**, *94*, 509–519. [CrossRef]

26. Derero, A.; Coe, R.; Muthuri, C.; Hadgu, K.M.; Sinclair, F. Farmer-led approaches to increasing tree diversity in fields and farmed landscapes in Ethiopia. *Agrofor. Syst.* **2020**, 1–18. [CrossRef]

27. Sinclair, F.; Coe, R.I.C. The options by context approach: A paradidm shift in agronomy. *Exp. Agric.* **2019**, *55*, 1–13. [CrossRef]

28. Chambers, R. The origins and practice of participatory rural appraisal. *World Dev.* **1994**. [CrossRef]

29. Rogers, E.M. *Diffusion of Innovations*, 5th ed.; Free Press: New York, NY, USA, 2003; ISBN 9780743222099.

30. Stern, P.C.; Dietz, T.; Abel, T.; Guagnano, G.A.; Kalof, L. A value-belief-norm theory of support for social movements: The case of environmentalism. *Hum. Ecol. Rev.* **1999**, *6*, 81–97.

31. Ajzen, I. The theory of planned behavior. *Organ. Behav. Hum. Decis. Process.* **1991**. [CrossRef]

32. Rodriguez, J.M.; Molnar, J.J.; Fazio, R.A.; Sydnor, E.; Lowe, M.J. Barriers to adoption of sustainable agriculture practices: Change agent perspectives. *Renew. Agric. Food Syst.* **2009**. [CrossRef]

33. Läpple, D.; Kelley, H. Understanding the uptake of organic farming: Accounting for heterogeneities among Irish farmers. *Ecol. Econ.* **2013**, *88*, 11–19. [CrossRef]

34. Lalani, B.; Dorward, P.; Holloway, G.; Wauters, E. Smallholder farmers' motivations for using Conservation Agriculture and the roles of yield, labour and soil fertility in decision making. *Agric. Syst.* **2016**, *146*, 80–90. [CrossRef]

35. Daxini, A.; O'Donoghue, C.; Ryan, M.; Buckley, C.; Barnes, A.P.; Daly, K. Which factors influence farmers' intentions to adopt nutrient management planning? *J. Environ. Manag.* **2018**, *224*, 350–360. [CrossRef]
36. Liu, T.; Bruins, R.J.F.; Heberling, M.T. Factors influencing farmers' adoption of best management practices: A review and synthesis. *Sustainability* **2018**, *10*, 432. [CrossRef]
37. Warren, C.R.; Burton, R.; Buchanan, O.; Birnie, R.V. Limited adoption of short rotation coppice: The role of farmers' socio-cultural identity in influencing practice. *J. Rural Stud.* **2016**, *45*, 175–183. [CrossRef]
38. Mercer, D.E.; Pattanayak, S.K. Agroforestry Adoption by Smallholders. In *Forests in a Market Economy*; Sills, E.O., Abt, K.L., Eds.; Springer: Dordrecht, The Netherlands, 2003; Volume 72, pp. 283–299.
39. Vanslembrouck, I.; Van Huylenbroeck, G.; Verbeke, W. Determinants of the willingness of Belgian farmers to participate in agri-environmental measures. *J. Agric. Econ.* **2002**, *53*, 489–511. [CrossRef]
40. Wilson, G.A.; Hart, K. Farmer participation in agri-environmental schemes: Towards conservation-oriented thinking? *Sociol. Ruralis* **2001**, *41*, 254–274. [CrossRef]
41. Nyaga, J.; Barrios, E.; Muthuri, C.W.; Öborn, I.; Matiru, V.; Sinclair, F.L. Evaluating factors influencing heterogeneity in agroforestry adoption and practices within smallholder farms in Rift Valley, Kenya. *Agric. Ecosyst. Environ.* **2015**. [CrossRef]
42. Ajayi, O.C.; Place, F. Policy Support for Large-Scale Adoption of Agroforestry Practices: Experience from Africa and Asia. In *Agroforestry—The Future of Global Land Use*; Nair, P.R., Garity, D., Eds.; Springer Science & Business Media: Dordrecht, The Netherlands, 2012; pp. 175–201. ISBN 9789400746763.
43. McGinty, M.M.; Swisher, M.E.; Alavalapati, J. Agroforestry adoption and maintenance: Self-efficacy, attitudes and socio-economic factors. *Agrofor. Syst.* **2008**, *73*, 99–108. [CrossRef]
44. Ghosh, M.K.; Sohel, M.H.; Ara, N.; Zahara, F.T.; Nur, S.B.; Hasan, M.M. Farmers Attitude towards Organic Farming: A Case Study in Chapainawabganj District. *Asian J. Adv. Agric. Res.* **2019**. [CrossRef]
45. Pontara, G. Analysing Farmers' Perceptions towards Agroforestry Adoption in Southern Belize. Master's Thesis, Wageningen University, Wageningen, The Netherlands, 2019.
46. Tadesse, S.A. Views and Attitudes of Local Farmers towards Planting, Growing and Managing Trees in Agroforestry System in Basona Worena District, Ethiopia. *J. Agric. Sci. Food Res.* **2019**, *10*, 1–10. [CrossRef]
47. Oduro, K.A.; Arts, B.; Kyereh, B.; Mohren, G. Farmers' Motivations to Plant and Manage On-Farm Trees in Ghana. *Small-Scale For.* **2018**, *17*, 393–410. [CrossRef]
48. Nouman, W.; Riaz, A. Farmer's attitude towardsagroforestry in district Faisalabad. *Pakistan J. Agric. Sci.* **2008**, *45*, 60–64.
49. Kiyani, P.; Andoh, J.; Lee, Y.; Lee, D.K. Benefits and challenges of agroforestry adoption: A case of Musebeya sector, Nyamagabe District in southern province of Rwanda. *Forest Sci. Technol.* **2017**, *13*, 174–180. [CrossRef]
50. Smith, D.E.; Gassner, A.; Agaba, G.; Nansamba, R.; Sinclair, F. The utility of farmer ranking of tree attributes for selecting companion trees in coffee production systems. *Agrofor. Syst.* **2019**, *93*, 1469–1483. [CrossRef]
51. Inwood, S. Social Forces and Cultural Factors Influencing Farm Transition. *Choices Mag. Food Farm Resour. Issues* **2013**, *28*, 1–5.
52. Chomba, S.W.; Nathan, I.; Minang, P.A.; Sinclair, F. Illusions of empowerment? Questioning policy and practice of community forestry in Kenya. *Ecol. Soc.* **2015**, *28*, 1–5. [CrossRef]
53. Madge, C. Ethnography and agroforestry research: A case study from the Gambia. *Agrofor. Syst.* **1995**, *32*, 127–146. [CrossRef]
54. Xu, J.; Ma, E.T.; Tashi, D.; Fu, Y.; Lu, Z.; Melick, D. Integrating sacred knowledge for conservation: Cultures and landscapes in Southwest China. *Ecol. Soc.* **2005**, *10*. [CrossRef]
55. Weber, R.; Faust, H.; Schippers, B.; Mamar, S.; Sutarto, E.; Kreisel, W. Migration and ethnicity as cultural impact factors on land use change in the rainforest margins of Central Sulawesi, Indonesia. In *Stability of Tropical Rainforest Margins*; Tscharntke, T., Leuschner, C., Zeller, M., Guhardja, E., Bidin, A., Eds.; Springer: Berlin/Heidelberg, Germany, 2007.
56. Ayantunde, A.A.; Briejer, M.; Hiernaux, P.; Udo, H.M.J.; Tabo, R. Botanical knowledge and its differentiation by age, gender and ethnicity in Southwestern Niger. *Hum. Ecol.* **2008**, *36*, 881–889. [CrossRef]
57. Coe, R.; Sinclair, F.; Barrios, E. Scaling up agroforestry requires research "in" rather than "for" development. *Curr. Opin. Environ. Sustain.* **2014**, *6*, 73–77. [CrossRef]
58. Adesina, A.A.; Chianu, J. Determinants of farmers' adoption and adaptation of alley farming technology in Nigeria. *Agrofor. Syst.* **2002**, *55*, 99–112. [CrossRef]
59. Bannister, M.E.; Nair, P.K.R. Agroforestry adoption in Haiti: The importance of household and farm characteristics. *Agrofor. Syst.* **2003**, *57*, 149–157. [CrossRef]
60. Meijer, S.S.; Catacutan, D.; Ajayi, O.C.; Sileshi, G.W.; Nieuwenhuis, M. The role of knowledge, attitudes and perceptions in the uptake of agricultural and agroforestry innovations among smallholder farmers in sub-Saharan Africa. *Int. J. Agric. Sustain.* **2015**, *13*, 40–54. [CrossRef]
61. Mills, J.; Gaskell, P.; Ingram, J.; Dwyer, J.; Reed, M.; Short, C. Engaging farmers in environmental management through a better understanding of behaviour. *Agric. Hum. Values* **2017**, *34*, 283–299. [CrossRef]
62. Sinclair, F.L.; Walker, D.H. Acquiring qualitative knowledge about complex agroecosystems. Part 1: Representation as natural language. *Agric. Syst.* **1998**, *39*, 223–252. [CrossRef]
63. Walker, D.H.; Sinclair, F.L. Acquiring qualitative knowledge about complex agroecosystems. Part 2: Formal representation. *Agric. Syst.* **1998**, *56*, 341–363. [CrossRef]
64. Sillitoe, P. The development of indigenous knowledge: A new applied anthropology. *Curr. Anthropol.* **1998**, *39*, 223–252. [CrossRef]

65. Dien Bien Statistic Office. *Dien Bien Statistical Yearbook 2018*; Statistical Publishing House: Dien Bien, Vietnam, 2019.
66. Son La Statistic Office. *Son La Statistical Yearbook 2018*; Statistical Publishing House: Son La, Vietnam, 2019.
67. Yen Bai Statistic Office. *Yen Bai Statistical Yearbook 2018*; Statistical Publishing House: Yen Bai, Vietnam, 2019.
68. Miles, M.B.; Huberman, A.M. *An Expanded Sourcebook: Qualitative Data Analysis*, 2nd ed.; Sage Publications, Inc.: Thousand Oaks, CA, USA, 1994; ISBN 0803955405.
69. De Jong, W.; Sam, D.D.; Trieu, V.H. *Forest Rehabilitation in Vietnam: Histories, Realities, and Future*; Center for International Forestry Research: Jakarta, Indonesia, 2006; ISBN 979-24-4652-4.
70. Gonzalez Gamboa, V.; Barkmann, J.; Marggraf, R. Social network effects on the adoption of agroforestry species: Preliminary results of a study on differences on adoption patterns in Southern Ecuador. In *Proceedings of the Procedia—Social and Behavioral Sciences*; Elsevier: Amsterdam, The Netherlands, 2010.
71. Neupane, R.P.; Sharma, K.R.; Thapa, G.B. Adoption of agroforestry in the hills of Nepal: A logistic regression analysis. *Agric. Syst.* **2002**, *72*, 177–196. [CrossRef]
72. Hernández-Morcillo, M.; Burgess, P.; Mirck, J.; Pantera, A.; Plieninger, T. Scanning agroforestry-based solutions for climate change mitigation and adaptation in Europe. *Environ. Sci. Policy* **2018**, *80*, 44–52. [CrossRef]
73. Bacon, C.M.; Getz, C.; Kraus, S.; Montenegro, M.; Holland, K. The social dimensions of sustainability and change in diversified farming systems. *Ecol. Soc.* **2012**. [CrossRef]
74. Scherr, S.J. Economic factors in farmer adoption of agroforestry: Patterns observed in Western Kenya. *World Dev.* **1995**, *23*, 787–804. [CrossRef]
75. Coulibaly, J.Y.; Chiputwa, B.; Nakelse, T.; Kundhlande, G. Adoption of agroforestry and the impact on household food security among farmers in Malawi. *Agric. Syst.* **2017**, *155*, 52–69. [CrossRef]
76. Assefa, E.; Hans-Rudolf, B. Farmers' Perception of Land Degradation and Traditional Knowledge in Southern Ethiopia—Resilience and Stability. *Land Degrad. Dev.* **2016**, *27*, 1552–1561. [CrossRef]
77. Kloppenburg, J.; Lezberg, S.; De Master, K.; Stevenson, G.W.; Hendrickson, J. Tasting food, tasting sustainability: Defining the attributes of an alternative food system with competent, ordinary people. *Hum. Organ.* **2000**, 177–186. [CrossRef]
78. Thangata, P.H.; Alavalapati, J.R.R. Agroforestry adoption in southern Malawi: The case of mixed intercropping of Gliricidia sepium and maize. *Agric. Syst.* **2003**, *78*, 57–71. [CrossRef]
79. Place, F.; Ajayi, O.C.; Torquebiau, E.; Detlefsen, G.; Gauthier, M.; Buttoud, G. *Improved Policies for Facilitating the Adoption of Agroforestry. In Agroforestry for Biodiversity and Ecosystem Services—Science and Practice*; Kaonga, M.L., Ed.; IntechOpen: London, UK, 2012; pp. 113–128.
80. Kabwe, G.; Bigsby, H.; Cullen, R. Factors Influencing Adoption of Agroforestry among Smallholder Farmers in Zambia. Ph.D. Thesis, Lincoln University, Oxford, PA, USA, 2010.
81. Rai, A.; Srivastava, A.K.; Singh, M. A socio-economic study on agroforestry in Chhachhrauli block of Yamunanagar District of Haryana. *Indian J. Agrofor.* **2001**, *3*, 148–152.

 *forests*

Article

Carbon Sequestration Potential of Agroforestry Systems in Degraded Landscapes in West Java, Indonesia

Mohamad Siarudin [1], Syed Ajijur Rahman [2,*], Yustina Artati [3,*], Yonky Indrajaya [4], Sari Narulita [3], Muhammad Juan Ardha [5] and Markku Larjavaara [2,*]

[1] Agroforestry Research and Development Center (ARDC), Ministry of Environment and Forestry, Jalan Raya Ciamis-Banjar Km 4, Ciamis 46201, Indonesia; msiarudin@yahoo.com
[2] Institute of Ecology and Key Laboratory for Earth Surface Processes of the Ministry of Education, College of Urban and Environmental Science, Peking University, Beijing 100871, China
[3] Center for International Forestry Research (CIFOR), Bogor 16115, Indonesia; s.narulita@cgiar.org
[4] Watershed Management Technology Center (WMTC), Ministry of Environment and Forestry, Surakarta 57169, Indonesia; yonky_indrajaya@yahoo.com
[5] Forest Carbon, Jakarta 10210, Indonesia; j.ardha@forestcarbon.com
* Correspondence: sumonsociology@yahoo.com (S.A.R.); y.artati@cgiar.org (Y.A.); markku.larjavaara@gmail.com (M.L.)

Abstract: When restoring degraded landscapes, approaches capable of striking a balance between improving environmental services and enhancing human wellbeing need to be considered. Agroforestry is an important option for restoring degraded land and associated ecosystem functions. Using survey, key informant interview and rapid carbon stock appraisal (RaCSA) methods, this study was conducted in five districts in West Java province to examine potential carbon stock in agroforestry systems practiced by smallholder farmers on degraded landscapes. Six agroforestry systems with differing carbon stocks were identified: gmelina (*Gmelina arborea* Roxb.) + cardamom (*Amomum compactum*); manglid (*Magnolia champaca* (L.) Baill. ex Pierre) + cardamom; caddam (*Neolamarckiacadamba* (Roxb.) Bosser) + cardamom; caddam + elephant grass (*Pennisetum purpureum* Schumach.); mixed-tree + fishpond; and mixed-tree lots. Compared to other systems, mixed-tree lots had the highest carbon stock at 108.9 Mg ha^{-1}. Carbon stock variations related to species density and diversity. Farmers from research sites said these systems also prevent soil erosion and help to restore degraded land. Farmers' adoption of agroforestry can be enhanced by the implementation of supportive policies and measures, backed by scientific research.

Keywords: agroforestry; land restoration; carbon sequestration; smallholder farmers; system adoption

Citation: Siarudin, M.; Rahman, S.A.; Artati, Y.; Indrajaya, Y.; Narulita, S.; Ardha, M.J.; Larjavaara, M. Carbon Sequestration Potential of Agroforestry Systems in Degraded Landscapes in West Java, Indonesia. *Forests* **2021**, *12*, 714. https://doi.org/10.3390/f12060714

Academic Editor: Julien Fortier

Received: 25 March 2021
Accepted: 26 May 2021
Published: 31 May 2021

Publisher's Note: MDPI stays neutral with regard to jurisdictional claims in published maps and institutional affiliations.

1. Introduction

Land degradation often causes severe environmental and socioeconomic problems [1]. It leads to declining environmental services, such as biodiversity, climate regulation, water supply, carbon stock, and other services [2]. Degradation also causes decreased land productivity, with serious cascading effects for human wellbeing, e.g., poverty, malnutrition, disease, forced migration, cultural damage, and even war [3]. Indonesia has approximately 14 million hectares (ha) of degraded land providing limited benefits for humans and nature [4]. Causes of land degradation can be categorized into direct and indirect ones. Direct causes in Indonesia include agricultural land expansion, overgrazing, commercial logging, and urbanization [5]. Indirect causes are typically population pressures coupled with poverty, as people often have to convert forests for agricultural land use to meet food demands beyond the capacity of existing farmlands [6]. The exploitation of forests to meet increasing demands for timber, fuel wood, and other products associated with population growth can also lead to land degradation [7]. Land degradation also increases emissions of greenhouse gases (GHGs) into the atmosphere, thereby contributing to climate change.

Indonesia is the world's third largest emitter of GHGs, with land-use change contributing most of its emissions [8]. The country has committed to unconditional and conditional emissions reductions of 29% and 41% respectively by 2030, as part of the global climate action following the Paris Agreement in 2015 [9]. Promoting climate resilience in food, water, and energy is the main pathway for the country's efforts to mitigate climate change. This includes the restoration of 12 million ha of degraded land by 2030 for landscape resilience [9,10].

The forest landscape restoration (FLR) approach is gaining interest for the restoration of degraded land. It is a comprehensive approach aimed at improving ecological functions and enhancing human wellbeing [11]. Agroforestry can be an important FLR approach with the potential to restore degraded land by reestablishing ecological processes, structures and ecosystem functions, while also enabling economic returns and the maintenance of livelihoods, local knowledge, and culture [12]. Agroforestry-based restoration projects across the tropics have been documented vividly by a number of scientific studies from Brazil to Mozambique and Indonesia [13–17].

By diversifying and enhancing farm products and services, agroforestry systems can meet the financial and social objectives crucial for local communities [18,19]. As an alternative source of important forest products, such as timber, fuelwood, fruits, and vegetables, agroforestry can reduce encroachment on local forests [20–22]. It can also improve soil fertility by enhancing nutrients, conserving soil moisture, protecting soil from erosion, moderating microclimates, sequestrating carbon, and diversifying habitats for wildlife and humans [23–26].

People across the Indonesian archipelago practice various agroforestry systems for their livelihood needs. These include spice and nut agroforests in the Moluccas; sugar palm and salak agroforests in Bali and Lombok; illipe-nut forests (tembawang) in West Kalimantan; benzoin gardens in North Sumatra; fruit and timber agroforests in Maninjau, West Sumatra; damar agroforests in Pesisir, Lampung; and fruit agroforests in Jambi and Palembang [21]. Notable examples of farmers managing to restore degraded landscapes are the planting of damar (*Agathis dammara* (Lamb.) Rich.) trees in degraded swidden areas in Jambi, Sumatra, and tamanu (*Calophyllum inophyllum* L.) trees on barren land in Wonogiri, Central Java [27,28]. Other studies provide evidence of agroforestry systems delivering carbon stock and sequestration [29,30]. The ability of agroforestry to store carbon varies from 37.7 Mg ha^{-1} at 1–10 years and 72.6 Mg ha^{-1} at 11–30 years [31]. A study of a 30-year-old home garden system in Lampung, Indonesia indicated above-ground biomass of 35.3 Mg C ha^{-1} [32]. In Indonesia's Bengkulu province, mixed tree agroforestry systems store around 95.2 tons ha^{-1} regardless of age [33]. Carbon stock storage in agroforestry systems with a combination of oil palm and agar wood vary between 78.28 and 79.13 Mg C ha^{-1} in regard to different levels of soil bulk density [34]. While simple and complex agroforestry systems adjacent to the Lore Lindu National Park buffer zones in Palu, Indonesia show significant differences in carbon stock. Simple agroforestry systems (combining individual trees and cash crops) store an average 37.30 Mg C ha^{-1} ranging from 30.32–45.05 Mg C ha^{-1}, while complex agroforestry systems (combining multiple tree species, shrubs, bushes, and crops) store an average 80.05 Mg C ha^{-1} of carbon stock ranging from 71.99–85.45 Mg C ha^{-1} [35]. As agroforestry systems are crucial for local people and nature, to corroborate agroforestry knowledge, the specific objective of this paper is to reinvestigate the type, component, management, and carbon sequestration potential of six common agroforestry systems, practiced by local communities in degraded landscapes in five administrative regions in West Java province. Although there are various definitions of agroforestry, we have defined it as a land-use system where woody perennials are used deliberately in the same land management units as other agricultural crops and/or animals, in some form of spatial arrangement or temporal sequence. We have hypothesized that mixed tree-based agroforestry systems have high carbon sequestration potential as they consist of various trees with understory.

2. Materials and Methods

2.1. Study Sites

Five districts in West Java province, i.e., Banjar, Ciamis, Garut, Tasikmalaya, and Pangandaran were selected as study areas (see Figure 1). These districts were selected as they represent 21% of all degraded land in West Java province (West Java province has the largest area of degraded land among all provinces in Java, i.e., 900,000 ha [4]) (see Table 1), and because prior information showed local farmers practicing various agroforestry systems on degraded land, allowing investigations of system types and their carbon sequestration potential. Conditions in these study areas can also be representative of many other parts of Indonesian and tropical Asian agricultural landscapes in general.

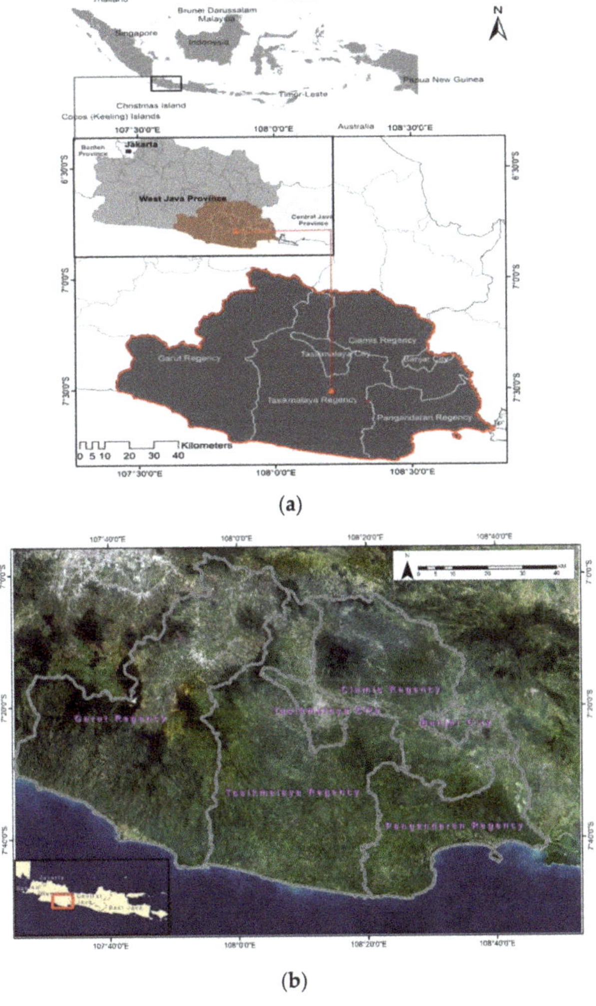

(a)

(b)

Figure 1. Locations of study sites in West Java (**a**), Landsat 8 imagery (2020) of study sites (**b**).

The five districts had a total area of 8850 km^2 and a human population of 6,198,824 in 2020, and therefore a high population density of 700 inhabitants per km^2 [36]. The climate is equatorial with two distinct seasons: dry from April to October, and rainy from

November to March. Average daily minimum air humidity varies from 50%–89% (see Table 1). Soils are dominated by volcanic sedimentary rocks and are highly fertile. Given the proximity of standing volcanoes, the area is considered highly seismic.

Table 1. Basic characteristics of the research sites.

District	Location	Average Temperature (°C)	Average Rainfall (mm)	Humidity (%)	Elevation (m)	Total Land Area (ha)/Total Degraded Land Area (ha)
Banjar	7.37° S 108.53° E	28.5	1600	69–85	31–79	13,195/620 (4.70%)
Ciamis	6.75° S 108.38° E	27	2800	68–87	31–768	159,763/24,259 (15.19%)
Garut	7.22° S 107.9° E	25.5	3294	73–85	100–1500	309,601/96,730 (31.24%)
Tasikmalaya	7.33° S 108.2° E	26	2171	50–73	0–2000	289,203/61,110 (21.13%)
Pangandaran	7.54° S 108.50° E	26.5	2750	85–89	0–2500	112,765/13,167 (11.68%)

Source: Badan Pusat Statistik [37].

Agriculture plays an important role in district economies, and contributes 43% of total income in Tasikmalaya [38]. Agriculture also dominates land use in Banjar district, covering 75% of the district's total land area [39]. These districts are also important providers of agricultural products to other areas in West Java. Garut, for example, is a major source of food for Bandung, the provincial capital of West Java, which is around 68 km away. These districts also provide important ecosystem services to other areas in West Java. Mountainous areas in Garut, Pangandaran, and Tasikmalaya are home to sources of important watersheds in the province. The highlands of Pangandaran district, for example, have three major rivers: the Cijulang, Citanduy, and Cimedang, which support water supplies, climate regulation, and aesthetics for lowland areas [40].

2.2. Data Collection and Analysis

This study combined quantitative and qualitative approaches, applying field surveys and interviews with key informants. Four farmer group heads, two forestry extension officers, and twenty landowners were selected purposively based on their experience in agroforestry and other land-use systems to obtain a general overview of major agroforestry practices in the study sites. Based on the field surveys and interviews, 50 plots (18 plots for the manglid-based system, 17 for the gmelina-based system, 6 for the caddam-based system, 3 for the sylvofishery system, and 6 for the mixed-tree lot system) from five districts were chosen purposively and observed to secure data on stand structure and composition. Primary data was collected during surveys by taking measurements directly in the field. The data collected covered the diameter at breast height (D) of trees using diameter tapes, tree height using digital dendrometers, tree species with local farmers identifying local names for expert translation to scientific names, and numbers of individual trees. Interviews were held with four key informants to secure qualitative information on socio-economic importance and management level, including product uses, economic purposes of product use (commercial, semi-commercial, subsistence), financial investment, and labor intensity. The data was analyzed comparatively between the identified systems. The agroforestry systems identified in the study locations were classified based on structure, function, socio-economic importance, and management level [41].

Carbon stock (C)estimation data was based on previous publications, i.e., Indrajaya et al. [29], Siarudin and Indrajaya [42,43], and Siarudin [44] with additional data measurements at the same research locations. The RaCSA (Rapid Carbon Stock Appraisal) method developed by Hairiah et al. [45] was used in this study to estimate C stocks in above-ground biomass (AGB). RACSA can reduce time and expense, and is considered more effective [46,47]. Several previous studies have used this method to estimate carbon stock in agroforestry systems [46,48]. The components measured were tree biomass, understory biomass, woody necromass, and non-woody necromass (litter). Within 10 m × 20 m or 5 m × 40 m plots, tree biomass measurements were collected in diameter at breast height (5–30 cm), and tree-height specifically for Arecaceae (palms). If trees with D greater than 30 cm were present in a plot, the main plot width was expanded to 20 m × 100 m to measure trees with D greater than 30 cm. The weight estimation of above-ground biomass of each tree was measured using the generic allometric equation for biomass calculations by Chave et al. [49].

$$AGB = \rho \times \left(-1.499 + 2.148 \ln(D) + 0.207 \left(\ln(D)^2 \right) - 0.028 \left(\ln(D)^3 \right) \right) \qquad (1)$$

where AGB represents the weight of above-ground biomass (kg), ρ is wood density, and D is the diameter at breast height (cm). The generic allometric equation used in this study [49] is suggested by Hairiah et al. [45] to estimate the weight of aboveground biomass. Research conducted by Hairiah et al. [45] shows that the generic equation suggested by Chave et al. [49] is valid in Indonesia, especially if diameter (D) is less than 30 cm. The generic allometric Chave equation has been applied to estimate C stocks in various forest types in the Gunung Halimun National Park (GHSNP), West Java, Indonesia [50]. However, as the local allometric equation for the caddam species in our research area is available [51], we used this local equation to get a more accurate estimation, specifically for the caddam-based agroforestry system, i.e.,

$$AGB = 0.014D^{2.958} \qquad (2)$$

The wood density value of each tree species refers to the Global Wood Density Database [52]. The content is assumed to be 0.47 of the biomass weight [53]. Measurements of woody necromass were collected in the same plots as tree measurements. These included standing dead trees, stumps of felled trees/fallen trees, or fallen dead tree trunks. Using RaCSA procedures for measuring necromass from Hairiah et al. [45], diameters of 5–30 cm were measured in plots of 5 m × 40 m or 10 m × 20 m, while necromasses with diameters of more than 30 cm were measured in 20 m × 100 m plots. The data measured covered volume and degree of decay. Total dry (biomass) weight was estimated by using wood density secondary data, volume and degree of decay. The degree of decay was measured subjectively in the field based on the degree of biomass intactness from decay, ranging from 50% (for half decayed) to 100% (undecayed) [45].

Data on understory biomass and non-woody necromass (litter) was collected by gathering biomass samples in plots of 0.5 m × 0.5 m. Fresh weight was measured, and 100 to 300 g samples were taken to measure sample dry weight and extrapolate total dry weight estimations. These data were analyzed descriptively. We have also analyzed the relations between the D, stand density and tree basal area (BA) to the C-stock using a simple linear regression.

3. Results

3.1. System Characteristics

In the study sites in West Java, agroforestry systems practiced by local farmers appear in several sub systems: intercropping combining trees and crops, trees and pasture, trees and aquaculture, and multipurpose tree lots (see Table 2 and Figure 2). Trees in the observed systems are fast-growing commercial species, i.e., manglid, gmelina, and caddam. Farmers sell timber from these species mainly to wood processing industries for veneer, plywood, furniture, and woodworking. Cardamom is the most popular crop for cultivation in the understory of these tree-based systems.

Table 2. Description of common agroforestry systems across the study sites in West Java.

System	Major Component	Species	Structure	Products	Uses and Management Level
Agri silviculture	Trees + crops	Manglid + cardamom (*Amomum compactum*)	Regular spacing (initial tree spacing 4 m × 2 m or 2 m × 2 m)	Timber (Manglid), food and medicine (cardamom)	Commercial. Medium—high input
		Gmelina + cardamom (*Amomum compactum*)	Regular spacing (initial tree spacing 2 m × 2 m or 2 m × 1 m)	Timber (Gmelina), food, and medicine (cardamom)	Commercial. Medium—high input
		Caddam + Cardamom (*Amomum compactum*)	Regular spacing (initial tree spacing 4 m × 2 m)	Timber (Caddam), food and medicine (cardamom)	Commercial. Medium—high input
Silvopasture	Trees + pastures	Caddam + elephant grass (*Pennisetum purpureum*)	Regular spacing (initial tree spacing 4 m × 2 m)	Timber (Caddam), fodder (elephant grass)	Commercial, subsistence. Medium—high input
Silvofishery	Trees + freshwater fishpond	Various trees + fish	Irregular spacing, trees along the embankment	Timber (trees), food (fish, fruits), soil conservation (trees)	Commercial, subsistence. Medium—high input
Mixed-tree lots	Trees + natural undergrowth	Various trees + weed	Irregular spacing	Timber (trees), food (fruits), fuelwood (trees)	Subsistence, semi-commercial. Low input

Figure 2. Common agroforestry systems in the study sites in West Java.

Some farmers in the study sites apply complex agroforestry cropping systems with mixed-tree species. Generally, they do not follow any specific spacing patterns in such systems, and allocate less space for understory crops. As a result, stands in such systems take the form of uneven-aged multispecies tree lots. Species found in these systems include teak (*Tectona grandis* Linn. F.), durian (*Durio zibethinus* Murr.), mahogany (*Swietenia* sp.), langsat (*Lansium domesticum* Corr.), large leaf rose mallow (*Hibiscus macrophyllus* Roxb. ex Hornem.), dog fruit (*Archidendron pauciflorum* (Benth.) I.C. Nielsen), coconut (*Cocos nucifera* L.), rambutan (*Nephelium lappaceum* L.), jackfruit (*Artocarpus heterophyllus* Lam.), and bitter bean (*Parkia speciosa* Hassk.). Such trees are also present in silvofishery systems where farmers breed various freshwater fish species in fishponds. In both mixed-tree lot and silvofishery systems, farmers allow trees to grow without intensive maintenance, relying on natural regeneration or planting a few species in any spaces. Based on observations and information from key informants, the above systems practiced by farmers in study sites help prevent landslides, topsoil erosion, and restore land through the regeneration of trees on the land which is already degraded.

The agroforestry systems studied in the study sites are often temporal in nature. Manglid has been planted in the eastern part of West Java province since the late 1980s, and the species persists on private forest land to this day through several rotations. Gmelina and caddam trees were planted more recently, around the early 2000s, and are now partially harvested. Based on observations in Ciamis and Garut districts, some farmers apply clear cutting, while others practice selective cutting. Some farmers maintain the same species either by replanting or relying on natural regeneration.

Farmers commonly apply selective cutting in mixed tree lot systems. They harvest trees that have reached diameters agreed with traders and leave others to continue growing. However, some landowners apply clear-cutting, mainly when they want to cultivate new tree species. Meanwhile, tree species composition in silvofishery systems is generally longer lasting as they generally comprise multipurpose tree species that do not require cutting within a certain period.

Seasonal crops (e.g., cardamom, elephant grass) are planted while trees are still young. Farmers carry out maintenance in the form of regular weeding and fertilizing, depending on their capacity to provide financial capital and labor. The understory is often damaged during tree harvesting, which provides an opportune moment for farmers to regenerate after planting the next rotation of trees. In mixed tree lot systems, farmers do not plant specific understory crops, as the ground is most commonly covered by weeds, thus requiring less input in its management than gmelina- and caddam-based agroforestry systems, which involve intensive crop cultivation.

3.2. Carbon Stock

C stock in agroforestry systems in the West Java study sites varied from 37–108.6 Mg ha^{-1} (see Table 3). The trees in each system comprise the majority of carbon stock in comparison to understory and necromass. Manglid- and caddam-based agroforestry systems in Garut and Tasikmalaya have 37 Mg ha^{-1} and 44 Mg ha^{-1} of C stock respectively, due to their maintaining regular tree density to support understory species growth. Silvofishery systems in Ciamis show the lowest basal area of trees (i.e., 12.9 m^2 ha^{-1}), due to the spatial arrangements of such systems providing dominant pond areas with fewer and younger trees (see Figure 2). In such silvofishery systems farmers often do not follow fixed spacing for planting trees along bunds. Commonly planted tree species are coconut and areca nut, which can strengthen embankments while producing fruits. Farmers also plant bitter bean, langsat, and jackfruit as multipurpose tree species around their fishponds.

The gmelina-based agroforestry system, with its stand density of 3794 trees ha^{-1}, has a relatively high basal area and C stock at 32.4 m^2 ha^{-1} and 63.7 Mg ha^{-1}, respectively. Some farmers plant gmelina with dense initial spacings of 2 m × 1 m and/or 2 m × 2 m, and allow several pre-existing tree species (e.g., *Swietenia* sp., *Acacia mangium*) to continue growing between the gmelina.

The mixed-tree lot system in the study site shows the highest C stock (i.e., 108.9 Mg ha^{-1}), due to the density and variety of trees planted. In this system, farmers also allow natural regeneration of certain shade-tolerant species (e.g., mahogany, large leaf rosemallow) to populate empty spaces between the planted trees. The high density of species in this mixed-tree-based system causes high biomass and C stock per unit area. As shown in Figure 3, a simple regression analysis suggests that the BA is a good predictor for C stock value with a coefficient of determination (R^2) value of 0.62. The BA value describes the accumulation of trees in the stand, which is a function of D and stand density, although the variables D and stand density themselves cannot be predictor variables for the value of C stock individually.

Table 3. C sequestration capacity of various agroforestry systems in study sites in West Java province.

Agroforestry System	Site	C Stock (Mg ha⁻¹)				Stand Density (Tree ha⁻¹)	Diameter (D) (cm)	Tree BA (m² ha⁻¹)	Ref
		Trees	Understory	Necromass	Total				
Manglid + cardamom	Tasikmalaya	42.3 (14.8–106.2)	0.6 (0.1–0.7)	1.1 (0.5-2)	44 (16.7–108)	1247 (500–2250)	14.5 (9.3–28.1)	20.1 (8.8–41.6)	[42]
Gmelina + cardamom	Tasikmalaya, Banjar, Pangandaran	61.7 (19–112.8)	0.8 (0.5–1.1)	1.4 (0.3–7.7)	63.7 (20.3–114.4)	3794 (1550–5850)	10.1 (5.3–16.3)	32.4 (13.2–53.9)	[43]
Caddam + Cardamom *	Garut	36.8	0.1	0.06	37.0	340.0	27.2	19.8	[29]
Caddam + elephant grass *	Garut	36.8	0.2	-	37.0	340.0	27.2	19.8	[29]
Mixed-tree species and freshwater fishpond	Ciamis	53.9 (12.7–89.1)	0.1	ND	54.0 (12.8–89.2)	704.0 (437–1200)	15.5 (9.4–18.5)	15.7 (14.8–17.5)	[44]
Mixed tree lots	Ciamis	108.6 (85.9–123.2)	0.3 (0.1–0.5)	ND	108.9 (86.3–123.4)	1633 (1000–2000)	11.9 (8.8–17.1)	30.2 (24.1–40.6)	[44]

Note: Numbers in parentheses are the range value; ND = no data; * Data developed from estimation model without range data.

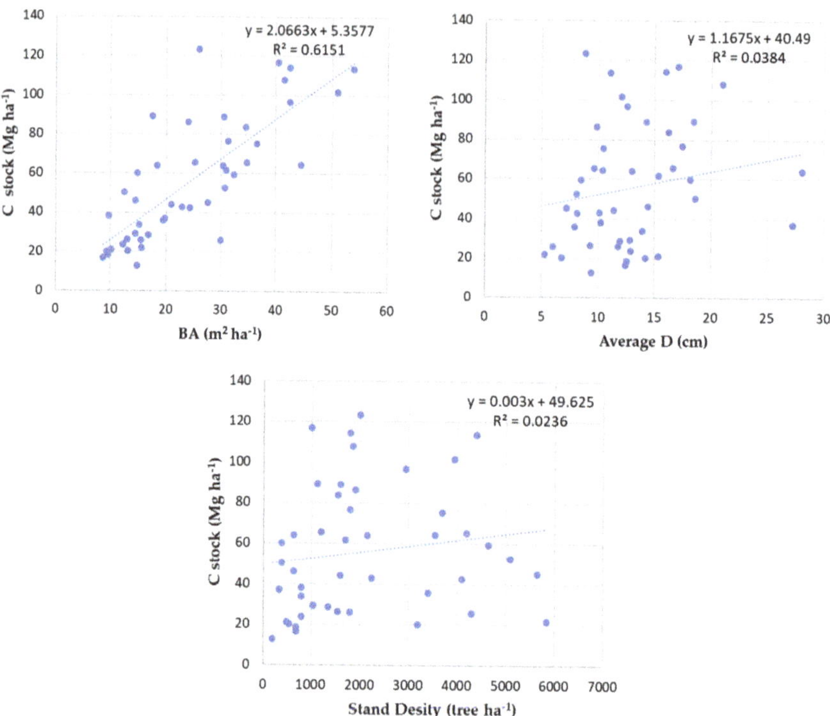

Figure 3. Correlation between C stock and average diameter (D), stand density and basal area (BA) in each measured plots.

4. Discussion

The agroforestry systems observed in the research sites in West Java are crucial for the livelihoods of local farmers, as they have both commercial and subsistence production value. In addition to timber from tree species, cardamom as an understory also has good market prospects, as it is widely used for food flavoring and herbal medicines [54,55]. Cardamom is also a suitable plant for growing under the tree canopy in a low light environment (29% to 82%) [56]. In managing these agroforestry systems, which are less labor-intensive than monoculture agriculture (e.g., rice, corn, beans), farmers mostly use family labor to enhance farm production and profits. However, maximizing the economic benefits of these systems may depend on the availability of labor, land, stable markets, and credit [57,58].

Farmers in the study sites reported that practicing agroforestry helps conserve soil and restore degraded land. Several studies have demonstrated that agroforestry systems can control runoff and soil erosion better than open cropland, thereby reducing losses of water, soil material, organic matter, and nutrients. They can also be employed to reclaim eroded and degraded land [59–61]. Local communities in the Indonesian archipelago possess the knowledge to use many traditional local tree garden—repong damar, simpukng and tembawang agroforestry systems as climate-smart farming, to manage ecosystems and restore degraded land [15]. Considering the high costs involved in land restoration (approximately 260 USD to 2880 USD ha^{-1}, depending on the restoration method used and the condition of land) and global concern for large-scale restoration [27,62,63], involving communities through their agroforestry practices can have low-cost potential (approximately \$181 US to \$402 US ha^{-1}, see also Rahman et al. [21]). Carbon sequestration across agricultural landscapes is important to minimize net C emissions from agriculture and mitigate climate change [64]. The agroforestry systems observed in our research sites have stored 37–108.9 Mg C ha^{-1}. The carbon stocks of the simple systems (i.e., combination of single trees with cash crop or grass) and mixed-tree system are slightly similar to C stock on agroforestry systems reported by other studies in Indonesia [31,32]. The C stock is found relatively higher in oil palm and agarwood-based systems in Sabah, Malaysia [34]. The mix tree plot systems in our study shows higher potential carbon stock than complex agroforestry system in the buffer zones of a national park in Sulawesi, and mixed tree systems in Bengkulu [33,35]. The variation in C stocks in these systems depends on the amount of biomass in the stands, particularly in tree components. Tree biomass accumulation representing the value of tree BA has a correlation of C stock value (see Figure 3). The higher the BA, the higher the biomass accumulation and C stocks. This is what we have found in the mixed-tree lot systems and the gmelina-based agroforestry system (i.e., relatively higher BA and C stock value compared to other three systems of our study).

As the agroforestry systems consisting of various tree species have high C stock potential (see Table 3, mixed tree lots), in order for agricultural land areas to absorb more C, increasing the adoption rate of agroforestry systems consisting of dense mixed tree species is crucial. The accumulation of soil C also improves soil quality and therefore productivity, and keeping soil in good condition supports farmers in adapting to climate change and extreme weather events [65]. However, targeting to have more C stock by having more trees in the agroforestry systems, farmers may need to compromise understory crop production due to increased tree canopy cover. Therefore, there is a trade-off between number of trees and understory crop production. To maximize overall benefit, farmers may consider planting fast growing timber and/or fruit trees with shade tolerant understory crops, e.g., pineapple (*Ananas comosus*) and ginger (*Zingiber officinale*), that also have high market value.

Farmers' adoption of agroforestry can be enhanced through implementation of supportive policies and measures by government and non-government organizations, such as capital support, the establishment of tailored market systems, secured land tenure, and technical assistance [57]. Effective policies should be propagated not by temporary projects but by long-term, government-backed institutions focused on agroforestry practices that may need adaptation to meet new opportunities and constraints. Our classification of agroforestry, i.e., agrisilviculture, silvopasture, silvofishery, and mixed tree lots (see Table 2), is based on the structures, components, and management practices of the systems, and this result can contribute to the documentation of agroforestry systems practiced in West Java. As there are gaps in information documenting the multitude of agroforestry systems practiced in different landscapes, such information is crucial for the development of supportive policy frameworks to implement climate smart agroforestry in other parts of Indonesia and elsewhere in the tropics, specifically for benefiting livelihoods and nature through land restoration [23,57,66]. The restoration of land using agroforestry can bring win-win solutions in attaining environmental and development objectives, which is also supported by other studies, e.g., degraded land restoration through agroforestry in Wonogiri, Central Java [27], and a burned and degraded peat land restoration project in Pulang Pisau, Central Kalimantan [16].

5. Conclusions

This study investigated the C sequestration potential of six agroforestry systems practiced by smallholder farmers in degraded landscapes in West Java province. The mixed-tree system practiced in the Ciamis area was found to have the highest C stock potential, i.e., 108.9 Mg ha^{-1}, while the caddam-based system with cardamom or elephant grass practiced in Garut provides the lowest C stock (37 Mg ha^{-1}). Tree stand density and diversity are key to having comparatively high or low C stock. Although the observed systems have variations in C stock, they are all crucial for local farmers to prevent topsoil erosion and landslides, and for the restoration of degraded land through tree regeneration. With this empirical evidence, this study contributes to agroforestry system science, specifically in its documentation of the characteristics of systems developed by farmers (i.e., species, structure, products, management) and their capacity for C sequestration, for practicing the restoration of degraded landscapes. However, further studies are required to investigate factors affecting farmers' selection of species, management intensity, market linkages, different landowners' interest in adopting such systems, and testing new systems (considering various fruit, spice, timber, and ornamental trees with various understory crops) targeted at enhancing and diversifying ecological functions in different types of degraded lands in West Java province and elsewhere that could also contribute to climate change mitigation and adaptation.

Author Contributions: Conceptualization, Y.A. and S.A.R.; methodology, M.S., Y.I., M.J.A. and M.L.; software, M.S. and Y.I.; validation, M.S., Y.I., S.A.R., Y.A. and M.L.; formal analysis, M.S. and Y.I.; investigation, M.S. and Y.I.; resources, M.S., S.A.R., Y.I., Y.A. and S.N.; data curation, M.S., S.A.R. and Y.I.; writing—original draft preparation, M.S., S.A.R. and Y.A.; writing—review and editing, S.A.R., Y.A. and M.L.; visualization, Y.I. and S.N.; supervision, Y.A.; funding acquisition, Y.A. All authors have read and agreed to the published version of the manuscript.

Funding: The publication of this research was funded by the National Institute of Forest Science (NIFoS), Republic of Korea.

Data Availability Statement: Data supported the result of this paper can be found by direct request to the authors.

Acknowledgments: The authors would like to thank a number of individuals at the Indonesian Ministry of Environment and Forestry's Research, Development and Innovation Agency (FORDA); CIFOR-ICRAF; and Peking University for their help with the publication. Many thanks are also extended to farmers in the study sites where field surveys were undertaken, who shared their precious time, thoughts and concerns. Our gratitude goes out to the National Institute of Forest Science (NIFoS) in Korea for funding the publication.

Conflicts of Interest: The authors declare no conflict of interest.

References

1. Abu Hammad, A.; Tumeizi, A. Land degradation: Socioeconomic and environmental causes and consequences in the eastern Mediterranean. *Land Degrad. Dev.* **2012**, *23*, 216–226. [CrossRef]
2. Dubois, O. *The State of the World's Land and Water Resources for Food and Agriculture: Managing Systems at Risk*; Earthscan: London, UK, 2011; p. 285.
3. Olsson, L.; Barbosa, H.; Bhadwal, S.; Cowie, A.; Delusca, K.; Flores-Renteria, D.; Hermans, K.; Jobbagy, E.; Kurz, W.; Li, D. *Land Degradation: IPCC Special Report on Climate Change, Desertification, Land 5 Degradation, Sustainable Land Management, Food Security, and 6 Greenhouse Gas fluxes in Terrestrial Ecosystems*; IPCC: Geneva, Switzerland, 2019.
4. Kementerian Lingkungan Hidup dan Kehutanan. *Statistik Lingkungan Hidup dan Kehutanan 2018*; Kementerian Lingkungan Hidup dan Kehutanan: Jakarta, Indonesia, 2019.
5. MoEF & UNCCD. *Indonesia—Land Degradation Neutrality National Report*; Ministry of Environment and Forestry & UNCCD: Jakarta, Indonesia, 2015.
6. FAO. *State of the World's Forests 2016: Forests and Agriculture: Land-Use Challenges and Opportunities*; Food Agriculture Organization of the United Nations: Rome, Italy, 2016.
7. Santoro, A.; Venturi, M.; Bertani, R.; Agnoletti, M. A Review of the Role of Forests and Agroforestry Systems in the FAO Globally Important Agricultural Heritage Systems (GIAHS) Programme. *Forests* **2020**, *11*. [CrossRef]
8. Baumert, K.A.; Herzog, T.; Pershing, J. *Navigating the Numbers: Greenhouse Gas Data and International Climate Policy*; World Resources Institute: Washington, DC, USA, 2005.
9. Government of Indonesia. *First Nationally Determined Contribution Republic of Indonesia*; Government of Indonesia: Jakarta, Indonesia, 2016.
10. Kementerian Lingkungan Hidup dan Kehutanan. *Rencana Strategis Direktorat Jenderal Pengendalian DAS and Hutan Lindung Tahun 2020–2024*; Kementerian Lingkungan Hidup dan Kehutanan: Jakarta, Indonesia, 2020.
11. Reed, J.; van Vianen, J.; Barlow, J.; Sunderland, T. Have integrated landscape approaches reconciled societal and environmental issues in the tropics? *Land Use Policy* **2017**, *63*, 481–492. [CrossRef]
12. ICRAF. *Restoring Land with Agroforestry: New Guide Published*; ICRAF: Bogor, Indonesia, 2021; Volume 2021.
13. Hillbrand, A. *What is the Potential of Agroforestry to Restore Degraded Land in Guatemala?* FAO: Rome, Italy, 2021.
14. Rahman, S.A.; Baral, H. Nature-based solution for balancing the food, energy, and environment trilemma: Lessons from Indonesia. In *Nature-Based Solutions for Resilient Ecosystems and Societies*; Dhyani, S., Gupta, A., Karki, M., Eds.; Springer: Singapore, 2020; pp. 69–82. [CrossRef]
15. Samsudin, Y.B.; Puspitaloka, D.; Rahman, S.A.; Chandran, A.; Baral, H. Community-Based Peat Swamp Restoration Through Agroforestry in Indonesia. In *Agroforestry for Degraded Landscapes*; Dagar, J.C., Gupta, S.R., Teketay, D., Eds.; Springer: Singapore, 2020; pp. 349–365. [CrossRef]
16. Maimunah, S.; Rahman, S.A.; Samsudin, Y.B.; Artati, Y.; Simamora, T.I.; Andini, S.; Lee, S.M.; Baral, H. Assessment of Suitability of Tree Species for Bioenergy Production on Burned and Degraded Peatlands in Central Kalimantan, Indonesia. *Land* **2018**, *7*. [CrossRef]
17. Miccolis, A.; Peneireiro, F.; Marques, H.; Vieira, D.; Arcoverde, M.; Hoffmann, M.; Rehder, T.; Pereira, A. *Agroforestry Systems for Ecological Restoration: How to Reconcile Conservation and Production. Options for Brazil's Cerrado and Caatinga Biomes. Instituto Sociedade, População e Natureza–ISPN/World Agroforestry Centre (ICRAF)*; Instituto Sociedade, População e Natureza: Brasilia, Brazil, 2016.
18. Rahman, S.A.; Sunderland, T.; Kshatriya, M.; Roshetko, J.M.; Pagella, T.; Healey, J.R. Towards productive landscapes: Trade-offs in tree-cover and income across a matrix of smallholder agricultural land-use systems. *Land Use Policy* **2016**, *58*, 152–164. [CrossRef]
19. Garrity, D.P. Agroforestry and the achievement of the Millennium Development Goals. *Agrofor. Syst.* **2004**, *61*, 5–17. [CrossRef]
20. Garrity, D.P.; Amoroso, V.B.; Koffa, S.; Catacutan, D.; Buenavista, G.; Fay, P.; Dar, W. Landcare on the Poverty-Protection Interface in an Asian Watershed. *Conserv. Ecol.* **2002**, *6*, 1–12. [CrossRef]
21. Rahman, S.A.; Jacobsen, J.B.; Healey, J.R.; Roshetko, J.M.; Sunderland, T. Finding alternatives to swidden agriculture: Does agroforestry improve livelihood options and reduce pressure on existing forest? *Agrofor. Syst.* **2017**, *91*, 185–199. [CrossRef]
22. Murniati, D.; Garrity, D.P.; Gintings, A.N. The contribution of agroforestry systems to reducing farmers' dependence on the resources of adjacent national parks: A case study from Sumatra, Indonesia. *Agrofor. Syst.* **2001**, *52*, 171–184. [CrossRef]
23. Snelder, D.J.; Lasco, R.D. *Smallholder Tree Growing for Rural Development and Environmental Services. Lessons from Asia*; Springer Science & Business Media: Berlin, Germany, 2008; Volume 5.
24. Jose, S. Agroforestry for ecosystem services and environmental benefits: An overview. *Agrofor. Syst.* **2009**, *76*, 1–10. [CrossRef]
25. Idol, T.; Haggar, J.; Cox, L. Ecosystem Services from Smallholder Forestry and Agroforestry in the Tropics. In *Integrating Agriculture, Conservation and Ecotourism: Examples from the Field*; Campbell, W.B., Lopez Ortiz, S., Eds.; Springer: Dordrecht, The Netherlands, 2011; pp. 209–270. [CrossRef]
26. Lasco, R.D.; Delfino, R.J.P.; Espaldon, M.L.O. Agroforestry systems: Helping smallholders adapt to climate risks while mitigating climate change. *WIREs Clim. Chang.* **2014**, *5*, 825–833. [CrossRef]
27. Rahman, S.A.; Baral, H.; Sharma, R.; Samsudin, Y.B.; Meyer, M.; Lo, M.; Artati, Y.; Simamora, T.I.; Andini, S.; Leksono, B.; et al. Integrating bioenergy and food production on degraded landscapes in Indonesia for improved socioeconomic and environmental outcomes. *Food Energy Secur.* **2019**, *8*, e00165. [CrossRef]

28. Michon, G. *Domesticating Forests: How Farmers Manage Forest Resources*; CIFOR: West Java, Indonesia, 2005.
29. Indrajaya, Y.; Siarudin, M.; Handayani, W. Karbon tersimpan dalam biomassa agroforestry jabon-kapulaga dan rumput gajah di Kecamatan Pakenjeng, Garut, Jawa Barat. *J. Penelit. Agrofor.* **2014**, *2*, 67–74.
30. Takimoto, A.; Nair, P.K.R.; Nair, V.D. Carbon stock and sequestration potential of traditional and improved agroforestry systems in the West African Sahel. *Agric. Ecosyst. Environ.* **2008**, *125*, 159–166. [CrossRef]
31. Rahayu, S.; Lusiana, B.; van Noordwijk, M. Above ground carbon stock assessment for various land use systems in Nunukan, East Kalimantan. In *Carbon Stock Monitoring in Nunukan, East Kalimantan: A Spatial and Modelling Approach. Report from the Carbon Monitoring Team of the Forest Resource Management for Carbon Sequestration (FORMACS) Project*; Lusiana, B., van Noordwijk, M., Rahayu, S., Eds.; World Agroforestry Center: Bogor, Indonesia, 2005; pp. 21–34.
32. Roshetko, J.M.; Delaney, M.; Hairiah, K.; Purnomosidhi, P. Carbon stocks in Indonesian homegarden systems: Can smallholder systems be targeted for increased carbon storage? *Am. J. Altern. Agric.* **2002**, *17*, 138–148.
33. Wiryono, W.; Puteri, V.N.U.; Senoaji, G. The diversity of plant species, the types of plant uses and the estimate of carbon stock in agroforestry system in Harapan Makmur Village, Bengkulu, Indonesia. *Biodivers. J. Biol. Divers.* **2016**, *17*. [CrossRef]
34. Besar, N.A.; Suardi, H.; Phua, M.-H.; James, D.; Mokhtar, M.B.; Ahmed, M.F. Carbon Stock and Sequestration Potential of an Agroforestry System in Sabah, Malaysia. *Forests* **2020**, *11*. [CrossRef]
35. Wardah; Toknok, B.; Zulkhaidah. Carbon stock of agroforestry systems at adjacent buffer zone of Lore Lindu National Park, Central Sulawesi. *J. Trop. Soils* **2013**, *16*, 123–128. [CrossRef]
36. Badan Pusat Statistik. *Jumlah Penduduk Menurut Kabupaten/Kota (Jiwa) 2020: Badan Pusat Statistik Provinsi Jawa Barat*; Badan Pusat Statistik: Jakarta, Indonesia, 2021.
37. Badan Pusat Statistik. Kabupaten Dalam Angka. Available online: https://jabar.bps.go.id/ (accessed on 13 March 2021).
38. Kabupaten Tasikmalaya Dalam Angka. 2021. Available online: https://tasikmalayakab.bps.go.id/publication/2021/02/26/88bd050d058c24b3fde94581/kabupaten-tasikmalaya-dalam-angka-2021.html (accessed on 26 February 2021).
39. Kota Banjar Dalam Angka. 2020. Available online: https://banjarkota.bps.go.id/publication/2020/04/27/d433ab1e85911408c2d935a8/kota-banjar-dalam-angka-2020.html (accessed on 27 April 2020).
40. Kabupaten Pangandaran Dalam Angka. 2020. Available online: https://pangandarankab.bps.go.id/publication/2020/04/27/3aa1afc640f74adeb926edd7/kabupaten-pangandaran-dalam-angka-2020.html (accessed on 27 April 2020).
41. Nair, P.R. *An Introduction to Agroforestry*; Springer Science & Business Media: Berlin, Germany, 1993.
42. Siarudin, M.; Indrajaya, Y. Struktur tegakan dan cadangan karbon hutan rakyat pola agroforestry manglid (Manglieta glauca Bl) di Tasikmalaya, Jawa Barat. *J. Penelit. Agrofor.* **2014**, *2*, 45–56.
43. Siarudin, M.; Indrajaya, Y. Dinamika cadangan karbon sistem agroforestri gmelina (gmelina arborea roxb.) pada hutan rakyat di Tasikmalaya Dan Banjar, Jawa Barat. *J. Penelit. Agrofor.* **2014**, *4*, 37–46.
44. Siarudin, M. Traditional cropping pattern and management of home garden: Lesson learnt from Ciamis Regency, West Java Province, Indonesia. *IOP Conf. Ser. Earth Environ. Sci.* **2019**, *250*. [CrossRef]
45. Hairiah, K.; Dewi, S.; Agus, F.; Velarde, S.; Ekadinata, A.; Rahayu, S.; Van Noodwijk, M. *Measuring Carbon Stocks Accross Land Use Systems*; ICRAF: Bogor, Indonesia, 2011.
46. Reyes, T.D., Jr.; Ludevese, E.G. Rapid Carbon Stock Appraisal (RACSA) Implementation in Wahig-Inabanga Watershed, Bohol, Philippines. *Int. J. Environ. Rural Dev.* **2015**, *6*, 102–109. [CrossRef]
47. Van Noordwijk, M. *World Agroforestry Centre (ICRAF) Southeast Asia Program*; Rapid Carbon Stock Appraisal (RaCSA): Bogor, Indonesia, 2007.
48. Gusli, S.; Sumeni, S.; Sabodin, R.; Muqfi, I.H.; Nur, M.; Hairiah, K.; Useng, D.; van Noordwijk, M. Soil Organic Matter, Mitigation of and Adaptation to Climate Change in Cocoa–Based Agroforestry Systems. *Land* **2020**, *9*, 323. [CrossRef]
49. Chave, J.; Andalo, C.; Brown, S.; Cairns, M.A.; Chambers, J.Q.; Eamus, D.; Fölster, H.; Fromard, F.; Higuchi, N.; Kira, T.; et al. Tree allometry and improved estimation of carbon stocks and balance in tropical forests. *Oecologia* **2005**, *145*, 87–99. [CrossRef]
50. Arifanti, V.; Dharmawan, I.; Wicaksono, D. Potensi cadangan karbon tegakan hutan sub montana di Taman Nasional Gunung Halimun Salak. *J. Penelit. Sos. Ekon. Kehutan.* **2014**, *11*, 13–31. [CrossRef]
51. Siarudin, M.; Indrajaya, Y. Persamaan allometrik jabon (Neolamarckia cadamba Miq) untuk pendugaan biomassa di atas tanah pada Hutan Rakyat Kecamatan Pakenjeng Kabupaten Garut. *J. Penelit. Hutan Tanam.* **2014**, *11*, 1–9. [CrossRef]
52. Zanne, A.E.; Lopez-Gonzalez, G.; Coomes, D.A.; Ilic, J.; Jansen, S.; Lewis, S.L.; Miller, R.B.; Swenson, N.G.; Wiemann, M.C.; Chave, J. Global Wood Density Database. 2009. Available online: https://datadryad.org/stash/dataset/doi:10.5061/dryad.234 (accessed on 30 June 2015).
53. IPCC. *IPCC Guideline 2006 Guidelines for National Green House Gas Inventories*; IPCC: Geneva, Switzerland, 2006.
54. Ashokkumar, K.; Murugan, M.; Dhanya, M.K.; Warkentin, T.D. Botany, traditional uses, phytochemistry and biological activities of cardamom [Elettaria cardamomum (L.) Maton]—A critical review. *J. Ethnopharmacol.* **2020**, *246*. [CrossRef]
55. Wang, X.; Chen, H.; Chang, C.; Jiang, M.; Wang, X.; Xu, L. Study the therapeutic mechanism of Amomum compactum in gentamicin-induced acute kidney injury rat based on a back propagation neural network algorithm. *J. Chromatogr. B* **2017**, *1040*, 81–88. [CrossRef] [PubMed]
56. Sudomo, A.; Handayani, W. Karakteristik tanah pada empat jenis tegakan penyusun agroforestry berbasis kapulaga (Amomum compactum Soland ex Maton). *J. Penelit. Agrofor.* **2013**, *1*, 1–11.

57. Rahman, S.A. *Incorporation of Trees in Smallholder Land Use Systems: Farm Characteristics, Rates of Return and Policy Issues Influencing Farmer Adoption*; Prifysgol Bangor University: Bangor, UK, 2017.
58. Mercer, D.E. Adoption of agroforestry innovations in the tropics: A review. *Agrofor. Syst.* **2004**, *61*, 311–328. [CrossRef]
59. Muchane, M.N.; Sileshi, G.W.; Gripenberg, S.; Jonsson, M.; Pumariño, L.; Barrios, E. Agroforestry boosts soil health in the humid and sub-humid tropics: A meta-analysis. *Agric. Ecosyst. Environ.* **2020**, *295*. [CrossRef]
60. Wolde, Z. *The Role of Agroforestry in Soil and Water Conservation*; LAP Lambert Academic Publishing: Sunnyvale, CA, USA, 2015.
61. Kiepe, P.; Rao, M.R. Management of Agroforestry for the Conservation and Utilization of Land and Water Resources. *Outlook Agric.* **1994**, *23*, 17–25. [CrossRef]
62. Brown, L.R. Restoring the Earth: The Earth Restoration Budget. Available online: http://www.earth-policy.org/mobile/books/pb4/PB4ch8_ss7?phpMyAdmin%20=%201d6bec-%201fea35111307d869d19bcd2ce7 (accessed on 20 March 2021).
63. Strassburg, B.B.N.; Latawiec, A.E. The Economics of Restoration: Costs, Benefits, Scale and Spatial Aspects. Available online: https://www.slideshare.net/CIFOR/the-economics-of-restoration-costs-benefits-scale-and-spatial-aspects (accessed on 18 June 2018).
64. Schahczenski, J.; Hill, H. *Agriculture, Climate Change and Carbon Sequestration*; ATTRA, Ed.; NCAT: Greensboro, NC, USA, 2009.
65. World Bank. *Carbon Sequestration in Agricultural Soils*; The World Bank: Washington, DC, USA, 2012.
66. Mbow, C.; Van Noordwijk, M.; Luedeling, E.; Neufeldt, H.; Minang, P.A.; Kowero, G. Agroforestry solutions to address food security and climate change challenges in Africa. *Curr. Opin. Environ. Sustain.* **2014**, *6*, 61–67. [CrossRef]

Article

Assessing Tree Coverage and the Direct and Mediation Effect of Tree Diversity on Carbon Storage through Stand Structure in Homegardens of Southwestern Bangladesh

Md Mizanur Rahman [1,2,†] , **Gauranga Kumar Kundu** [3,†], **Md Enamul Kabir** [3], **Heera Ahmed** [3] and **Ming Xu** [1,2,*]

1   Henan Key Laboratory of Earth System Observation and Modeling, Henan University, Kaifeng 475004, China; mizanfwt04@gmail.com
2   College of Geography and Environmental Science, Henan University, Kaifeng 475004, China
3   Forestry and Wood Technology Discipline, Khulna University, Khulna 9208, Bangladesh; gauranga.fwt@gmail.com (G.K.K.); menamulkabir@fwt.ku.ac.bd (M.E.K.); heera.fwt09@gmail.com (H.A.)
*   Correspondence: mingxu@henu.edu.cn
†   Authors contributed equally to this work.

Abstract: Dealing with two major challenges, climate change mitigation and biodiversity loss, under the same management program, is more noteworthy than addressing these two separately. Homegardens, a sustainable agroforestry system and a home of diverse species, can be a possible choice to address these two issues. In this study, we assessed tree coverage, and the direct and indirect effects of tree diversity on carbon storage in different carbon pools through stand structure in homegardens of southwestern Bangladesh, using Sentinel 2 and field inventory data from 40 homesteads in eight villages. An unsupervised classification method was followed to assess homegardens' tree coverage. We found a high tree coverage (24.34% of total area of Dighalia) in homesteads, with a high overall accuracy of 96.52%. The biomass and soil organic carbon ($p < 0.05$) varied significantly among the eight villages, while total carbon stock did not vary significantly ($p > 0.05$). Shannon diversity had both direct and indirect effects on biomass carbon, upper layer soil organic carbon and total carbon storage, while basal area mediated the indirect effect. Both basal area and tree height had positive effects on biomass carbon and total carbon storage, with basal area having the strongest effect. These findings suggest that we must maintain higher diversity and tree height in order to maximize and sustain carbon storage, where tree diversity increases stand basal area and improves total carbon storage (including soil organic) in homegardens. Therefore, privately managed homegardens could be a potential nature-based solution for biodiversity conservation and climate change mitigation in Bangladesh.

Keywords: traditional homegardens; agroforestry system; biodiversity and carbon; optical remote sensing

Citation: Rahman, M.M.; Kundu, G.K.; Kabir, M.E.; Ahmed, H.; Xu, M. Assessing Tree Coverage and the Direct and Mediation Effect of Tree Diversity on Carbon Storage through Stand Structure in Homegardens of Southwestern Bangladesh. *Forests* **2021**, *12*, 1661. https://doi.org/10.3390/f12121661

Academic Editors: Himlal Baral and Mi Sun Park

Received: 1 October 2021
Accepted: 23 November 2021
Published: 30 November 2021

Publisher's Note: MDPI stays neutral with regard to jurisdictional claims in published maps and institutional affiliations.

1. Introduction

Nature-based solutions are one of the best strategies for adapting to global climate change and biodiversity loss [1]. To combat global warming, world leaders committed in the Paris Agreement to keep atmospheric temperature below 2 °C by end of the century [2]. Each signatory country to the Paris Agreement has pledged to cut a part of its greenhouse gas emissions by 2030 [2]. Under the framework of Nationally Determined Contributions (NDCs), each country is meeting their obligation by lowering either low carbon sectors (transport, industry, power, etc.) or by increasing forestry activities (afforestation or restoration) [3]. Homegardens, a sustainable and integrated agroforestry system in tropical and subtropical countries, has huge carbon sequestration potential [4,5]. This climate regulatory role of homegardens can be a potential NDC component for many countries in meeting its carbon emission commitment [3].

Sustainably managed lands that have the ability to supply food and other domestic products to conserve biodiversity and regulate climate are of high demand in the current changing climate [6]. Homegardens are a tiny piece of land just in and around the homestead in rural areas of the tropical and subtropical regions. This unique type of agroforestry system provides multiple ecosystem services over the year to members of the household [7]. It is an intensive cropping system of domesticated annual and perennial plants and/or animals, which are the primary source of household demand, such as food, fruits, fodder, fuel wood and furniture [7,8]. Different coexisting species of trees, shrubs and herbs in homegardens with vertical differentiation make this traditional agroforestry system a forest-like structure [7]. Within human-dominated landscapes, homegardens are considered as highly biodiverse agroforestry systems which play an important role in biodiversity conservation and carbon sequestration [4].

Studies on homegardens' carbon storage have shown a high capacity for minimizing atmospheric carbon through photosynthesis compared to other agroforestry systems [4,9,10]. Findings on homegardens carbon stocks have revealed that aboveground carbon is mainly contributed by tree and palm species in homegardens, and that soil carbon occupies the largest percentage of the total carbon stored in homegardens [9]. Homegardens with high species diversity and stem density lead to a more basal area and biomass (above and below ground), and therefore, they store more biomass and soil carbon compared to homegardens with less species density and stem density [5,11,12]. In the forest ecosystem, the positive role of species diversity in carbon storage is underpinned by the niche complementarity hypothesis, which believes that higher numbers of species in a stand with greater trait variation have used more resources, and therefore, store more carbon [13]. Some studies have, however, documented no particular relationship between species diversity and carbon stock, which can be explained by selection effect. The selection effect believes that among the coexisting species, the most dominant species with its key traits such as size vigor (height and diameter) regulates carbon storage [12]. Both of these hypotheses can be a driver for homegarden agroforestry carbon regulation, because it is a multi-storied-species landscape. Furthermore, inconsistent results of species diversity and carbon storage relationships have also been reported in different studies in agroforestry systems [5,14]. Thus, for understanding the species diversity and carbon storage and their relationship in a specific study area, we need to conduct direct field investigations.

Bangladesh committed to the Paris Agreement to reduce its greenhouse gas emission from power, transport and industry sectors with an equivalent to 12 Mt CO_{2e} by 2030, which is 5% below the total business as usual emission level from those sectors. However, Bangladesh is a country of villages where almost every household has a homegarden, which are well-established land use systems [7]. A recent study reported that the total tree coverage in non-forest area is about 4.5 million ha, where homegardens have dominant contributions [15]. The homegardens in Bangladesh are biologically diverse and most of the studies on homegardens in Bangladesh remained descriptive on the floristic characteristics, structure, uses and the relation between household and homegarden characters [7,16]. Homegardens in Bangladesh can be a potential component of NDCs, similar to other developing countries. However, few studies focused on the total carbon storage capacity and carbon and biodiversity relationship in Bangladesh [17,18]. This knowledge gap in homegardens agroforestry system in Bangladesh might be one of the most important reasons for excluding agroforestry as a key NDCs component by policy makers, despite its huge coverage and high potential for climate change mitigation.

In this study, we assessed homegarden tree coverage, stand structure, species diversity and carbon storage in southwestern Bangladesh. We set three specific objectives in this study. First, to assess the total tree coverage in homesteads of Dighalia upazila (administrative unit) in the Khulna district, Bangladesh by using Sentinel 2 imagery. Second, to quantify the direct and indirect effects of species diversity on biomass and soil organic carbon stocks through different stand structures in homegardens. Third, to examine how results in line with carbon emission targets have contributed to homegardens' management system.

2. Materials and Methods

2.1. Study Site

The study was conducted in Dighalia Upazilla (administrative unit) in the Khulna district of Bangladesh (Figure 1). Dighalia Upazilla is located in the southwestern part of Bangladesh. The study area is primarily a flood plain land mass between 22.50° to 22.60° N latitude and 89.31° to 89.37° E longitude. The deltaic landscape of this region is mainly low (<10 m above average sea level), flat and fertile [19]. The average area of the homegarden over the study site is 0.05 ha and rich in biodiversity [19]. It generally enjoys a tropical to subtropical monsoon climate with an average annual temperature of 26 °C. January is the coolest month (average temperature: 12.4 °C), while April is the hottest month (average temperature: 34.6 °C) in this region. The average annual rainfall for this region is 1986 mm (range: 1400–2600 mm). This upazila is enclosed by three main rivers, i.e., the Bhairab, Citra and Naboganga, and is adjacent to Khulna City, the largest city in southwestern Bangladesh [20].

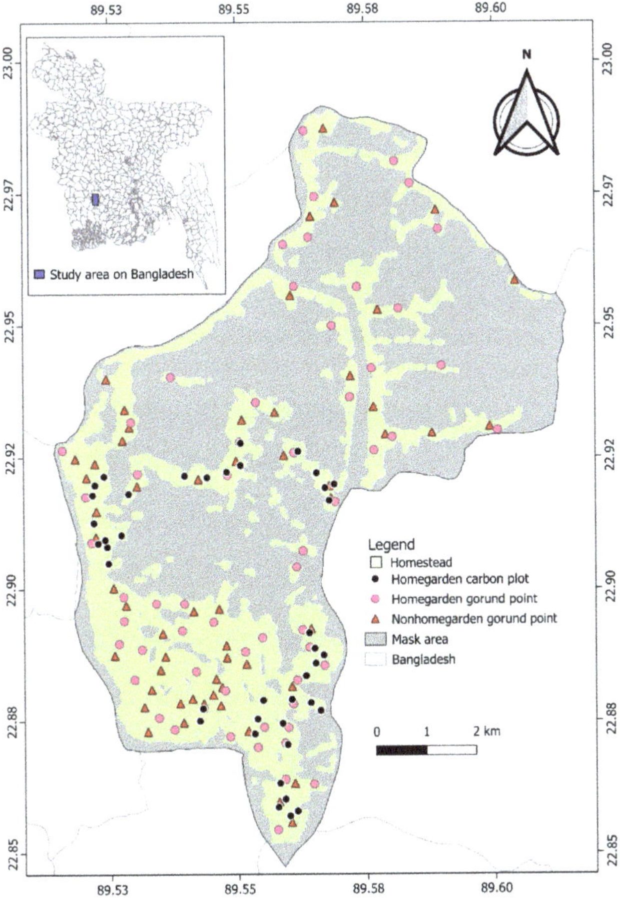

Figure 1. Map showing carbon sampling plots and ground truthing points in Dighalia Upazilla.

2.2. Sampling Design and Field Inventory

A multi-stage random sampling procedure was followed to select homegardens. First, two unions (smallest administrative unit), namely Senhati and Barakpur, from the total of six unions of Dighalia Upazila were selected randomly. Then, eight villages were selected randomly from the two randomly selected unions (four villages from each union). A total of 40 plots of 10 × 10 m (five from each randomly selected village) were taken purposively (homestead with homegarden), as shown in Figure 1. We measured the diameter at breast height (DBH) and canopy height of all species with a DBH ≥ 3 cm. All species in each sampled plot were identified and recorded to species level or by local name and were later confirmed from an authentic source [19]. We collected soil samples for soil organic carbon assessment from the center of each plot. We pulled out soil cores using an open face peat augur. From the soil core, we took a total of 80 (40 homegardens × 1–plot × 2 depths) soil samples; two from each plot using a five-centimeter-long steel core at the midpoint of 0–15 cm depth and 15–30 cm depth. All samples were air-dried and analyzed for estimating bulk density and soil organic carbon. We analyzed all the soil samples at the Nutrient Dynamics Laboratory of Khulna University, Bangladesh.

2.3. Remote Sensing Data and Processing

Sentinel -2 satellite image that was acquired in 13 December 2016 was used in this study for assessing the vegetation coverage. The Level-1C product of Sentinel 2 is a version of Top-Of-Atmosphere reflectance image, which is radiometrically and geometrically corrected. Sentinel 2 is a multispectral imager launched on 23 June 2015 by the European Space Agency, which has 13 bands of different resolutions (10–60 m) [21]. In this study, we used the 10 m resolution bands (Red and Near Infrared) for the Normalized Difference Vegetation Index (NDVI). We masked non-homestead areas (water body, crop field and fishpond) in order to distinguish rural settlement with homesteads. Then, we applied an ISODATA algorithm, an iteration method that collapses clusters of similar groups into one by measuring the Euclidean distance between the cluster centers [22,23]. We set the number of classes in the classification to 10, the total class size to 10 and the sample interval to 2. Using Google Earth visual interpretation and field data, we assigned homegardens and non-homegardens classes (houses, paddy fields, fish ponds, waterbodies and built-up areas) to the resultant one band image of ISODATA classification with 10 clusters [24]. We collected 115 points randomly from homegardens (59) and non-homegardens (56) within homestead areas (Figure 1) by using Google Earth visual interpretation for accuracy assessment of our classification result [23–25]. A confusion matrix was used for assessing the accuracy of our classification [25]. We used ENVI 5.00 and QGIS 3.14 for classification and mapping.

2.4. Data Analysis

2.4.1. Biomass Carbon

Common allometric equations were used (Table 1) to calculate the biomass [26,27], as it was developed from wide graphical (tropical countries) data and has diameter range (Table 1). The wood density data were obtained from the Global Wood Density Database [28]. For species with missing wood density in this database, we used the average wood density. Biomass carbon was calculated by multiplying 0.5, as it is assumed that wood biomass contains 50% carbon [28].

Table 1. Allometric equations used for biomass estimation in this study.

Equation(s)	Notes	Reference
$AGB = \rho \times \exp(-1.499 + 2.148 \times \ln(DBH) + 0.207 \times (\ln(DBH))^2 - 0.0281(\ln(DBH))^3)$	For all dicot trees	Chave et al. [26]
$AGB = 6.666 + 12.826 \times ht0.5 \times \ln(ht)$	For palm, coconut and date trees	Pearson et al. [29]
$BGB = \exp(-1.0587 + 0.8836 \times \ln AGB)$	For root biomass	Cairns et al. [27]

AGB = Aboveground biomass, ρ = Wood density (gcm^{-3}), DBH = Diameter at breast height, ht = Height, BGB = Belowground Biomass.

2.4.2. Soil Carbon Calculation

Soil carbon storage was estimated by the product of soil organic carbon concentration, soil bulk density and soil depth range. Soil bulk density was determined for each soil layer by dividing the oven-dried soil sample mass by the volume of the sample (Equation (1)) [30]:

$$Bulk\ Density = \frac{(Wt_{105°C})}{V_{core}}$$

$$V_{core} = \frac{\left(\pi D_{core}{}^2 L_{core}\right)}{4}$$

(1)

where $Wt_{105°C}$ is the weight of oven dried soil, V_{core} is the volume of the core, D_{core} is the inner diameter of the core and L_{core} is the length of the core.

The Loss of Ignition method was used for calculating soil organic matter [30]. Generally, soil organic matter contains 58% organic carbon [31]. Thus, we divided the organic matter by 1.724 (a universal conversion factor (called the van Bemmelen factor) for converting the soil organic carbon percentage [31].

After getting the soil bulk density and organic carbon percentage, we calculated the soil carbon stock for each layer using Equation (2):

$$Soil\ C\ (Mg\ ha^{-1}) = bulk\ density\ (g\ m^{-3}) \times soil\ depth\ interval\ (m) \times \%OC \times 0.01 \quad (2)$$

where Soil depth interval = 0.15 for 0–15 cm depth and 15–30 cm depth, %OC is expressed as a decimal fraction (e.g., 5% is expressed as 0.05) and 0.01 is a conversion factor to convert units to Mg ha^{-1}.

2.4.3. Total Carbon Stock

We calculated total carbon stocks (Equation (3)) per hectare by summing the biomass carbon stocks (aboveground and belowground biomass carbon) and soil organic carbon stocks (soil carbon, 0–15 cm and soil carbon, 15–30 cm):

$$Total\ C\ stock\ (Mg\ ha^{-1}) = Biomass\ carbon_{(AGC + BGC)} + TSOC_{SOC\ (0–15cm)} + SOC_{(15–30)} \quad (3)$$

2.4.4. Woody Species Diversity Calculation

The Shannon–Wiener index (H; Equation (4)) was used for species diversity [32], while for species richness, the Margalef index (D; Equation (5)) was used [33]:

$$H = \sum_{i=1}^{S} (P_i \times \ln(P_i)) \tag{4}$$

$$D = \frac{(S - 1)}{\ln(N)} \tag{5}$$

where H is the Shannon diversity for a plot, S is the number of species (species richness), P_i is the proportional of individuals of species i in the plot, N is the total number of individuals in a plot and "ln" is the natural logarithm.

One-way analysis of variance (ANOVA) was used to test the significant difference of carbon pools across the eight villages and Duncan's Multiple Range Test was used for multiple comparisons. Pearson's correlation tests were performed using the "rstatix" package in R environment (Version 3.6.3) [34], to explore the correlation between stand structure, woody species diversity and carbon pools. Structural equation model (SEM) was used to quantify the direct and indirect effects of tree diversity on carbon pools through stand structure [35]. We used the Lavaan package for employing SEM in R environment (Version 3.6.3) [34]. As recommended, we log transformed and standardized all the covariables before applying the structural equation model. Multiple linear regression was used to quantify the variance inflation factors (VIF) with a threshold value of 2 and assuming that covariables having a value below this threshold have no multicollinearity [36]. In the final steps of the variance inflation factors test, mean tree height, basal area, stem density and Shannon diversity were retained (MDBH had a VIF value that exceeded the threshold value, Supplementary Table S1). Then, we employed SEM for quantifying the direct, indirect and total effect of the Shannon diversity on different carbon pools through stem density, canopy height and basal area. A total 14 SEMs were tested for seven carbon pools, where highly insignificant covariables were removed until the model satisfied fit statistics [37] (Supplementary Table S2). The indirect effects of Shannon diversity were assessed by the path value through its mediator to different carbon pools (response variable) [35]. We assessed the total effect of the Shannon diversity on a carbon pool by adding the direct and indirect effect through different mediators [35,37].

3. Results

3.1. Homegardens Cover

The spatial distribution of homegardens and non-homegardens based on the unsupervised classification of Sentinel 2 images in Dighalia Upazilla showed a good match with the Google Earth map (same year image as of Sentinel 2; Figure 2). Our classification separated homegardens and non-homegardens in homestead areas in Dighalia with a higher overall accuracy and kappa coefficient (overall accuracy = 96.52% and Kappa = 0.93; Table 2). The overall coverage of the homegardens was 17.32 km^2 and the total percentage of homegardens in Dighalia was 24.34 (the total area of Dighalia is 71.16 km^2).

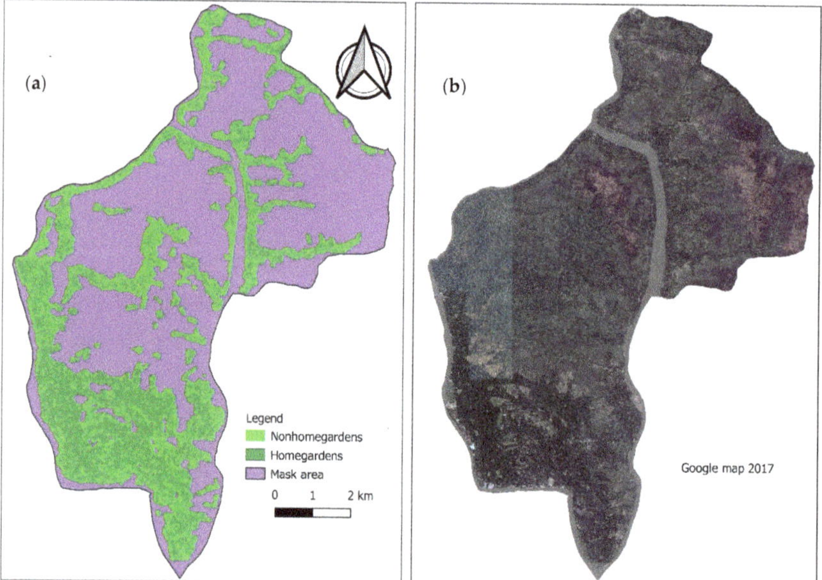

Figure 2. Extent of homegardens and non-homegardens in Dighalia Upazila. (**a**) Based on unsupervised classification of Sentinel 2 imagery; (**b**) Google Earth satellite image® shows homegardens covered by dark area.

Table 2. Result of the classification accuracy based on the Confusion Matrix. Column represents the Google Earth vegetation class, while the row represents the Sentinel 2-based class.

Pixel Based	Google Earth Based				
	Homegardens	Non-Homegardens	Total	Users (%)	Commission (%)
Homegardens	56	1	57	98.25	1.75
Non-homegardens	3	55	58	95.16	5.17
Total	59	56	115		
Producers (%)	94.92	98.21		Overall	96.52
Commission (%)	5.08	1.79		Kappa	0.93

3.2. Diversity and Stand Structure

The mean Shannon diversity, Margalef diversity and stand structures exhibited a spatial variation across the 40 plots (Table 3). The range of Shannon diversity was 1.91 to 3.01 with a mean of 2.56 ± 0.04 in the 40 plots (Table 3). The Margalef diversity index had a mean value of 2.31 ± 06 (range: 1.44–3.03). The estimated mean stem density, DBH, tree height and basal area were 1055.00 ± 35.62 treesha^{-1}, 13.30 ± 0.49 cm, 7.88 ± 0.17 m and 13.76 ± 0.96 m^2 ha^{-1}, respectively, across the 40 sample plots (Table 3).

Table 3. Descriptive statistics of stand structures and diversity indices across the 40 plots in Dighalia.

Variables	Minimum	Maximum	Mean	SE
Shannon–Wiener index (H')	1.91	3.01	2.56	0.04
Margalef index (R)	1.44	3.03	2.31	0.06
Stem density (trees ha^{-1})	600	1600	1055	35.61
Mean DBH (cm)	8.77	24.45	14.49	0.49
Tree height (m)	5.84	11.00	7.88	0.17
Basal area (m^2 ha^{-1})	4.04	29.31	13.76	0.96

3.3. Carbon Stocks

High variations of above- and belowground biomass carbon were observed (aboveground: 45.01 ± 6.95 to 88.00 ± 6.39 Mg ha^{-1}; belowground: 9.98 ± 1.36 to 18.11 ± 2.6 Mg ha^{-1}). We found similar biomass carbon stocks in Senhati, Baracpur and Lakhoati villages, but these three villages have significantly lower biomass carbon than in the Chandani mahal village ($p < 0.05$; Table 4). However, the biomass carbon stocks in these four villages were not significantly different from the other four villages ($p > 0.05$). Overall, above- and belowground carbon stocks across the 40 plots in Dighalia were 60.42 ± 3.93 and 12.89 ± 0.75 Mg ha^{-1}, respectively. Although the upper layer (0–15 cm) soil organic carbon stocks were insignificant across the different villages ($p > 0.05$; Table 4), the lower layer (15–30 cm) soil organic carbon stocks were found to be significantly different ($p < 0.05$; Table 4). In the Chandani mahal and Hagigram villages, soil organic carbon stocks at lower layers were significantly lower than at the Bativita village ($p < 0.05$; Table 4). Although we found a significant difference in biomass and lower layer soil organic carbon stocks, the differences in overall carbon stocks were insignificant across the villages ($p > 0.05$). The total carbon stock (biomass and soil carbon pools) in the eight villages ranged from 109.10 ± 10.89 Mg ha^{-1} (in Senhati) to 152.22 ± 10.15 Mg ha^{-1} (in Chandani mahal) with an overall average total carbon stock of 129.47 ± 5.10 Mg ha^{-1} (Table 4). The overall biomass carbon stock (73.31 ± 4.67 Mg ha^{-1}) contributes 56.62 percent of the total carbon stock, while soil carbon (56.16 ± 1.71 Mg ha^{-1}) contributes 43.38 percent of the total carbon stock. As we mapped the total area of homegardens and mean total carbon per hectare in Dighalia, we also assessed the total carbon stock in Dighalia by multiplying the mean carbon stocks by the total area. The total carbon stocks in Dighalia homegardens was 224,242.2 Mg C, which is equivalent to 822,968.29 Mg CO_2.

Table 4. Mean ($\pm$SE) difference of different carbon pools between villages. According to the Duncan's Multiple Range Test, the columns with a similar letter suggest a non-significant difference ($p > 0.05$).

Village	Aboveground Biomass Carbon (Mg ha^{-1})	Belowground Biomass Carbon (Mg ha^{-1})	Soil Organic Carbon (0–15 cm) (Mg ha^{-1})	Soil Organic Carbon (15–30 cm) (Mg ha^{-1})	Total Biomass Carbon (Mg ha^{-1})	Total Soil Organic Carbon (Mg ha^{-1})	Total Carbon (Mg ha^{-1})
Senhati	45.01 ± 6.95 b	9.98 ± 1.36 b	26.18 ± 2.25 a	27.94 ± 3.38 ab	54.99 ± 8.30 b	54.12 ± 3.99 ab	109.10 ± 10.89 a
Chandani mahal	88.00 ± 6.39 a	18.11 ± 1.16 a	26.04 ± 2.18 a	20.08 ± 3.29 b	106.11 ± 7.56 a	46.12 ± 4.63 b	152.22 ± 10.15 a
Bativita	55.05 ± 10.07 ab	11.89 ± 1.94 ab	29.44 ± 3.02 a	34.79 ± 2.00 a	66.94 ± 12.01 ab	64.23 ± 3.85 a	131.17 ± 12.14 a
Hagigram	61.47 ± 10.47 ab	13.12 ± 1.99 ab	27.48 ± 3.63 a	22.21 ± 3.35 b	74.59 ± 12.46 ab	49.70 ± 6.28 ab	124.30 ± 16.22 a
Baracpur	49.96 ± 14.79 b	10.82 ± 2.81 b	31.28 ± 2.67 a	27.04 ± 3.85 ab	60.78 ± 17.59 b	58.32 ± 5.66 ab	119.10 ± 19.06 a
Ghosgati	66.83 ± 7.28 ab	14.18 ± 1.38 ab	30.80 ± 2.20 a	29.23 ± 3.27 ab	81.01 ± 8.67 ab	60.03 ± 4.21 ab	141.03 ± 10.83 a
Kamargati	67.82 ± 10.86 ab	14.34 ± 2.00 ab	30.03 ± 6.46 a	29.32 ± 4.39 ab	82.16 ± 12.86 ab	59.35 ± 3.74 ab	141.51 ± 14.14 a
Lakhoati	49.20 ± 13.18 b	10.70 ± 2.51 b	30.30 ± 2.27 a	27.10 ± 2.06 ab	59.91 ± 15.69 b	57.40 ± 3.38 ab	117.31 ± 18.40 a
Average	60.42 ± 3.93	12.89 ± 0.75	28.94 ± 1.12	27.21 ± 1.25	73.31 ± 4.67	56.16 ± 1.71	129.47 ± 5.10

3.4. Bivariate Relationship between Stand Structure, Diversity and Carbon Pools

All four stand structure parameters were positively related to biomass carbon (aboveground and belowground), where basal area had a stronger relation compared to stem density, mean tree height and mean diameter ($p < 0.05$; Figure 3). However, these four-stand structure parameters had no significant relation with any of the soil organic carbon layers ($p > 0.05$; Figure 3). When we considered all the carbon pools together as total mean carbon stocks, all four stand structure parameters except mean tree height had a significant relation and basal area also had a stronger relation ($p < 0.05$, Figure 3). Shannon diversity index also had positive effects on all carbon pools (except lower layer soil organic carbon; $p > 0.05$; Figure 3) and total carbon stocks as well as stem density and basal area ($p < 0.05$; Figure 3).

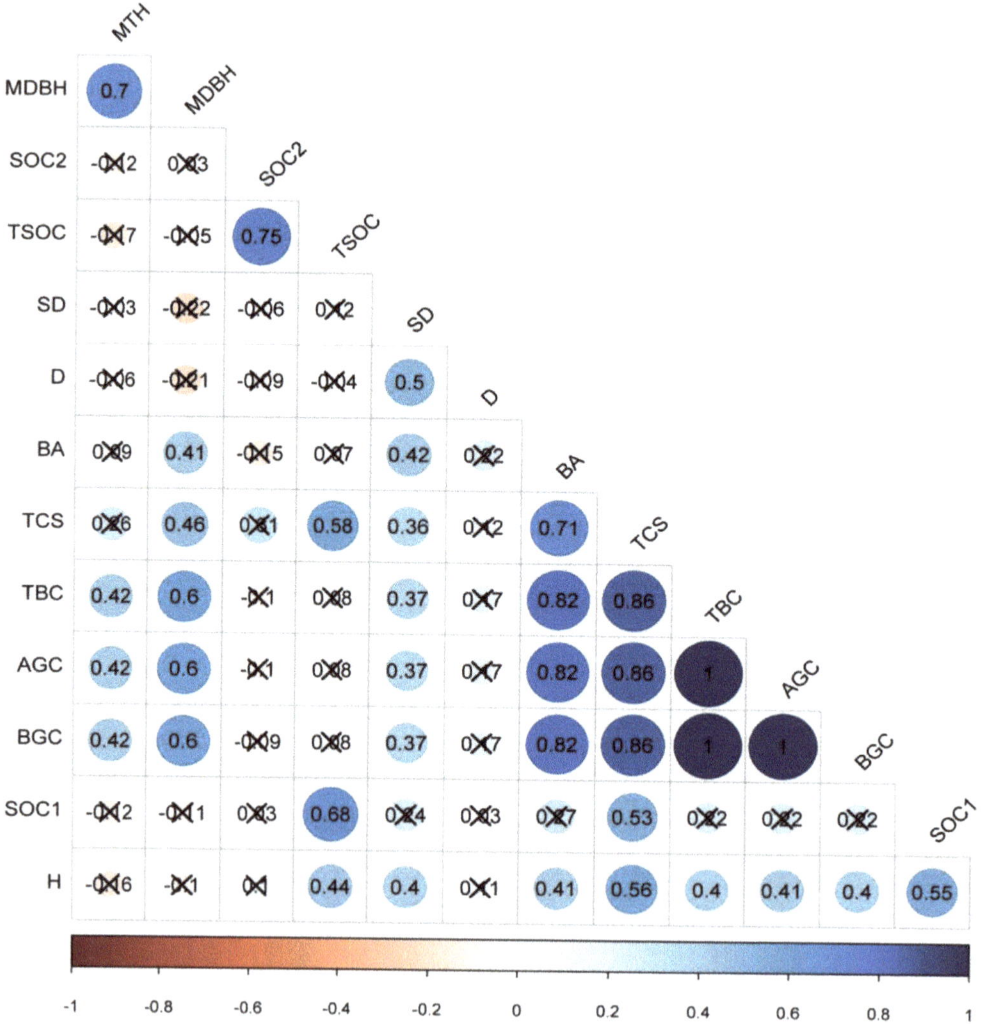

Figure 3. Correlation matrix between stand structure, diversity and different carbon pools. Block with cross sign indicates insignificant relation ($p > 0.05$). MDBH: Mean Diameter at Breast Height (1.37 m), MTH: Mean tree height, SD: Stem Density, BA: Basal Area, D = Margalef Diversity index, H = Shannon Diversity Index, AGC: Aboveground Carbon, BGC: Belowground Carbon, SOC1 = Soil Organic Carbon at 0–15 cm depth, SOC2 = Soil Organic Carbon at 15–30 cm, TBC = Total Biomass Carbon (AGC + BGC), TSOC = Total Soil Organic Carbon (0–30 cm).

3.5. Direct, Indirect and Total Effects of Biodiversity and Stand Structure on Carbon Pools

Shannon diversity had significant direct and indirect effects on aboveground biomass carbon, belowground biomass carbon total biomass carbon, and total carbon storage, but had only direct effects on the upper layer soil organic carbon and total soil organic carbon (Figure 4a–g). The basal area only significantly mediated the Shannon diversity effect on biomass carbon (above- and belowground) and total carbon stocks (Figure 4a,b,e,g). Basal area had the strongest direct effect on biomass and total carbon storage (Figure 4a,b,e,g), while Shannon diversity had the strongest effect on upper layer and total soil organic carbon storage (Figure 4c,f). Tree height also had positive effects on biomass carbon and total carbon storage (Figure 4a,b,e,g). As the Shannon diversity has indirect effects on the biomass and total carbon storage and, which were mediated by the basal area, the total effect of the Shannon diversity on the aboveground biomass carbon, belowground biomass carbon, total biomass carbon and total carbon storage were 0.393, 0.393, 0.513, 0.393, 0.411 and 0.544, respectively (Table 5). The Shannon diversity, basal area and mean tree height together explained 80%, 80%, 27%, 04%, 80%, 19% and 61% ariations in aboveground biomass carbon, belowground biomass carbon, upper layer soil organic carbon, total biomass carbon, total soil organic carbon and total carbon storage in homegardens, respectively (Figure 4a–g).

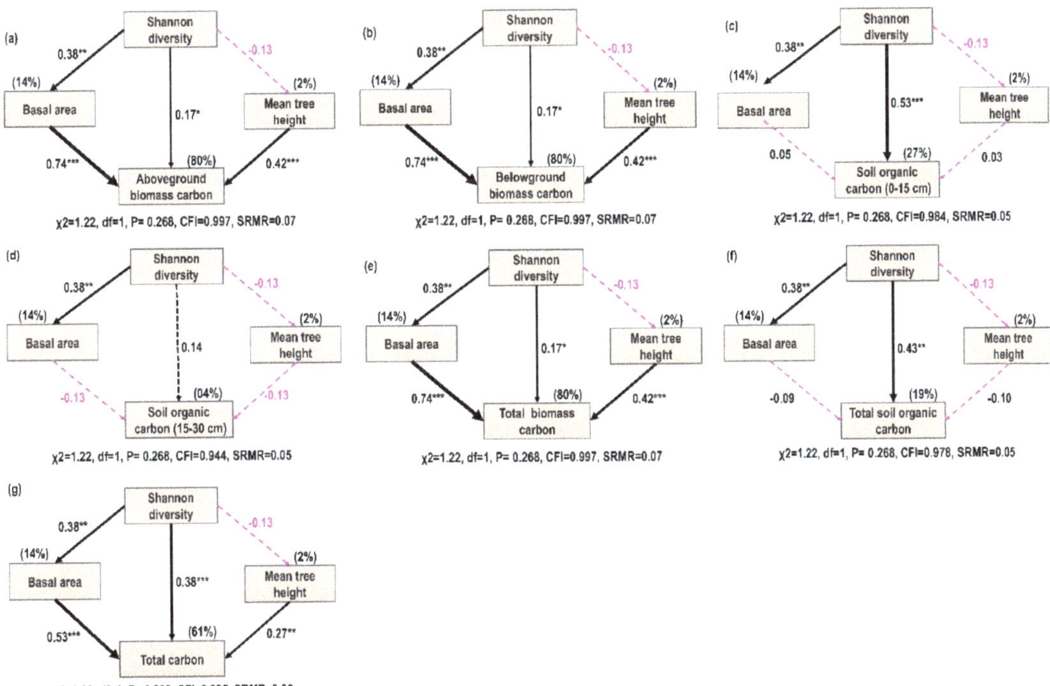

Figure 4. Structural equation models. (**a–g**) Aboveground biomass carbon (**a**), belowground biomass carbon (**b**), soil organic carbon (0–15 cm) (**c**), soil organic carbon (15–30 cm) (**d**), total biomass carbon (**e**), total soil organic carbon (**f**) and total carbon (**g**). All SEMs had a similar insignificant χ^2 (Chi-Square) of 1.22 ($p = 0.268$), with a comparative fit index close to one (CFI; 0.98–0.99) and a standardized root mean square residual close to zero (0.08), indicating no significant deviation from model and datasets at one degree of freedom. The lines with pink and black indicate a negative and positive association between the two covariables. Arrows with numbers indicate the standardized association of predictors with dependent variables. Numbers with percentages above boxes independent variables indicate their explained variance (Coefficient of determinant: R squared) by all the predictors. The path values with asterisks indicate their significance level (*** $p < 0.001$, ** $p < 0.01$, * $p < 0.05$), while the insignificant paths were indicated as dotted lines ($p > 0.05$).

Table 5. Indirect and total standardized effect of the Shannon diversity on different carbon pools. The path values with asterisks indicate their significance level (*** $p < 0.001$, ** $p < 0.01$, * $p < 0.05$).

Indirect and Total Effects	AGC	BGC	SOC1	SOC2	TBC	TSC	TCS
	Std. eff.	Std. eff.	Std. eff.	Std. eff.	Std. eff.	Std. eff.	Std. eff.
Indirect effect of the Shannon diversity through basal area	0.282 *	0.282 *	−0.017	−0.049	0.282 *	−0.035	0.201 *
Indirect effect of the Shannon diversity through tree canopy height	−0.056	−0.056	0.004	0.017	−0.056	0.014	−0.035
Total effect of the Shannon diversity	0.393 **	0.393 **	0.513 ***	0.106	0.393 **	0.411 **	0.544 ***

4. Discussion

Homegardens' agroforestry system is rich in biodiversity in the tropics and provides a variety of ecosystem services, including carbon sequestration. We quantified homegardens' agroforestry vegetation coverage and the direct and mediation effects of tree diversity on carbon storage via stand structure in Dighalia, Khulna, Bangladesh. We found that homegardens had high vegetation coverage and carbon stocks where tree diversity promotes carbon storage directly and indirectly through basal area. The stand basal area had a stronger effect on carbon storage than other stand structural parameters.

4.1. Homegardens Vegetation Coverage

We found a high vegetation coverage (as homegardens) in homesteads, which accounted for 24% of the total area of Dighalia Upazila. Two recent studies reported that the percentage of tree coverage outside forests (woodlots, homegardens and other plantations) in Bangladesh is 13.01 (1,920,700 ha/country area) by Landsat Satellite with 30 m × 30 m pixel [38] and 15.1% by Sentinel 1 and 2 with 10 m × 10 m pixel [15]). While the other two studies focus on all types of trees outside forests, we focused only on the homegardens that were found in villages, which play a key role in biodiversity conservation and carbon sequestration. The difference in percentage of tree coverage in our study could be due to the spatial resolution of satellite data and the specific types of trees outside forests. The forest area percentage within the total land area is one of the key indicators of the Sustainable Development Goals (15.1.1), Achi biodiversity targets of the convention on biological diversity (Target 5; related to forest coverage or extent) and a key activity data for estimating countries' emission factor [6]. Higher accuracy in the estimation of homegardens' coverage with high-resolution imagery (Sentinel 2 10 m band), thus, contributes to assessing the accurate uncertainty in the carbon emission factor.

4.2. Stand Structure

Forest stand structures (tree height, DBH, stand density and basal area) are the key indicators of growth and productivity modeling, and recently, the stand level means of these parameters have been used in stand level biomass carbon modeling [39]. Higher amounts of those parameters at the stand level reveals standing resources and, thereby, growth and biomass carbon stocks [18,39]. The mean stem density in our study (1055.00 ± 35.62 trees ha^{-1}) was very high compared with the natural forests in the Chittagong Hill Tracts (South) Forest Division (381 trees ha^{-1}; Nath et al., 1998) in the Chunati Wildlife Sanctuary, Cox's Bazar (459 trees ha^{-1}) [40]. Higher stem density is also reported in homegardens in Ethiopia (1125.23 ± 334.6) [41]. Homegardens are an intensive land use system, where low planting space is being maintained, which may be the cause of the higher stem density in homegardens compared to natural and secondary forests in Bangladesh. These traditional homegardens with high stem density, thus, play a major role in meeting the timber demand in a sustainable manner while reducing pressure on natural forests [7,18].

4.3. Carbon Stocks

We found a high carbon stock in the studied homegardens, which indicates a huge potential for homegardens to mitigate climate change through sequestering atmospheric CO_2.

The total average carbon stock in our study (129.47 ± 5.10 Mg ha^{-1}; biomass + soil 0–30 cm) was within the range of total carbon stocks in Ethiopian homegardens (91.75 ± 4.31 to 156.17 ± 13.78 Mg ha^{-1}) [14], but higher than Indonesian homegardens (107 Mg ha^{-1}) [9]. Our result was also higher than that of other agroforestry system's carbon stocks, such as agroforest (120 Mg ha^{-1}; soil carbon at 40 cm only) in Panama and traditional agroforestry systems (14–70.08 Mg ha^{-1}) in Mali [42]. The variations in total carbon stocks in different studies are mainly due to the difference in biomass and soil carbon pools that depend on higher stem density, basal area and tree size (height; Table 4 and Figure 3) as well as homestead age [9,11]. For example, in our study, the stem density (1055 stem/ha) in homegardens was higher than that of Indonesian homegardens (624.4 stem/ha). Our findings also showed that across villages, some had higher biomass carbon and lower soil carbon stocks than other villages and vice versa, where the biomass carbon varies mainly due to the difference in mean stem density, basal area and canopy height (Table 3; Figure 3). The positive correlation between these stand structures and different carbon pools indicates that high carbon storage is present in homegardens, and thus, villages are associated with higher stand structure (Figure 3).

4.4. Effects of Tree Diversity and Stand Structure on Carbon Pools

Dealing with two major challenges, climate change mitigation and biodiversity loss under the same management program is more noteworthy than addressing these two separately [43]. In order to practice this type of management, we need to show the evidence that species diversity and carbon storage are positively linked. However, conflicting results (positive, negative or neutral) in different ecosystems (or forest types), sparked debate among scientists and posed a barrier to the adoption of a unified management plan [5,11,18,44]. In our study, we found that tree diversity (Shannon) promotes all carbon pools (except for lower layer soil organic carbon) including total carbon stock (Figures 3 and 4). Tree diversity also had a mediation effect on all carbon pools (except soil organic carbon) through increasing stand basal area. These findings indicate that homegardens associated with high diversity had higher carbon stocks. Our result, therefore, supports the complimentary niche hypothesis which believes that higher niche differences between coexisting species can promote more efficient resource acquisition, leading to higher productivity and higher carbon stocks. The basal area and tree height are both good indicators of biomass and productivity, and hence, carbon storage, at the stand level [18,39,45]. Both of these stand structural features had a positive influence on biomass and total carbon storage in homegardens, according to bivariate relationships and subsequently verified by structural equation models, with the basal area having the strongest effect (Figures 3 and 4a,b,e,g). As a result, in order to maximize and sustain both above- and belowground carbon storage (including soil organic) in homegardens, we must maintain higher diversity and stand tree height, where tree diversity increases stand basal area and improves total carbon storage in homegardens. Recent studies in tropical forests have shown that that both selection and complementarity directed the biomass carbon stock together [13]. Potential work should consider different components of diversity, e.g., species richness, functional trait diversity and functional trait composition, to investigate the mechanism between diversity and carbon relationship in homegardens.

4.5. Implication for Nationally Determined Contributions

Agricultural activities are among the main sources of carbon emissions in developing countries. Integrated agroforestry systems, however, are considered to be sustainable land-use systems, such as traditional homegardens, which can be a crucial tool for achieving carbon mitigation goals because it has high potential for sequestering atmospheric carbon. Nevertheless, some countries have not included homegardens as possible NDCs, such as Bangladesh, whose homegardens coverage is so rich. Here, we showed that in the Digholia sub-district, each village had high tree coverage (Figure 2) with high diversity (Table 3) and carbon storage (Table 4). In its initial forest reference level, homegardens have not been listed as forests by the FAO (Food and Agricultural Organization) definition in Bangladesh. Nevertheless, our study shows that both tree coverage (24%) and mean tree canopy height (7.87 m) in homegardens within homesteads satisfied the forest definition by the UNFCCC (United Nations Framework Convention on Climate Change; tree cover: 10–30%; Canopy height: 2–5 m) [46]. Through considering many homegardens as a unit, we can view it as continuous forests (Figure 2). Therefore, homegardens also follow the concept of minimum area of land (0.05–1.0 ha) [42]. Thus, districts with homegardens in Bangladesh satisfy the requirements of REDD+ or the IPCC (Intergovernmental Panel on Climate Change) definition of forests, which has been used in other forest-based mitigation alternatives such as the clean development mechanism (CDM). Nonetheless, since homegardens are not currently included in the REDD+ scheme, we should explicitly enlist homegardens as a choice for prospective NDCs (although a large-scale study is needed in other districts of Bangladesh). Therefore, by setting up a Village Forestry Committee or a group (composed of several homegardens owner), we can manage it like community-based forestry program [18], e.g., social forestry (managed by the Social Forestry Committee with Gender Equity under direct oversight of the Forestry Department of Bangladesh). In this way, it can retain some of the contributions to the Government's climate change adaptation and mitigation measures promised by the Paris Agreement (NDCs; 5% reduction in greenhouse gas emissions by 2030).

5. Conclusions

In this study, we assessed the tree coverage and direct and indirect effects of tree diversity on carbon storage through stand structure in homestead agroforestry systems (homegardens) in Dighalia, using both remote-sensing and field data. Remote sensing-based tree coverage and consistency of extent of homegardens and field-based canopy height met the forest definition as per IPCC or UNFCCC protocol. High tree diversity with a higher stand structure and carbon storage both above- and belowground indicates that homegardens in Bangladesh play a key role in the conservation of biodiversity and climate change mitigation. We also found that tree diversity promotes carbon storage directly and indirectly via basal area. Therefore, we suggest that for maximizing and sustaining carbon storage, we need to maintain diverse tree species which would lead to a higher stand basal area and, hence, high carbon storage in homegardens. In this way, districts with high homegardens in Bangladesh could contribute to mitigating carbon emissions, and thereby, be a prospective choice (as an agroforestry system) for the NDC.

Supplementary Materials: The following are available online at https://www.mdpi.com/article/10.3390/f12121661/s1, Table S1: Result of multicollinearity test of different covariates using variance inflation factors analysis. Table S2: Fit statistics of the rejected and accepted structural equation models for predicting different carbon pools in combination of plant diversity and stand structure.

Author Contributions: Conceptualization, M.M.R. and G.K.K.; methodology, M.M.R. and G.K.K.; software, M.M.R.; formal analysis, M.M.R. and G.K.K.; investigation, G.K.K. and H.A.; writing—original draft preparation, M.M.R. and G.K.K.; writing—review and editing, M.M.R. and M.E.K.; visualization, M.M.R.; supervision, M.E.K. and M.X.; funding acquisition, G.K.K. and M.M.R. All authors have read and agreed to the published version of the manuscript.

Funding: Gauranga Kundu provided funding for field data collection and soil organic carbon assessment. This research was also supported by the National Key Research and Development Program of China (2017YFA0604300, 2018YFA0606500).

Institutional Review Board Statement: Not applicable.

Informed Consent Statement: Not applicable.

Data Availability Statement: Data analyzed in this manuscript will be provided by the corresponding author upon reasonable request.

Acknowledgments: We would like to thank Mahmood Hossain, of Khulna University, Khulna, Bangladesh, for providing laboratory support in Nutrient Dynamic Lab. We want to acknowledge the European Space Agency for providing the image data. We also want to thank Steven Xu from the Johns Hopkins University for improving the English writing of the paper.

Conflicts of Interest: The authors declare no conflict of interest.

References

1. Fargione, J.E.; Bassett, S.; Boucher, T.; Bridgham, S.D.; Conant, R.T.; Cook-Patton, S.C.; Ellis, P.W.; Falcucci, A.; Fourqurean, J.W.; Gopalakrishna, T.; et al. Natural climate solutions for the United States. *Sci. Adv.* **2018**, *4*, eaat1869. [CrossRef]
2. Gao, Y.; Gao, X.; Zhang, X. The 2 °C Global Temperature Target and the Evolution of the Long-Term Goal of Addressing Climate Change—From the United Nations Framework Convention on Climate Change to the Paris Agreement. *Engineering* **2017**, *3*, 272–278. [CrossRef]
3. Duguma, L.A.; Nzyoka, J.; Minang, P.A.; Bernard, F. How agroforestry propels achievement of nationally determined contributions. *ICRAF Policy Br.* **2017**, *34*, 1–8.
4. Nair, P.K.R.; Kumar, B.M.; Nair, V.D. Agroforestry as a strategy for carbon sequestration. *J. Plant Nutr. Soil Sci.* **2009**, *172*, 10–23. [CrossRef]
5. Kirby, K.R.; Potvin, C. Variation in carbon storage among tree species: Implications for the management of a small-scale carbon sink project. *For. Ecol. Manag.* **2007**, *246*, 208–221. [CrossRef]
6. Cardinael, R.; Umulisa, V.; Toudert, A.; Olivier, A.; Bockel, L.; Bernoux, M. Revisiting IPCC Tier 1 coefficients for soil organic and biomass carbon storage in agroforestry systems. *Environ. Res. Lett.* **2018**, *13*, 124020. [CrossRef]
7. Kabir, E.; Webb, E.L. Floristics and structure of southwestern Bangladesh homegardens. *Int. J. Biodivers. Sci. Manag.* **2008**, *4*, 54–64. [CrossRef]
8. Mohri, H.; Lahoti, S.; Saito, O.; Mahalingam, A.; Gunatilleke, N.; Irham, I.; Hoang, V.T.; Hitinayake, G.; Takeuchi, K.; Herath, S. Assessment of ecosystem services in homegarden systems in Indonesia, Sri Lanka, and Vietnam. *Ecosyst. Serv.* **2013**, *5*, 124–136. [CrossRef]
9. Delaney, M.; Hairiah, K.; Purnomosidhi, P. Carbon stocks in Indonesian homegarden systems: Can smallholder systems be targeted for increased carbon storage? *Am. J. Altern. Agric.* **2002**, *17*, 138–148. [CrossRef]
10. Saha, S.K.; Nair, P.K.R.; Nair, V.D.; Kumar, B.M. Soil carbon stock in relation to plant diversity of homegardens in Kerala, India. *Agrofor. Syst.* **2009**, *76*, 53–65. [CrossRef]
11. Kumar, B.M. Species richness and aboveground carbon stocks in the homegardens of central Kerala, India. *Agric. Ecosyst. Environ.* **2011**, *140*, 430–440. [CrossRef]
12. Mattsson, E.; Ostwald, M.; Nissanka, S.P.; Pushpakumara, D.K.N.G. Quantification of carbon stock and tree diversity of homegardens in a dry zone area of Moneragala district, Sri Lanka. *Agrofor. Syst.* **2014**, *89*, 435–445. [CrossRef]
13. Mensah, S.; Veldtman, R.; Assogbadjo, A.E.; Kakaï, R.G.; Seifert, T. Tree species diversity promotes aboveground carbon storage through functional diversity and functional dominance. *Ecol. Evol.* **2016**, *6*, 7546–7557. [CrossRef]
14. Birhane, E.; Ahmed, S.; Hailemariam, M.; Negash, M.; Rannestad, M.M.; Norgrove, L. Carbon stock and woody species diversity in homegarden agroforestry along an elevation gradient in southern Ethiopia. *Agrofor. Syst.* **2020**, *94*, 1099–1110. [CrossRef]
15. Thomas, N.; Baltezar, P.; Lagomasino, D.; Stovall, A.; Iqbal, Z.; Fatoyinbo, L. Trees outside forests are an underestimated resource in a country with low forest cover. *Sci. Rep.* **2021**, *11*, 7919. [CrossRef]
16. Bardhan, S.; Jose, S.; Biswas, S.; Kabir, K.; Rogers, W. Homegarden agroforestry systems: An intermediary for biodiversity conservation in Bangladesh. *Agrofor. Syst.* **2012**, *85*, 29–34. [CrossRef]
17. Islam, M.; Dey, A.; Rahman, M. Effect of Tree Diversity on Soil Organic Carbon Content in the Homegarden Agroforestry System of North-Eastern Bangladesh. *Small-Scale For.* **2014**, *14*, 91–101. [CrossRef]
18. Rahman, M.; Kabir, E.; Akon, A.J.U.; Ando, K. High carbon stocks in roadside plantations under participatory management in Bangladesh. *Glob. Ecol. Conserv.* **2015**, *3*, 412–423. [CrossRef]
19. Kabir, E.; Webb, E.L. Household and homegarden characteristics in southwestern Bangladesh. *Agrofor. Syst.* **2008**, *75*, 129–145. [CrossRef]
20. Statistic Division, Ministry of Planning, Bangladesh Secretariat. *BBS Statistical Year Book of Bangladesh.Bangladesh Bureau of Statistics (BSS)*; The Government of the People's Republic of Bangladesh: Dhaka, Bangladesh, 2012.
21. Drusch, M.; Del Bello, U.; Carlier, S.; Colin, O.; Fernandez, V.; Gascon, F.; Hoersch, B.; Isola, C.; Laberinti, P.; Martimort, P.; et al. Sentinel-2: ESA's Optical High-Resolution Mission for GMES Operational Services. *Remote Sens. Environ.* **2012**, *120*, 25–36. [CrossRef]
22. Dhodhi, M.K.; Saghri, J.A.; Ahmad, I.; Ul-Mustafa, R. D-ISODATA: A Distributed Algorithm for Unsupervised Classification of Remotely Sensed Data on Network of Workstations. *J. Parallel Distrib. Comput.* **1999**, *59*, 280–301. [CrossRef]
23. Rahman, M.; Lagomasino, D.; Lee, S.; Fatoyinbo, T.; Ahmed, I.; Kanzaki, M. Improved assessment of mangrove forests in Sundarbans East Wildlife Sanctuary using WorldView 2 and Tan DEM -X high resolution imagery. *Remote Sens. Ecol. Conserv.* **2019**, *5*, 136–149. [CrossRef]
24. Yu, L.; Gong, P. Google Earth as a virtual globe tool for Earth science applications at the global scale: Progress and perspectives. *Int. J. Remote Sens.* **2011**, *33*, 3966–3986. [CrossRef]

25. Congalton, R.G.; Green, K. *Assessing the Accuracy of Remotely Sensed Data: Principles and Practices*, 2nd ed.; CRC Press: Boca Raton, FL, USA, 2009; ISBN 978-1-4200-5512-2.
26. Chave, J.; Andalo, C.; Brown, S.; Cairns, M.A.; Chambers, J.; Eamus, D.; Fölster, H.; Fromard, F.; Higuchi, N.; Kira, T.; et al. Tree allometry and improved estimation of carbon stocks and balance in tropical forests. *Oecologia* **2005**, *145*, 87–99. [CrossRef] [PubMed]
27. Cairns, M.A.; Brown, S.; Helmer, E.H.; Baumgardner, G.A. Root biomass allocation in the world's upland forests. *Oecologia* **1997**, *111*, 1–11. [CrossRef]
28. Zanne, A.E.; Lopez-Gonzalez, G.; Coomes, D.A.; Ilic, J.; Jansen, S.; Lewis, S.L.; Miller, R.B.; Swenson, N.G.; Wiemann, M.C.; Chave, J. Data from: Towards a worldwide wood economics spectrum. *Ecol. Lett.* **2009**, *12*, 351–366. [CrossRef]
29. Pearson, T. *Sourcebook for Land Use, Land-Use Change and Forestry Projects*; Winrock International and the BioCarbon Fund of the World Bank: Washington, DC, USA, 2005; pp. 56–64.
30. Maynard, D.G.; Curran, M.P. Bulk density measurement in forest soils. In *Soil Sampling and Methods of Analysis*; Carter, M.R., Gregorich, E.G., Eds.; Taylor & Francis: London, UK, 2008; pp. 863–869.
31. Sparks, D.L.; Page, A.L.; Helmke, P.A.; Loeppert, R.H. *Methods of Soil Analysis Part 3—Chemical Methods*; SSSA Book Series; Soil Science Society of America, American Society of Agronomy: Madison, WI, USA, 1996; ISBN 978-0-89118-866-7.
32. Shannon, C.E. A Mathematical Theory of Communication. *Bell Syst. Tech. J.* **1948**, *27*, 379–423. [CrossRef]
33. Margalef, R. Information theory in biology. *Gen. Syst. Yearb.* **1958**, *3*, 36–71.
34. R Core Team. *R: A Language and Environment for Statistical Computing*; R Foundation for Statistical Computing: Vienna, Austria, 2020.
35. Grace, J.B.; Bollen, K.A. Interpreting the Results from Multiple Regression and Structural Equation Models. *Bull. Ecol. Soc. Am.* **2005**, *86*, 283–295. [CrossRef]
36. Zuur, A.F.; Ieno, E.N.; Elphick, C.S. A protocol for data exploration to avoid common statistical problems: Data exploration. *Methods Ecol. Evol.* **2010**, *1*, 3–14. [CrossRef]
37. Rahman, M.; Zimmer, M.; Ahmed, I.; Donato, D.; Kanzaki, M.; Xu, M. Co-benefits of protecting mangroves for biodiversity conservation and carbon storage. *Nat. Commun.* **2021**, *12*, 3875. [CrossRef] [PubMed]
38. Potapov, P.; Siddiqui, B.N.; Iqbal, Z.; Aziz, T.; Zzaman, B.; Islam, A.; Pickens, A.; Talero, Y.; Tyukavina, A.; Turubanova, S.; et al. Comprehensive monitoring of Bangladesh tree cover inside and outside of forests, 2000–2014. *Environ. Res. Lett.* **2017**, *12*, 104015. [CrossRef]
39. Khan, N.I.; Islam, R.; Rahman, A.; Azad, S.; Mollick, A.S.; Kamruzzaman, M.; Sadath, N.; Feroz, S.; Rakkibu, G.; Knohl, A. Allometric relationships of stand level carbon stocks to basal area, tree height and wood density of nine tree species in Bangladesh. *Glob. Ecol. Conserv.* **2020**, *22*, e01025. [CrossRef]
40. Rahman, M.L.; Hossain, M.K. Status of fodder and non-fodder tree species in Chunati wildlife sanctuary of Chittagong forest division, Bangladesh. *Int. J. For. Usufructs Manag.* **2003**, *4*, 9–14.
41. Bajigo, A.; Tadesse, M. Woody species diversity of traditional agroforestry practices in Gununo Watershed in Wolayitta Zone, Ethiopia. *For. Res.* **2016**, *4*, 155. [CrossRef]
42. Takimoto, A.; Nair, P.R.; Nair, V.D. Carbon stock and sequestration potential of traditional and improved agroforestry systems in the West African Sahel. *Agric. Ecosyst. Environ.* **2008**, *125*, 159–166. [CrossRef]
43. Phelps, J.; Webb, E.; Adams, W.M. Biodiversity co-benefits of policies to reduce forest-carbon emissions. *Nat. Clim. Chang.* **2012**, *2*, 497–503. [CrossRef]
44. Ruiz-Jaen, M.C.; Potvin, C. Can we predict carbon stocks in tropical ecosystems from tree diversity? Comparing species and functional diversity in a plantation and a natural forest. *New Phytol.* **2010**, *189*, 978–987. [CrossRef] [PubMed]
45. Rahman, M.; Khan, N.I.; Hoque, A.K.F.; Ahmed, I. Carbon stock in the Sundarbans mangrove forest: Spatial variations in vegetation types and salinity zones. *Wetl. Ecol. Manag.* **2014**, *23*, 269–283. [CrossRef]
46. UNFCCC. Report of the Conference of the Parties on its Seventh Session. In Proceedings of the Conference of the Parties, Seventh Session, Marrakesh, Morocco, 29 October–10 November 2001.

*forests*

MDPI

Article

Agroforestry Systems and Their Contribution to Supplying Forest Products to Communities in the Chure Range, Central Nepal

Deepa Khadka [1,*], Anisha Aryal [2], Kishor Prasad Bhatta [1], Bed Prakash Dhakal [1] and Himlal Baral [3,4]

[1] Faculty of Forest Science and Forest Ecology, Georg-August-Universität, Busgenweg 5, 37077 Goettingen, Germany; k.bhatta@stud.uni-goettingen.de (K.P.B.); bed.dhakal@stud.uni-goettingen.de (B.P.D.)

[2] Faculty of Environmental Sciences, Technische Universität Dresden, 01737 Dresden, Germany; anisha.aryal@mailbox.tu-dresden.de

[3] Center for International Forestry Research (CIFOR), Jalan CIFOR, Situ Gede, Bogor 16115, Indonesia; h.baral@cgiar.org

[4] School of Ecosystem and Forest Sciences, University of Melbourne, Parkville, VIC 3010, Australia

* Correspondence: deepa.khadka@stud.uni-goettingen.de

Abstract: Agroforestry (AF), an integration of agricultural and/or pastureland and trees, is a powerful tool for the maximization of profit from a small unit of land; however, it has been less well explored and recognized by existing policies. AF could be the best approach to conserving the fragile soils of Chure and to supplying subsistence needs to the local people. This study endeavored to understand how the adoption of various AF practices contributed to people's livelihoods in the Bakaiya rural municipality of Makawanpur District. To achieve this, 5 focus group discussions, 10 key informant interviews and 100 household surveys were conducted. These were analyzed using various statistical analysis tools: Kruskal–Wallis test, Games–Howell post hoc comparison test and Wilcoxon test. Thematic analysis was employed to understand the status and growth process of AF in the study area. Of three different AF systems used in the area, agri-silviculture was found to be the dominant form. Local people derived forest products, especially fuelwood, fodder and leaf litter from AF, where agri-silvi-pasture was most common. The three AF systems studied here were in turn compared with community forestry (CF), which is a participatory forest management system overseen by the community. People derived almost 75% of fuelwood from CF, whereas in the case of fodder and leaf litter, contributions from CF and AF were almost equal. Despite the potentiality of AF in fulfilling the demands of local people, promotional and development activities were lacking. This study recommends a strong collaboration of local people and concerned stakeholders for the promotion and technical facilitation of AF systems.

Keywords: agroforestry systems; Chure conservation; livelihood; community forestry

Citation: Khadka, D.; Aryal, A.; Bhatta, K.P.; Dhakal, B.P.; Baral, H. Agroforestry Systems and Their Contribution to Supplying Forest Products to Communities in the Chure Range, Central Nepal. *Forests* **2021**, *12*, 358. https://doi.org/10.3390/f12030358

Academic Editor: Julien Fortier

Received: 30 January 2021
Accepted: 15 March 2021
Published: 18 March 2021

Publisher's Note: MDPI stays neutral with regard to jurisdictional claims in published maps and institutional affiliations.

1. Introduction

Improving the well-being of poor people along with the sustainable management of natural resources are global targets under the United Nation's sustainable development goals [1]. The exploitation of the world's natural resources is occurring at such an alarming rate that the livelihoods of poor people could become even more precarious in the near future, especially in the global south [2,3]. Analogous to other developing countries, Nepal is facing severe pressure on its natural resources, as more than 80% of the population depend on them [4,5]. Food insecurity and land degradation have grown as major problems impacting the livelihoods of the people [6–8]. About 52% of people in Nepal have differing levels of food insecurity [9], although about 68% of the population engage in agriculture [10]. Currently, community forestry (CF), a forest management approach where local communities are provided with a certain degree of responsibility and authority for the forest management is regarded as the most effective way of addressing the subsistence needs of local people [11]. Over the last few decades, CF has been a priority for policy makers [12] to address forest degradation and widespread rural poverty in a single package of programs by mobilizing local people [13]. However, the maximal use by local people has created a heavy pressure in CF thereby inducing land degradation and biodiversity loss [14]. To tackle these issues of degradation and to obtain a high and sustained level of production, agroforestry (AF) has been recognized as the most efficient land management system, as it integrates different land use practices on a single unit of land [15,16]. Recently, AF has received significant attention for its efficiency in conserving natural resources along with improving livelihoods [17].

AF is defined as a land use system in which woody perennials are included within the agricultural landscapes and where both ecological and economical interactions occur between the woody and non-woody components for various social, economic and environmental benefits [18]. AF incorporates the optimal use of the land for the woody and non-woody components and provides several effects on a sustainable basis that are beneficial, such as biodiversity conservation [19], soil erosion mitigation [20], protection of (ground)water quality [21,22] and household food security and income [23].

In Nepal, AF has been practiced in a traditional way where farmers use their agricultural land for propagating trees as an integral part of their farming system [5,24]. The promotion of AF as a means of generating income has been included under policies related to agriculture and forests [25,26]. Nepal is one of the few countries that has a separate policy for AF. National AF policy [27] aims to prioritize the commercial and collective farming system, facilitate farmers' access to market, promote industry-based AF, provide incentives to farmers adopting AF systems, develop AF on bare, fallow land and develop special area-based AF models [27]. To develop the commercial aspect of AF and site-specific AF models, it is necessary to understand the system adopted by local farmers and the products they derive from it.

Many studies have been conducted regarding AF practices in both the hill and the Terai regions of Nepal. Most of them focus on the linkage between the socioeconomic characteristics of the households and the adoption of AF practices e.g., [17,24,28]. The evolution of AF from subsistence to commercial enterprises has been studied by Dhakal et al. [29] for the Eastern Terai of Nepal, whereas other research has highlighted the contribution of AF to rural livelihoods e.g., [30,31]. Similarly, the effect of AF on the quality of soil has also been studied [32,33]. However, studies on the AF systems in the region of most interest, Chure, are still lacking. Because the land in the Chure region is fragile which means the soil in the Chure region is young with low resistance to erosion and vulnerable to degradation [34] and topographically complex in nature, finding the best land management strategy is important [35]. Further, lack of livelihood assets and inadequate food security due to increasing population necessitates the development of an integrated land use system to fulfill these requirements and for conservation purposes [36]. The Chure region has been receiving considerable attention from the government [37]; despite this, the ecological, geographical and biophysical conditions of the region are degrading rapidly [38]. This study was conceptualized to contribute to the sparse knowledge around the AF system in the Chure region.

The objectives of this study are: (i) to assess the major AF systems and farmers' preferences relating to AF species, (ii) to analyze the characteristics of the respondents practicing different AF systems, (iii) to estimate the supply of forest-based products from AF and compare those with forest-based products obtained from community forestry (CF) and (iv) to understand the status and growth process of AF in the study area.

2. Materials and Methods

2.1. Study Area

The study was conducted in the Chure region which is located between plain Terai and the mid-hills [35]. It lies within the geographical location of 80°9′25″ to 88°11′16″

longitudes and 26°37′47″ to 29°10′27″ latitudes and comprises the youngest mountains in the world [39]. It covers about 12.78% of the area of Nepal and is among the most fragile ecoregion of Nepal [40]. The Chure region is the most highlighted region considering its heavy vulnerability to mass erosion and landslides caused by heavy deforestation [41]. The Government of Nepal has therefore initiated the Chure conservation program through President Chure-Terai Madhesh Conservation Development Committee (hereafter, Chure conservation board) [42,43].

Within the Chure region, the study focused on Makawanpur District, Nepal (Figure 1) that lies between 27°10′ to 27°40′ N and 84°41′ to 85°31′ E, and covers an area of 2426 km^2 [44]. With a wide variation of altitude i.e., 166 m to 2584 m above mean sea level, this district accommodates both flatlands and hilly regions [45]. Almost half of the district has a slope inclination greater than 30°, which makes it more vulnerable to landslides and soil erosion [46]. This district experiences an annual mean precipitation of 2535 mm, where the average temperature ranges from 13.3 °C to 16.6 °C. From south to north, the climate varies from subtropical to temperate [47]. The livelihoods of the population in this district depend upon agriculture; however, the integration of agriculture, forest and pastures was adopted long ago [48,49]. The practice of shifting cultivation transitioned into sustainable AF systems due to lack of land ownership [49].

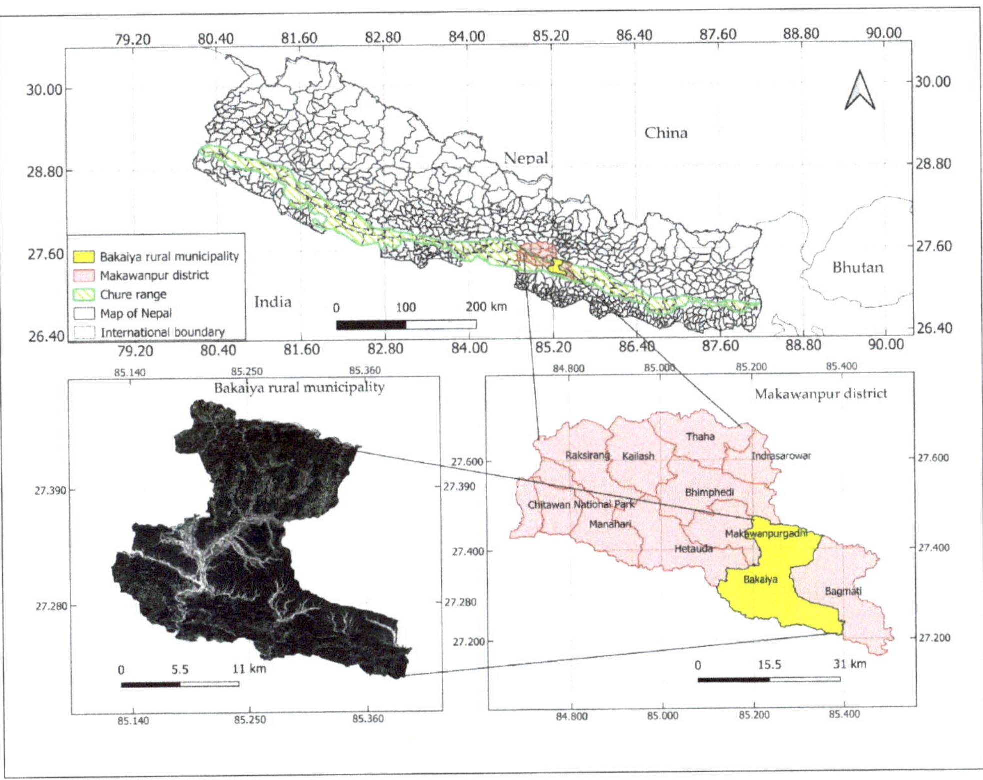

Figure 1. Map showing the study area in Central Nepal.

Studies related to food security and farm-based agroforestry have been conducted in different Village Development Committees (VDCs) of Makawanpur district [48,50]. The contribution of agroforestry to the rural livelihood of the Chepang ethnic group in central Nepal has also been documented [51]. Further, the transition of the farming system from slash-and-burn farming to permanent agroforestry in Makawanpur district has been studied, and a comparison of the cost and benefits between the two has been done which showed agroforestry as the financially profitable system [52]. Also, the role of agroforestry in conserving tree species was studied by Sharma and Vetaas [53]. However, studies focusing on the choice of agroforestry systems, the supply of forest-based products by farm-based agroforestry and the status of agroforestry in Makawanpur are lacking.

For the specific study site, we consulted with district forest officials and the members of the Chure conservation board. We were interested in the areas where people were more dependent on forest causing severe destruction. Bakaiya rural municipality was listed as a vulnerable region to landslides due to deforestation [54,55] but many people were still dependent on community forests for their livelihood. Agroforestry is regarded as a viable option in fulfilling both agricultural and forest-based products but only about 13% of total households adopted agroforestry in this area. To understand the socio-economic condition of the people adopting agroforestry, their choice of system, species and the amount of products they acquired from farm-based agroforestry, we selected Bakaiya rural municipality as the study area.

Bakaiya rural municipality is majorly inhabited by the ethnic group—Majhi. Most of the people were dependent on agriculture and labor activities for their livelihood. Very few people were engaged in business, government offices and remittance.

2.2. Data Collection

This paper is based on our quantitative research on AF systems and their contribution to livelihoods in Central Nepal. Out of 1000 households practicing farm-based agroforestry, purposive random sampling was conducted to select the 100 interviewee households. Three key instruments were used: focus group discussions (FGDs), key informant interviews (KII) and semi structured interviews. The semi structured interview was used because we wanted data according to our objective alongside providing the interviewee with open space allowing them to bring up new ideas. FGDs consisted of 25 open-ended questions (Appendix A) focused on the current status of AF systems, their growth processes and the support available for the growth of AF. KII consisted of 13 open-ended questions (Appendix B) with a focus on the contribution of AF to local livelihoods and the strengths and weaknesses of the AF practices. The interview guide consisted of 27 items (Appendix C) that focused on the major AF systems adopted by the local people and the contributions of AF to their livelihoods compared with those of CF. Prior to data collection, secondary data sources, specifically reports from government agencies such as the Chure Conservation Board the District Forest Office (DFO) and the Ministry of Forests and Environment (MOFE), were reviewed to enhance our understanding of the adoption, use and impact of AF at district and local levels. The data collection process started with FGD, where five main actors were identified: one community forestry user group, two women's groups, one youth group and one ethnic group (Majhi). Each group consisted of 7–12 individuals who were given the opportunity to express their views. From each focus group we identified one key informant and in addition, five other key informants were selected: one member from the district forest office, one member from Chure Conservation Board, two local leaders and one teacher at the primary school. The individuals from the FDG who had greater knowledge about agroforestry systems and policies and who couldn't discuss openly during focus group discussion were selected as key informants. Other key informants were selected based on their knowledge about agroforestry, governance, national policies and marketing sector. Thus, in total, 10 key informants were interviewed. Interviews with officers of the DFO and the Chure Conservation Board were conducted to gain insights into the role of government in facilitating the AF system at the study site. To understand the participation of local people in the AF system, their choice of system and the contribution of AF to livelihoods, a survey of 100 households was carried out. The interview guides were formulated in English, but the interview was carried out in Nepali language. We tried our best to make local people understand the technical terms which made the data interpretation process easier. Additionally, through direct observation, the major species used in the AF systems were determined and further validated with responses from households (Figure 2).

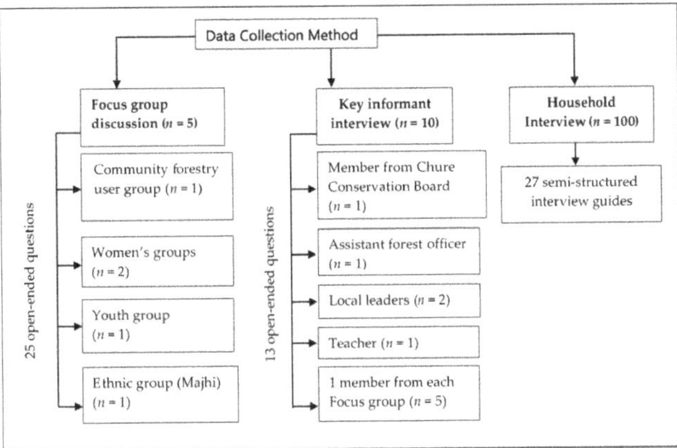

Figure 2. Data collection chart showing the distribution of focus group discussions, household survey and key informant interviews.

2.3. Data Analysis

The Shapiro–Wilk test [56] was used to check the normality of the data. Socioeconomic characteristics of respondents were analyzed through descriptive statistics. The Kruskal–Wallis Test [57] was used to assess the significant difference between the AF systems in terms of socioeconomic variables and the supply of forest products. Later, the variable resulting in significant difference was analyzed using the Games–Howell post hoc test (posthocTGH) in R-studio [58] to find the exact pair of AF systems with statistical difference. For the post hoc test, the package userfriendlyscience [59] was used. To analyze the difference in the supply of products between CF and AF, the Wilcoxon rank-sum test, also called the Mann–Whitney test [60], was employed in R-studio [58].

To assess if the difference in the supply of forest products between different AF systems is statistically significant, the Kruskal–Wallis test was employed with the following hypothesis:

Hypothesis 1.

Null hypothesis (H_0): There is no significant difference in the supply of forest products between different AF systems.

Alternative hypothesis (H_1): There is a significant difference in the supply of forest products between different AF systems.

To assess if the difference in the supply of forest products between AF and CF systems is statistically significant, a Wilcoxon rank-sum test was applied with the following hypothesis:

Hypothesis 2.

Null hypothesis (H_0): There is no significant difference in the supply of forest products between AF and CF systems.

Alternative hypothesis (H_1): There is a significant difference in the supply of forest products between AF and CF systems.

3. Results

3.1. Agroforestry Systems in the Study Area and the Choice of Plant Species

In the study area, three farm-based AF systems were adopted by the residents: agri-silvicultural (AS), horti-agri-silvicultural (HAS) and agri-silvi-pastoral (ASP) systems. AS is a system of AF in which tree species are grown and managed in the farmland together with agricultural crops, whereas HAS is an integrated system of land use, where fruit trees, agricultural crops and timber trees/fuel wood are grown together to meet householders' needs for food grains, fruits, timber and fuel wood [61]. Similarly, under the ASP system the same unit of land is managed to produce agricultural crops and trees/fuelwood and where farmers can also rear animals [61]. Out of 100 respondents, 66 respondents adopted the AS system whereas HAS and ASP systems were adopted by 29 and 5 respondents, respectively. The establishment of trees on farmland was through deliberate retention and management, plantation, or both. The major plant species used in three different AF systems are listed in Table 1.

Table 1. Major species used in different AF systems.

System	Crops/Grasses	Tree Species
Agri-silvicultural	*Oryza sativa* L. (paddy rice) *Zea mays* L. (maize) *Triticum aestivum* L. (wheat) *Brassica campestris* L. (mustard)	*Ficus semicordata* Buch.-Ham. ex Sm. (Dropping fig), *Ficus hispida* L. (Fig tree, *Litsea polyantha* Juss (Litsea), *Premna integrifolia* L. (Wind killer), *Ficus roxburghii* Steud. (Nevaro), *Terminalia elliptica* Willd. (Indian laurel), *Ficus religiosa* L. (Sacred fig), *Bixa orellana* L. (Annato)
Agri-silvi-pastoral	**Crops** *Oryza sativa* (paddy rice) *Triticum aestivum* (wheat) *Curcuma longa* L. (turmeric) **Grasses** *Pennisetum purpureum* cv. Mott (dwarf elephant grass) *Setaria splendida* Schum.Stapf & Hubb. (setaria) *Brachiaria brizantha* Hochst. Ex A. Rich. Stapf (Mulato) *Arachis pintoi* Krapov & Gregory (forage peanut) *Thysanolaena maxima* Roxburgh O. Kuntze (broom grass) *Bambusa tulda* Roxb. (bamboo)	*Ficus semicordata* (Dropping fig), *Terminalia bellirica* Roxb. (Belliric myrobalan), *Litsea polyantha* (Litsea), *Ficus hispida* (Fig tree), *Ficus infectoria* Miq. (White fig), *Leucaena leucocephala* Lam, de Wit. (Ipil-ipil), *Shorea robusta* Gaertn. (Sal)
Horti-agri-silvicultural	**Crops** *Zingiber officinale* Roscoe (ginger) *Curcuma longa* (turmeric) *Dioscorea alata* L. (yam) *Colocasia esculenta* L., Schott. (colocasia) **Vegetables** *Solanum tuberosum* L. (potato) *Brassica oleracea* var. *botrytis* L. (cauliflower) *Brassica oleracea* var. *capitata* L. (cabbage) *Solanum melongena* L. (aubergine)	**Fodder and firewood trees** *Dalbergia sissoo* Roxb. (Indian rosewood), *Eucalyptus Camaldulensis* Dehnh. (Eucalyptus), *Melia azedarach* L. (White cedar), *Artocarpus lakoocha* Roxb. (Monkey tree), *Bauhinia longifolia* Bong. (Bauhinia), *Leucaena leucocephala* (Ipil-ipil) **Fruit trees** *Mangifera indica* L. (Mango), *Litchi chinensis* Sonn. (Litchi), *Artocarpus heterophyllus* Lam. (Jack fruit), *Bauhinia longifolia* Bong. (Bauhinia), *Nyctanthes arbor-tristis* L., *Ziziphus mauritiana* Lam.(Jujube fruit), *Prunus persica* L. (Peach), *Musa acuminata* Colli.(Banana), *Citrus maxima* Burm. F, Merr (Pomelo), *Psidium guajava* L. (Guava), *Morus indica* L. (Black mulberry), *Carica papaya* L. (Papaya), *Ananas comosus* L, Merr. (Pineapple), *Annona reticulata* L. (Wild sweetsop), *Areca catechu* L. (Palm), *Saccharum officinarum* L. (Sugarcane)

3.2. Socioeconomic Characteristics of the Respondents Practicing Different AF Systems

Out of 100 households surveyed, the majority i.e., 66% were females. The average age of respondents was 40 years (range, 17–74 years) and the median age was 36 years. The majority of respondents were from the Tamang ethnic group constituting about 56% of total households surveyed. About 55% of total respondents were illiterate and only 6% were educated up to university level. Therefore, their knowledge about proper AF systems was poor. Agriculture dominated occupations in the study area, with about 52% of total respondents relying on agriculture for their subsistence needs. The average land holding size of the household was about 14 katha (0.48 ha), which is half of the national average of 0.8 ha [62]. However, households practicing the ASP system had the highest land holding of 21 katha (0.7 ha) on average. The amount of livestock owned by the household was measured in livestock standard units (LSU). Buffalo, cow, goat and poultry were considered while estimating the LSU. A value of 1 was assigned to mature cows and buffalos, whereas a value of 0.1 was given to goats and 0.03 to poultry [63]. On average, the LSU of households was 3.05. Households that were practicing HAS showed the highest average LSU i.e., 3.36. The accessibility of respondents to a forest was taken into account to understand its impact on the choice of AF system. The distance from the house to a nearby forest was measured in units of time (minutes). On average, respondents required about 19 min to reach a nearby CF and the minimum and maximum time was found to be 2 min and 60 min respectively. The average annual income of those respondents adopting the HAS system was higher in comparison with the other two AF systems (Table 2). When asking about respondents' interest in developing AF, the majority (about 67%) responded in a positive way, showing a profound interest in developing the AF systems.

Table 2. Characteristics of the respondents in the study area.

Socioeconomic Variables	Type of Measure	Mean Values of the Socioeconomic Variables (±SD)		
		AS (*n* = 66)	ASP (*n* = 5)	HAS (*n* = 29)
Age	Years	40.67 (13.65) [a]	39.00 (20.79) [a]	38.76 (12.74) [a]
Sex	Male = 1, Female = 2	1.65 (0.48) [a]	1.4 (0.55) [a]	1.72 (0.45) [a]
Education	Illiterate = 1, Primary = 2, Secondary = 3, University = 4	1.65 (0.91) [a]	2 (1.22) [a]	1.82 (0.92) [a]
Ethnic group	Tamang = 1, Brahmin = 2, Chettri = 3, Magar = 4, Others = 5	2.18 (1.54) [a]	1.8 (1.09) [a]	2.14 (1.61) [a]
Major occupation	1 = Agriculture, 2 = Government employee, 3= Non-Government employee, 4 = Business, 5 = Remittance, 6 = Wage Labour	2.59 (1.86) [a]	1.00 [b]	2.66 (2.04) [a]
Livestock unit	Livestock standard unit (LSU)	2.91 (1.84) [a]	3.11 (2.05) [a]	3.36 (2.22) [a]
Land holding	Katha (30 Katha = 1 hectare)	13.02 (7.79) [a]	21.6 (7.23) [a]	16.05 (9.00) [a]
Annual cash income	NRs (Nepalese Rupees)	150,455 (85,150) [a]	110,000 (82,158) [a]	161,035 (57,591) [a]
Time to reach forest	Minutes	19.94 (13.25) [a]	18 (7.58) [a]	16.69 (10.82) [a]
Interest in developing agroforestry	0 = No, 1 = Yes	0.65 (0.54) [a]	0.6 (0.55) [a]	0.79 (0.41) [a]

Values in parenthesis are the ± SD of the socioeconomic variables. The similar letter in superscript indicates the nonsignificant difference between agroforestry systems in pairwise comparison. AS = agri-silvicultural, ASP = agri-silvi-pastoral, HAS = horti-agri-silvicultural.
[a] = The similar letter in superscript indicates the nonsignificant difference between agroforestry systems in pairwise comparison.

3.3. Supply of Forest Products from Different Agroforestry Systems

Fodder, fuelwood and leaf litter were the major forest products derived from AF systems, and therefore, the difference between AF systems was estimated for these three forest products. The local people derived a higher amount of forest products from ASP. The average annual fodder supply from ASP was higher (6000 kg) in comparison with HAS (3600 kg) and AS (3222 kg) (Figure 3a) Similarly, ASP provided a higher average annual fuelwood supply (1875 kg) than AS (1048 kg) and HAS (760 kg) (Figure 3b). Moreover, ASP provided a higher average annual leaf litter supply (2900 kg) compared with AS (2610 kg) and HAS (1700 kg) (Figure 3c). The *p*-values obtained from Kruskal–Wallis test for all the products were greater than 0.05 at the 95% confidence interval, the null hypothesis was accepted, implying that the difference in the supply of forest products from different AF systems was not statistically significant (Table 3).

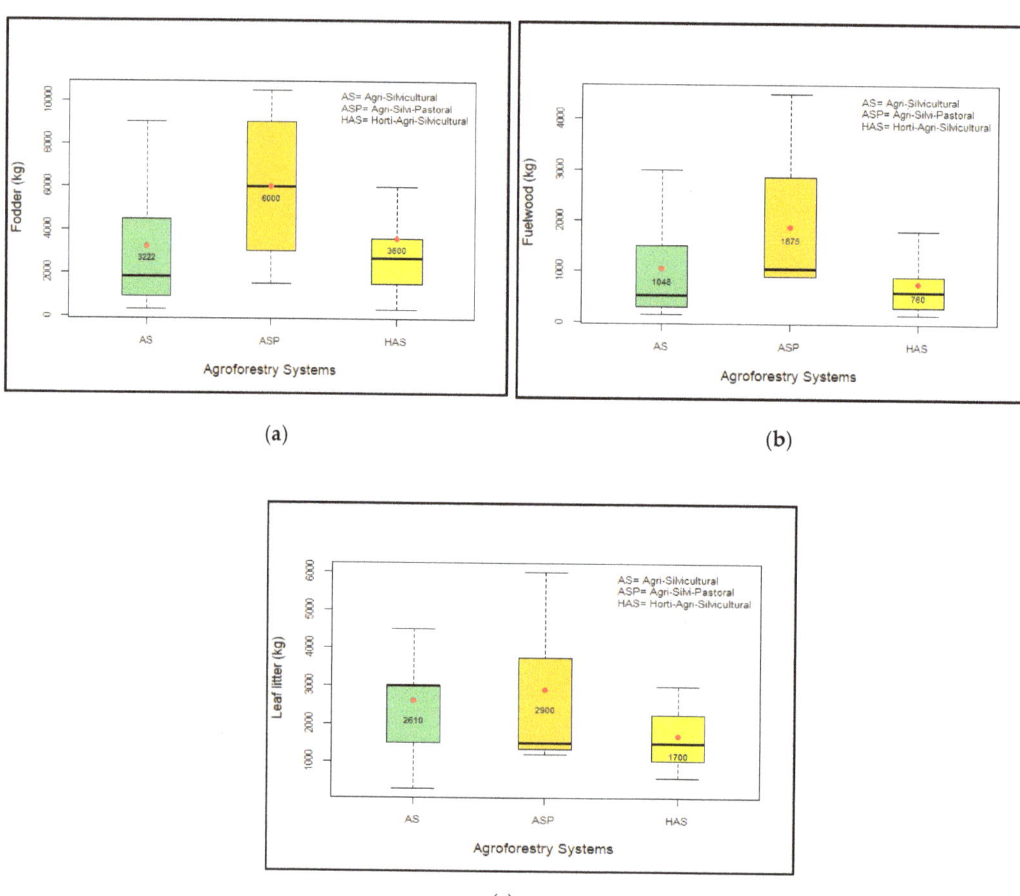

(a)   (b)   (c)

Figure 3. The average annual supply of forest products from different agroforestry systems in kilograms; (**a**) fodder, (**b**) fuelwood and (**c**) leaf litter where a red dot indicates the mean while a black line indicates the median.

Table 3. Kruskal–Wallis test result for Hypothesis 1 *.

Forest Products	Chi-Square	*p*-Value
Fodder	3.7	0.15
Fuelwood	2.93	0.23
Leaf litter	1.06	0.58

* = Null hypothesis (H$_0$): There is no significant difference in the supply of forest products between different AF systems.

Alternative hypothesis (H$_1$): There is a significant difference in the supply of forest products between different AF systems.

3.4. Contribution of Forest Products from AF and CF Systems

A comparison was made between the annual supply of forest products from AF and CF systems. The total supply of forest based products were measured in terms of the amount derived by local households in one year. The average annual supply of fodder from CF was higher (4791 kg) than that from the AF system (3483 kg) (Figure 4). The average annual supply of fuelwood from CF was higher than that from AF (i.e., 3085 > 1012). The average annual supply of leaf litter by CF was higher (3466 kg compared with 2494 kg from AF). Thus, it was evident that the local people relied mainly on CF for their forest products. About 75% of fuelwood demand was fulfilled through CF, whereas in the case of fodder and leaf litter, the contribution was around 57%. It can be hypothesized that people used the AF system for the fulfilment of fodder and leaf litter requirements, and the choice of species was similarly affected.

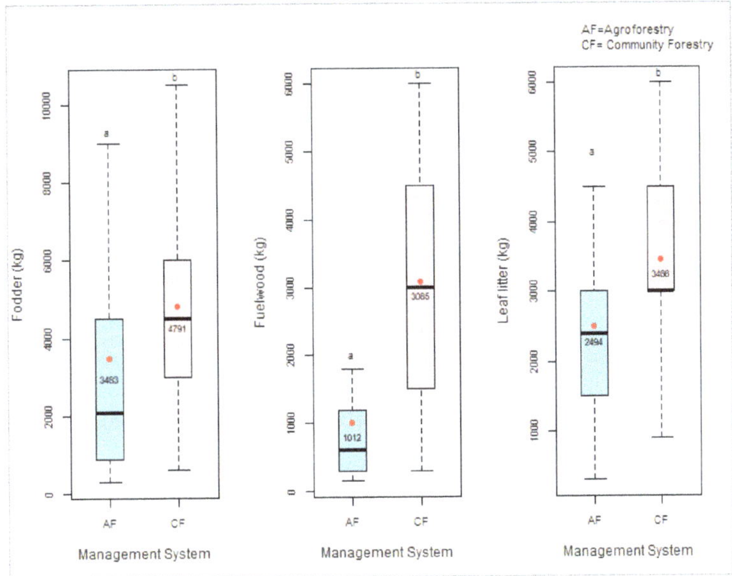

Figure 4. The average annual supply of forest products (fuelwood, fodder, leaf litter) from AF and CF systems where a red dot indicates the mean value while a black line indicates the median. The different letters (a and b) denotes the significance difference between the mean values.

Using Wilcoxon test for Hypothesis 2 at the 95% confidence interval, all the *p*-values were less than 0.05, null hypothesis was rejected, implying that the difference in the supply of all forest products from AF and CF systems was statistically significant (Table 4).

Table 4. Wilcoxon test result for Hypothesis 2 *.

Forest Product	Wilcoxon Test Statistic (W)	*p*-Value
Fodder	2242.5	<0.001
Fuelwood	827	0.001
Leaf litter	273.5	0.01

* = Null hypothesis (H_0): There is no significant difference in the supply of forest products between AF and CF systems.

Alternative hypothesis (H_1): There is a significant difference in the supply of forest products between AF and CF systems.

3.5. Status and Growth of Agroforestry in the Study Area

Through FGDs and KIIs, the status of AF in terms of its potency and limitations were assessed along with the advantages and disadvantages of further growth of this system at the study site. While the allocation of a small unit of land to provide multiple benefits is a very important aspect of AF, the local people were unaware of AF systems, and therefore no scientific techniques were employed at the study site. The lack of appropriate tools, irrigation facilities and high-yielding varieties of seed were major limitations in the promotion of AF systems in this region. Due to the fragile soils of the Chure region, the fact that planted seedlings and crops may be swept away by floods and destroyed by landslides was also a considerable hindering factor in the adoption of AF systems in this region. The majority of households had medium and small land holdings. Thus, the use of limited land for trees along with agricultural crops would reduce food production, which may be a serious problem for poor farmers. Those who retained tree species on their farmland were less motivated to embrace the notion of a plantation. Although there was high potential for the fruit tree species because of the suitable soil properties, people were not interested in planting fruit trees. The committee of the community forestry user group was active at the study site to manage and develop CF, but a similar committee for AF was lacking, showing there was greater interest in administering the growth of CF. As regards AF, there is a sufficiency of low-skilled manpower, little investment is required and it is simple to adopt the necessary practices. With less input than is needed for CF, AF provides maximum output. Because Chure falls in a critical zone of Nepal, several government and nongovernment organizations have shown an interest in Chure conservation, which may provide the greatest opportunity for the implementation and promotion of AF. However, the focus on conservation was at odds with the promotion of AF systems. No external or internal agencies were found to support the use of AF techniques in this region. Also, no funding was allocated for the development of AF in this region.

However, this lack of foresight on the part of the government was contrasted by some key informants who commended the efforts of local government in organizing training sessions on AF systems, facilitating the choice of species and distributing the required seedlings. The focus groups disagreed with this, explaining that it led to ineffectiveness caused by speculative guidance. The insufficiency of distributed seedlings demotivated the farmers from diversifying their land and planting local species. On a positive note, youths were interested in developing AF systems and were aware of potential markets, which led to the evolution of the choice of species from subsistence to cash crops and vegetables. Youth groups were motivated to commercialize their farm production and therefore, the development of markets and enterprises although slow, had begun. Since women were responsible for the collection of fuelwood and fodder from the forest, they showed greater interest in developing AF as it could enhance easy access to these needs. But women played a minimal role in the decision-making process, and therefore, AF in the study area was much less developed. Even among those taking part in the focus groups and KIIs, many did not know the details of this policy and how it would benefit the local farmers. However, the development of AF in the near future was anticipated.

4. Discussion

4.1. Choice of AF System and Plant Species

Out of 12 AF systems documented by the Forest Research and Training Center (FRTC) [64] in hilly and Terai regions of Nepal, the people of the study area adopted only three of those: AS, ASP and HAS. This may be because they do not have the proper level of knowledge about the various AF systems, which could be associated with the higher illiteracy level at the study site. Since 52% of local people were engaged in agriculture as the main occupation, any diversification of land use was minimal owing to the prioritization of subsistence crops for the fulfillment of basic needs. Similar reasons were mentioned by Kiptot and Franzel [65] in their study in Africa. Kassie [66], in Ethiopia, discovered that the people who were engaged in more diverse income sources would adopt various AF systems, which supported the findings of Dhakal and Rai [28]. The choice of crops worldwide has developed over time from food crops to cash crops and vegetables. As explained by Dhakal et al. [29], the pattern of the farming system has changed over time. The change was linked with the growth of the market in many cases [29,30,67], which was also observed at our study site.

AS was the most dominant AF system in the study area, possibly because of the small size of land holdings of the people. AS is the most common AF system in both hilly and Terai regions of Nepal, as also determined by FRTC [64] and Amatya et al. [68]. Since the landholding size of the respondents was below the national average, the integration of different systems on a small unit of land would restrict the space available for crop production, thereby inducing food insecurity. Therefore, most of the people relied on the simple AS system. The influence of landholding size on the adoption of AF systems was noted by different researchers [28,69,70].

As Dhakal et al. [29] mentioned, the choice of tree species is affected by the farmer's priority. Since the production of timber would take several years, people opted instead for fodder and fruit species. Which local fodder species are preferred by farmers in different countries has been analyzed in many studies, e.g., [71–73]. The findings of our study are in line with those of Kunwar et al. [74] for Nepal, as different *Ficus* species that are indigenous to Nepal were used by most of the farmers. Despite the higher potentiality of HAS to improve the livelihoods of the farmers by providing extra income, farmers still view that system as less profitable, which is a valid argument due to a lack of commercial prospects. The study by Snelder et al. [75] in the Philippines also discovered negative perceptions among farmers in regard to fruit species in comparison with seasonal cash crops.

4.2. Agroforestry for the Support of Rural Livelihoods in Comparison with Community Forestry

Since the focus of this study was on the supply of forest-based products from AF, the benefits of AF in regard to cash crops and fruit production were not studied. The major forest-based products used by local people in this study area were fuelwood, fodder and leaf litter. Even though more than 69% of people depend on fuelwood as their main source of energy in Nepal [76], the AF system in our study area contributed only 25% of the total demand; the remaining 75% was derived from CF. The dependency of people in CF for energy sources is well documented by many studies, e.g., [5,77,78]. However, the insufficiency of AF in meeting the fuelwood demand in the study area could be associated with the preference for palatable fodder species by the farmers for their livestock. As explained by Iiyama et al. [79], in the context of Africa, the adoption of AF with a focus on fuelwood is minimal and therefore pressure is placed on the natural forest. The use of multipurpose trees or of rotational woodlots with fast growing species could address this issue [79].

In the case of fodder and leaf litter, the contribution of AF systems and CF were almost equal, which implies that the dependency on the natural forest for these products is slowly decreasing. A study in Indonesia also showed a strong linkage between farm diversification and low dependency on natural forest [80]. Enabling farmers to integrate multiple species rather than monocultures could minimize the pressure on the adjacent forest [80]. The study by Rahman et al. [81] in Bogor, Indonesia stated that the use of natural forests by farmers adopting subsistence agriculture is much higher than that among farmers adopting AF systems.

Despite the wide opportunities provided by AF of sustaining subsistence and commercial livelihood aspects, the policy implications restrict the use of some trees as an income-generating source [82]. Even though AF makes a significant contribution to livelihoods, it does not have a strong policy base [5]. Therefore, CF is still the first choice of poor farmers with less land holding capacity [83]. While 44% of the country's total land area is covered by forest, there has been a gradual decrease in the extent of the forests of the Chure region [39,84], a trend that is even more pronounced in and around the agricultural areas of Nepal [31]. Research studies have shown there are direct connections between forest degradation and poverty [85]. Owing to the large population in poverty and unemployed in the Chure region [37], AF has a strong potential to Supplement CF in this region by reducing the pressure on CF. Along with the supportive roles they play in relation to each other, CF and AF make remarkable contributions to livelihoods.

As an integrated approach, agroforestry plays a multifunctional role in achieving the agenda 2030 by increasing the yields/farms output significantly and achieving food security towards SDG-2, and can reduce poverty (SDG-1) by enhancing the income with diversified income opportunities. The role played by AF in the study area to conserve the biodiversity and ecosystem services (SDG-15) while reducing the pressure on CF for fodder, fuelwood and litter is remarkable. Furthermore, AF acts as the source of energy (SDG-7) by providing fuelwood. Several studies highlight the role of agroforestry to achieve various goals under agenda 2030 e.g., [86–88]. The role of CF in making contributions to SDGs has been widely recognized in many research studies e.g., [89–91]. Along with the complementary role played to each other, AF and CF have strong potentiality to make remarkable contributions to SDGs.

4.3. Prospects for Growth of Agroforestry

The major restrictions in the proper development of AF at the study site were the lack of awareness of local people around greater diversification around AF systems and lack of technical knowledge. Dhakal and Rai [28] in their study in Dhanusha, Nepal also found that the improved accessibility of farmers to extension services had a positive influence on the adoption of a variety of AF systems. As in the case of Rwanda, where farmers declined the adoption of AF systems due to a lack of technical expertise, capital and quality seeds [92], this study also found the lack of scientific techniques, funding and availability of seedlings to be the major limitations for the adoption of AF systems. With the transfer of knowledge to farmers regarding the selection of appropriate species, and techniques of tree planting and harvesting, the growth of AF systems could be ensured. A study in Africa by Meijer et al. [69] has also explained the linkage of perceptions of local people about AF to the knowledge they have. While AF is practiced at the study site for subsistence needs, there are no well-developed markets for farmers' AF products such as fruits and non-timber forest products. Research conducted by Gilmour et al. [5] showed that the unavailability of a proper market has been a drawback that limits the development of AF systems.

There is a gap in the demand and supply chain of forest products from AF in the study area, so there is an immediate need to innovate and improve the current system of AF. There is great opportunity for the conservation of fragile Chure regions through AF by enhancing livelihood productivity; this would give higher economic returns and other related opportunities. Nepal is one of few countries with an AF policy that aims to prioritize the commercialization and collective advantages of AF systems and to facilitate farmers' access to the market [27]. The policy further aims to provide incentives to those farmers adopting an AF system and to generate specific area-based AF models. However, the implementation at rural ground level is still deficient. Lack of proper laws and directives and organizational incompetence has restricted the successful implementation of this policy. Furthermore, the provision of harvesting, transportation and marketing of commercial timber from agroforestry is completely contradictory with existing forest law [93]. The feasibility of the commercial use of forest products from community forests has caused a drawback in developing agroforestry. Even though much remains to be done around the development of AF systems in different parts of Nepal, the initiation of promotion through the formulation of policies has created an anticipation of better AF systems. To increase the adoption of AF among the stakeholders, positive perceptions around AF could be built through extension services and technical training.

5. Conclusions

In conclusion, the practice of farm-based agroforestry in the study area was more traditional, which has eventually grown from sustenance to commercial. The adoption of agroforestry in the study area was much limited as people lacked knowledge about the multiple uses of land and were not prepared to take risks with small land holdings. Only three farm-based agroforestry systems: AS, HAS and ASP were practiced by the local people. The supply of forest products from ASP was higher than other systems in spite of a low number of respondents practicing it. People still relied on the community forests for the products thereby creating pressure on natural forests. Agroforestry though has a significant potential to contribute to livelihoods with fewer inputs, the horizon of its practice was restricted by the lack of technological development and market availability. Proper extension services, supply of proper raw materials to the farmers and development of markets would help agroforestry to prosper in this region alongside reducing the heavy forest degradation. This study recommends further comprehensive study on the motivational drivers for agroforestry, competence of policies in developing agroforestry, quantitative measure of the contribution of agroforestry to the total income of local people and the pertinence of the market for agricultural and forest products in this area.

Author Contributions: Conceptualization, D.K.; Methodology, A.A. and D.K.; Software, K.P.B. and B.P.D.; Validation, A.A. and H.B.; Formal Analysis, K.P.B. and B.P.D.; Investigation, D.K.; Resources, D.K.; Data Curation, A.A. and D.K.; Writing—Original Draft Preparation, A.A. and B.P.D.; Writing—Review & Editing, H.B. and K.P.B.; Visualization, D.K.; Supervision, H.B.; Project Administration, B.P.D.; Funding Acquisition, D.K. All authors have read and agreed to the published version of the manuscript.

Funding: This research received no external funding.

Institutional Review Board Statement: Not applicable.

Informed Consent Statement: Informed consent was obtained from all subjects involved in the study.

Data Availability Statement: The data presented in this study are available on request from the corresponding author.

Acknowledgments: We acknowledge the financial contribution from the Nepal Agroforestry Foundation (NAF) to carry out this research and CGIAR research program on Forests, Trees and Agroforestry for the partial support to publish the work. The staff of Terai Private Forest Development Association (TPFDA) are also acknowledged for their support during data collection. The authors extend thanks to Nab Raj Subedi who helped produce the study area map. We appreciate the reviewers' comments and suggestions.

Conflicts of Interest: The authors declare no conflict of interest.

Appendix A

Questionnaire for focus group discussions

1. How well do you understand agroforestry?
2. Are you aware of different types of agroforestry systems?
3. Do you have any idea about the scientific techniques of agroforestry?
4. Which species do you think is beneficial (timber, fodder, fuelwood, fruit)?
5. How do you evaluate the benefit of agroforestry?
6. Do you think agroforestry can provide more benefits than the community forest?
7. Would you prefer agroforestry over subsistence agriculture?
8. Are you interested in developing agroforestry?
9. Which type of support are you receiving from the government?
10. Do you think that the support from the government is enough as well as inclusive?
11. Are there any other organizations supporting agroforestry in this area?
12. What kind of support do you expect from the government?

13. Is there any committee for agroforestry development in the area?
14. Are you planning to form any such committee?
15. What motivates you to adopt agroforestry?
16. What restricts you from adopting agroforestry?
17. Are there enough markets where you can sell the products from your farm?
18. If yes, are you satisfied with the market price of the agroforestry products?
19. Do you think increasing size of markets could help in developing agroforestry in this region?
20. Do you have any suggestions to improve agroforestry?
21. Do you think agroforestry is well-promoted in this region?
22. Are you willing to participate in any forms of training related to agroforestry?
23. Do you think equal attention is given to agroforestry and community forestry?
24. Are you aware of the recent agroforestry policy?
25. If yes, do you think this will help toward the development of an agroforestry system in your area?

Appendix B

Questionnaire for Key Informant Interview

1. Do people in the study area know about agroforestry practices?
2. Are you aware of any scientific techniques used for the practices of agroforestry?
3. Are villagers interested in private agroforestry? If yes, what is the trend for developing private agroforestry?
4. Is there any committee for the development of an agroforestry system? If yes, what is the name of the committee involved?
5. How do you evaluate the efforts of the central and local governments in promoting agroforestry?
6. Are NGOs and INGOs involved in the development of an agroforestry system?
7. If yes, do they provide any type of funding for the development of an agroforestry system?
8. What are the possible agroforestry techniques that could be practiced here?
9. What are the constraints for developing farm-based agroforestry?
10. What are the opportunities for agroforestry promotion on this site?
11. What are the agroforestry practices that can give maximum benefit and reduce pressure in natural forests?
12. Are you convinced that agroforestry could contribute to decreasing pressure in natural forests as they supply major forest products? If yes, why? If no, why?
13. At last, would you like to suggest any agroforestry practice that would be suitable for this area?

Appendix C

Interview Guide
Socioeconomic characteristics

1. Name of Respondent: ....

 i. Age: .... ii. Sex: .... iii. Address: ....
 iv. Ethnic group: (a) Tamang (b) Brahmin (c) Chettri (d) Magar (e) Others

2. Annual cash income of family............................NRs.
3. Educational status of family:

Number of Family Members with Their Educational Status									
Illiterate		Primary (1–5 class)		Secondary (6–10 class)		Higher Secondary (10–12 class)		University	
M	F	M	F	M	F	M	F	M	F

4.   What is your major occupation?

Types	Choice
Agriculture	
Government employee	
Non-gov. employee	
Business	
Pension	
Remittance	
Wage laboring	
Others	

5.   Livestock standard units

Types of Animal	LSU
Buffalo	
Cow	
Goat	
Poultry	
Others	

6.   Land holding:

Land Types	Registered Land Area (in ha)	Unregistered Land Area (in ha)
Khet (Large Productive Agricultural Land)		
Bari (Medium Productive Agricultural Land)		
Kharbari (Poor Productive Agricultural Land)		
Home garden		

7.   Area of home garden (Farm-based agroforestry) . . . . (in ha).
8.   Time to reach forest from household . . . . in minutes/hours.
9.   Time to reach the road head from household . . . . minutes/hours.
10.   Free grazing months in your farm-based agroforestry in the whole year.................months.
11.   Number of fodder collection months in your farm-based agroforestry in the whole year . . . . . . months.

Supply from agroforestry

12.   Do you think planting trees along with agricultural crops is a good idea?

(a) Yes   (b) No   (c) Do not know

13.   Have you planted trees with agricultural crops? (If no, go to Q.N. 15)

(a) Yes   (b) No   (c) Do not know

14.   If yes, what types of trees are more preferable to plant on your farmland?

(a) Fodder   (b) Fuel wood   (c) Timber   (d) Fruit tree

15.   In your opinion, why do you plant trees on your farmland? (Please Tick ☑ one or more)

(a) For fodder   (b) For timber   (c) For fuel wood   (d) For soil conservation   (e) For income
(f) For aesthetic value   (g) To increase land productivity   (h) Others (Specify..........)

16.   What are the methods of establishment of farm trees? {Please Tick ☑ one}

(a) Deliberate retention and management of natural regeneration
(b) Plantation
(c) Both a and b

17. What are your sources of firewood collection? (Tick √ one or more than one)

    (a) Community forest (b) Agroforestry (c) National forest (d) Private forest
    (e) Buy from market (f) Others.............

18. What are the sources of tree fodder for your cattle? (Tick √ one or more than one)

    (a) Community forest (b) Agroforestry (c) National forest (d) Private forest
    (e) Buy from market (f) Others.............

19. What are the sources of forage for your cattle? (Tick √ one or more than one)

    (a) Community forest (b) Agroforestry (c) National forest (d) Private forest
    (e) Buy from market (f) Others.............

20. What are the sources of timber for your household?

    (a) Community forest (b) Agroforestry (c) National forest (d) Private forest (e) Buy from market
    (f) Others.............

21. What are the sources of forest products for your household consumption?

    (a) Community forest (b) Agroforestry (c) National forest (d) Private forest (e) Buy from market
    (f) Others.............

22. Are you interested in developing an agroforestry system?

    (a) Yes   (b) No   (c) Do not know

23. If yes, what are the reasons behind developing agroforestry?

    (a) Increased crop yield   (d) Soil quality improvement
    (b) Increased biodiversity   (e) Healthy environment
    (c) Livestock benefits   (f) Soil conservation
    (g) Others..........

24. Estimate the contribution of different forest product needs to your **household/year.**

Forest Product	Agroforestry			Community Forest		
	Unit	Price/Unit	Qty	Unit	Price/Unit	Qty
Timber						
Pole						
Fuelwood						
Fodder						
Others (specify . . . .)						

25. How often do you visit to collect forest products from natural forest? (Tick ☑ one)

    (a) Daily (b) Weekly (c) Monthly (d) Yearly

26. How often do you visit to collect forest products from your farm-based agroforestry? (Tick ☑ one)

    (a) Daily (b) Weekly (c) Monthly (d) Yearly

27. List the major tree species available in your farm-based agroforestry.

S. N.	Name of Species	Uses	No. of Plants
1			
2			
3			
4			

References

1. UN General Assembly, Transforming Our World: The 2030 Agenda for Sustainable Development, 21 October 2015, A/RES/70/1. Available online: https://www.refworld.org/docid/57b6e3e44.html (accessed on 1 December 2020).
2. Fisher, E.; Bavinck, M.; Amsalu, A. Transforming asymmetrical conflicts over natural resources in the Global South. *Ecol. Soc.* **2018**, *23*, 234. [CrossRef]
3. Rauch, T.; Schmidt, M.; Segebart, D. New rural dynamics and challenges in the Global South. *Geogr. Helv.* **2014**, *69*, 225–226. [CrossRef]
4. Varughese, G. Coping with changes in population and forest resources: Institutional mediation in the middle hills of Nepal. In *Forest Resources and Institutions*; Chapter 8; Forests, Trees and People Programme, Working Paper; FAO: Rome, Italy, 1998.
5. Gilmour, D.; Pradhan, U.; Malla, Y.; Bartlett, T.; Finlayson, R.; Shah, R. *Enhancing Livelihoods and Food Security from Agro-Forestry and Community Forestry Systems in Nepal: Current Status, Trends, and Future Directions*; Southeast Asia Regional Program; World Agro-forestry Centre (ICRAF): Bogor, Indonesia, 2014.
6. Deshar, B.D. An overview of agricultural degradation in Nepal and its impact on economy and environment. *Glob. J. Econ. Soc. Dev.* **2013**, *3*, 1–20.
7. Chalise, D.; Kumar, L.; Kristiansen, P. Land Degradation by Soil Erosion in Nepal: A Review. *Soil Syst.* **2019**, *3*, 12. [CrossRef]
8. Paudel, P.P.; Devkota, B.D.; Kubota, T. Land Degradation in Nepal: A Review on its Status and Consequences. *J. Fac. Agric. Kyushu Univ.* **2009**, *54*, 477–479. [CrossRef]
9. Government of Nepal (GoN). *Nepal Demographic and Health Survey, Government of Nepal*; Ministry of Health and Population, Ramshah Path: Kathmandu, Nepal, 2016. Available online: https://www.dhsprogram.com/pubs/pdf/fr336/fr336.pdf (accessed on 2 December 2020).
10. United States Agency for International Development (USAID). Agriculture and Food Security. 2020. Available online: https://www.usaid.gov/nepal/agriculture-and-food-security (accessed on 5 December 2020).
11. Charnley, S.; Poe, M.R. Community Forestry in Theory and Practice: Where Are We Now? *Annu. Rev. Anthr.* **2007**, *36*, 301–336. [CrossRef]
12. Gauld, R. Maintaining Centralized Control in Community-based Forestry: Policy Construction in the Philippines. *Dev. Chang.* **2000**, *31*, 229–254. [CrossRef]
13. Maryudi, A.; Devkota, R.R.; Schusser, C.; Yufanyi, C.; Salla, M.; Aurenhammer, H.; Rotchanaphatharawit, R.; Krott, M. Back to basics: Considerations in evaluating the outcomes of community forestry. *For. Policy Econ.* **2012**, *14*, 1–5. [CrossRef]
14. Shrestha, K.; Cedamon, E.; Ojha, H.; Sharma, N.; Paudel, B.H.P.; Amatya, S.M.; Joshi, M.; Karki, R.; Dangal, S.; Shah, R. *Enhancing Livelihoods and Food Security from Agroforestry and Community Forestry in Nepal (EnLiFT1)*; ACIAR: Bruce, Australia, 2018.
15. Conacher, A. Land degradation: A global perspective. *N. Z. Geogr.* **2009**, *65*, 91–94. [CrossRef]
16. Sheppard, J.P.; Reckziegel, R.B.; Borrass, L.; Chirwa, P.W.; Cuaranhua, C.J.; Hassler, S.K.; Hoffmeister, S.; Kestel, F.; Maier, R.; Mälicke, M.; et al. Agroforestry: An Appropriate and Sustainable Response to a Changing Climate in Southern Africa? *Sustain. J. Rec.* **2020**, *12*, 6796. [CrossRef]
17. Cedamon, E.; Nuberg, I.; Pandit, B.H.; Shrestha, K.K. Adaptation factors and futures of agroforestry systems in Nepal. *Agrofor. Syst.* **2017**, *92*, 1437–1453. [CrossRef]
18. Oelbermann, M.; Voroney, R.P.; Gordon, A. Carbon sequestration in tropical and temperate agroforestry systems: A review with examples from Costa Rica and southern Canada. *Agric. Ecosyst. Environ.* **2004**, *104*, 359–377. [CrossRef]
19. Torralba, M.; Fagerholm, N.; Burgess, P.J.; Moreno, G.; Plieninger, T. Do European agroforestry systems enhance biodiversity and ecosystem services? A meta-analysis. *Agric. Ecol. Environ.* **2016**, *230*, 150–161. [CrossRef]
20. Muchane, M.N.; Sileshi, G.W.; Gripenberg, S.; Jonsson, M.; Pumariño, L.; Barrios, E. Agroforestry boosts soil health in the humid and sub-humid tropics: A meta-analysis. *Agric. Ecosyst. Environ.* **2020**, *295*, 106899. [CrossRef]
21. Allen, S.C.; Jose, S.; Nair, P.; Brecke, B.J.; Nkedi-Kizza, P.; Ramsey, C.L. Safety-net role of tree roots: Evidence from a pecan (*Carya illinoensis* K. Koch)–cotton (*Gossypium hirsutum* L.) alley cropping system in the southern United States. *For. Ecol. Manag.* **2004**, *192*, 395–407. [CrossRef]
22. Jose, S. Agroforestry for ecosystem services and environmental benefits: An overview. *Agrofor. Syst.* **2009**, *76*, 1–10. [CrossRef]
23. Barrios, E.; Valencia, V.; Jonsson, M.; Brauman, A.; Hairiah, K.; Mortimer, P.E.; Okubo, S. Contribution of trees to the conservation of biodiversity and ecosystem services in agricultural landscapes. *Int. J. Biodivers. Sci. Ecosyst. Serv. Manag.* **2018**, *14*, 1–16. [CrossRef]
24. Neupane, R.P.; Sharma, K.R.; Thapa, G.B. Adoption of agroforestry in the hills of Nepal: A logistic regression analysis. *Agric. Syst.* **2002**, *72*, 177–196. [CrossRef]
25. Government of Nepal (GoN). *Agriculture Development Strategy (ADS), Government of Nepal, Ministry of Agricultural Development*; Singhadurbar: Kathmandu, Nepal, 2014. Available online: http://nnfsp.gov.np/PublicationFiles/bf53f040-32cb-4407-a611-d891935d2e97.pdf (accessed on 1 November 2020).
26. Government of Nepal (GoN). *National Forest Policy. Government of Nepal, Ministry of Forest and Environment*; Singhadurbar: Kathmandu, Nepal, 2018. Available online: http://mofe.gov.np/downloadfile/Ban%20Niti_1554873640.pdf?fbclid=IwAR1kJMxO7q_8P1HV4QnD6puoGEUmFxtaPB2EGBMzdCppjRlBB3D1f1fG-EQ (accessed on 10 November 2020).
27. Government of Nepal (GoN). *National Agroforestry Policy, Ministry of Forest and Environment*; SinghaDurbar: Kathmandu, Nepal, 2019. Available online: https://www.mofe.gov.np/downloadsdetail/8/2018/36366627/ (accessed on 28 November 2020).

28. Dhakal, A.; Rai, R.K. Who Adopts Agroforestry in a Subsistence Economy?—Lessons from the Terai of Nepal. *Forests* **2020**, *11*, 565. [CrossRef]
29. Dhakal, A.; Cockfield, G.; Maraseni, T.N. Evolution of agroforestry based farming systems: A study of Dhanusha District, Nepal. *Agrofor. Syst.* **2012**, *86*, 17–33. [CrossRef]
30. Regmi, B.N. Contribution of agroforestry for rural livelihoods: A case of Dhading District, Nepal. In Proceedings of the International Conference on Rural Livelihoods, Forests and Biodiversity, Bonn, Germany, 19–23 May 2003.
31. Pandit, B.H.; Thapa, G.B. Poverty and Resource Degradation under Different Common Forest Resource Management Systems in the Mountains of Nepal. *Soc. Nat. Resour.* **2004**, *17*, 1–16. [CrossRef]
32. Neupane, R.P.; Thapa, G.B. Impact of agroforestry intervention on soil fertility and farm income under the subsistence farming system of the middle hills, Nepal. *Agric. Ecosyst. Environ.* **2001**, *84*, 157–167. [CrossRef]
33. Schwab, N.; Schickhoff, U.; Fischer, E. Transition to agroforestry significantly improves soil quality: A case study in the central mid-hills of Nepal. *Agric. Ecosyst. Environ.* **2015**, *205*, 57–69. [CrossRef]
34. USDA (United States Department of Agriculture). Fragile Soil Index Interpretation. 2021. Available online: https://www.nrcs.usda.gov/wps/PA_NRCSConsumption/download?cid=nrcseprd1298845&ext=pdf (accessed on 1 March 2021).
35. Singh, B.K. Land Tenure and Conservation in Chure. *J. For. Livelihood* **2017**, *15*, 87–102. [CrossRef]
36. Central Department of Environment Science (CDES). Hazard and Vulnerability Analysis of Nepal's Chure Area. Central Department of Environmental Science, Tribhuvan University and Government of Nepal. 2014. Available online: http://202.70.82.238/cgi-bin/koha/opac-detail.pl?biblionumber=73406 (accessed on 26 November 2020).
37. Ministry of Forest and Soil Conservation (MOFSC). *Sustainable Management of Chure: Efforts, Challenges and Potential, Proceedings of the National Seminar on Chure*; Ministry of Forest and Soil Conservation, Singhadurbar: Kathmandu, Nepal, 2012; pp. 1–262. Available online: https://chureboard.gov.np/wp-content/uploads/2015/07/National-Seminar-2012_Sustainable-Management-of-Chure_Efforts-Challenges-and-Potential.pdf (accessed on 10 December 2020).
38. Government of Nepal (GoN). Chure Terai Madhesh Conservation or Management Master Plan. (Nepali version). President Chure Terai Madhesh Conservation, Development Board, Kathmandu. 2018. Available online: https://chureboard.gov.np/wp-content/uploads/2017/07/Master-Plan_Chure_NEPALI_final_29th-_Shrawan_2074.pdf (accessed on 3 December 2020).
39. DFRS (Department of Forest Research and Survey). *Churia Forests of Nepal. Forest Resource Assessment Nepal Project/Department of Forest Research and Survey*; Babarmahal: Kathmandu, Nepal, 2014. Available online: http://frtc.gov.np/old/downloadfile/CHURE_REPORT_1453193322.pdf (accessed on 10 March 2021).
40. Bhattarai, B.; Bishwokarma, D.; Legras, M. Breaking the Bottleneck: Conflicts Metamorphosis of Chure Landscape Man-agement in Federal Nepal. *J. For. Livelihood* **2018**, *16*, 71–86. [CrossRef]
41. Bjärnlid, H. The Slippery Slopes of Forest Governance in the Chure Range. Master's Thesis, Sveriges Lantbruksuniversitet, Uppsala, Sweden, 2016.
42. Bishwokarma, D.; Thing, S.J.; Paudel, N.S. Political Ecology of the Chure Region in Nepal. *J. For. Livelihood* **2016**, *14*, 84–96. [CrossRef]
43. Giri, A.; Rana, R.; Devkota, B.; Yadav, B. *Intervention Status Mapping of Churia Region: Concept and Scope*; Biological Resources Unit, Faculty of Science Nepal Academy of Science and Technology (NAST): Kathmandu, Nepal, 2012.
44. District Development Committee (DDC). Profile of Makawanpur district. District Development Committee. Hetauda, Makawanpur. 2020; (Nepali Version). Available online: http://ddcMakawanpur.gov.np/en/ (accessed on 10 November 2020).
45. Maraseni, T.N.; Shivakoti, G.P.; Cockfield, G.; Apan, A. Nepalese non-timber forest products: An analysis of the equitability of profit distribution across a supply chain to India. *Small-Scale For.* **2006**, *5*, 191–206. [CrossRef]
46. Chhetri, M.B.P. Chapter 10: Disaster Management System in Nepal: Time for Policy to Action. In *Compendium of Environment Statistics Nepal 2015*; ADB Technical Assistance Consultant's Report; Government of Nepal National Planning Commission Secretariat Central Bureau of Statistics; Ramshah Path: Kathmandu, Nepal, 2016.
47. Department of Hydrology and Meteorology (DHM). *Study of Climate and Climatic Variation over Nepal*; Department of Hydrology and Meteorology: Kathmandu, Nepal, 2015. Available online: http://dhm.gov.np/climate/ (accessed on 26 December 2020).
48. Shrestha, R.P.; Nepal, N. An assessment by subsistence farmers of the risks to food security attributable to climate change in Makawanpur, Nepal. *Food Secur.* **2016**, *8*, 415–425. [CrossRef]
49. Ghimire, P.; Bolakhe, S. Agroforestry systems: Biodiversity, carbon stocks and contribution to rural livelihood. *J. Agric. For. Univ.* **2020**, *4*, 197.
50. Chetry, M.R.; Silwal, U.K.; Kanel, M.D. Perspective of Farm Gate Agroforestry Product in the Emergency of the Chepangs Community: A Case Study of Mega earthquake of Central Nepal. *Int. J. Sci. Res. Manag.* **2018**, *6*, 3.
51. Chhetri, R.; Saxena, S.; Silwal, U.K. Addressing Climate Change and Enriching the Livelihood through Agroforestry: A Case Study of Chepang Community in Raksirang Rural Municipality, Makawanpur. *Int. Res. J. Manag. Sociol. Humanit.* **2019**, *10*, 161–177.
52. Khadka, R. Transition from Slash-and-Burn (Khoriya) Farming to Permanent Agroforestry in the Middle Hills of Nepal: An Analysis of Costs, Benefits and Farmers' Adoption. Master's Thesis, Department of Noragric, Norwegian University of Life Sciences, As Municipality, Norway, 2010.
53. Sharma, L.N.; Vetaas, O.R. Does agroforestry conserve trees? A comparison of tree species diversity between farmland and forest in mid-hills of central Himalaya. *Biodivers. Conserv.* **2015**, *24*, 2047–2061. [CrossRef]

54. Kafle, G. An overview of shifting cultivation with reference to Nepal. *Int. J. Biodivers. Conserv.* **2011**, *3*, 147–154. [CrossRef]

55. Pandey, B.P.; Kafle, K.R. Landslide hazard assessment using UAV imagery and GIS for road planning and development in Chure area: Sindhuli-Hetauda Section. 2020. Available online: https://www.researchsquare.com/article/rs-72224/v1 (accessed on 18 March 2021).

56. Razali, N.M.; Wah, Y.B. Power comparisons of shapiro-wilk, kolmogorov-smirnov, lilliefors and anderson-darling tests. *J. Stat. Model. Anal.* **2011**, *2*, 21–33.

57. Kruskal, W.H.; Wallis, W.A. Use of Ranks in One-Criterion Variance Analysis. *J. Am. Stat. Assoc.* **1952**, *47*, 583–621. [CrossRef]

58. R Studio Team. *RStudio: Integrated Development for R*; R Studio: Boston, MA, USA, 2019; Available online: http://www.rstudio.com/ (accessed on 10 October 2020).

59. Peters, G.J.; Verbon, P.; Green, J. Userfriendlyscience: Quantitative Analysis Made Accessible. 2018. Available online: https://userfriendlyscience.com/ (accessed on 10 October 2020).

60. Wilcoxon, F. Individual comparisons by ranking methods. In *Breakthroughs in Statistics*; Springer: New York, NY, USA, 1992; pp. 196–202.

61. Gill, A.S.; Deb Roy, R. Agroforestry Research Situation in India. In *Managing Agriculture for a better Tomorrow: The Indian Experience*; Pande, D.C., Ed.; M.D. Publication: Punjab, India, 1998; pp. 385–400.

62. Rapsomanikis, G. *The Economic Lives of Smallholder Farmers: An Analysis Based on Household Data from Nine Countries*; Food and Agriculture Organization: Rome, Italy, 2015.

63. Eurostat. Eurostat's Concepts and Definitions Database. Glossary: Livestock Unit (LU). 2020. Available online: https://ec.europa.eu/eurostat/statistics-explained/index.php/Glossary:Livestock_unit_(LSU) (accessed on 10 December 2020).

64. Forest Research and Training Center (FRTC). *Agroforestry Systems and Practices in Terai and Mid-hills of Nepal*; Forest Research and Training Center (FRTC), Ministry of Forests and Environment: Kathmandu, Nepal, 2019. Available online: https://frtc.gov.np/downloadfile/finalReport%20agroforestry_1598333214.pdf (accessed on 1 December 2020).

65. Kiptot, E.; Franzel, S. Gender and Agroforestry in Africa: Are women participating. In *Gender and Agroforestry in Africa: Are Women Participating*; World Agroforestry Centre: Nairobi, Kenya, 2011; pp. 1–72. [CrossRef]

66. Kassie, G.W. Agroforestry and farm income diversification: Synergy or trade-off? The case of Ethiopia. *Environ. Syst. Res.* **2017**, *6*, 8. [CrossRef]

67. Brown, S.; Kennedy, G. A case study of cash cropping in Nepal: Poverty alleviation or inequity? *Agric. Hum. Values* **2005**, *22*, 105–116. [CrossRef]

68. Amatya, S.M.; Cedamon, E.; Nuberg, I. Classification of Agroforestry systems. In *Agroforestry Systems and Practices in Nepal*; Amatya, S.M., Cedamon, E., Nu-berg, I., Eds.; Agriculture and Forestry University (AFU): Rampur, Nepal, 2018; pp. 23–33.

69. Meijer, S.S.; Catacutan, D.; Ajayi, O.C.; Sileshi, G.W.; Nieuwenhuis, M. The role of knowledge, attitudes and perceptions in the uptake of agricultural and agroforestry innovations among smallholder farmers in sub-Saharan Africa. *Int. J. Agric. Sustain.* **2015**, *13*, 40–54. [CrossRef]

70. Fischer, A.; Vasseur, L. Smallholder perceptions of agroforestry projects in Panama. *Agrofor. Syst.* **2002**, *54*, 103–113. [CrossRef]

71. Roothaert, R.L.; Franzel, S. Farmers' preferences and use of local fodder trees and shrubs in Kenya. *Agrofor. Syst.* **2001**, *52*, 239–252. [CrossRef]

72. Roothaert, R.; Franzel, S.; Kiura, M. On-farm evaluation of fodder trees and shrubs preferred by farmers in central kenya. *Exp. Agric.* **2003**, *39*, 423–440. [CrossRef]

73. Mekoya, A.; Oosting, S.; Fernandez-Rivera, S.; Van Der Zijpp, A. Multipurpose fodder trees in the Ethiopian highlands: Farmers' preference and relationship of indigenous knowledge of feed value with laboratory indicators. *Agric. Syst.* **2008**, *96*, 184–194. [CrossRef]

74. Kunwar, R.M.; Bussmann, R.W. Ficus (Fig) species in Nepal: A review of diversity and indigenous uses. *Lyonia* **2006**, *11*, 85–97.

75. Snelder, D.J.; Klein, M.; Schuren, S.H.G. Farmers preferences, uncertainties and opportunities in fruit-tree cultivation in Northeast Luzon. *Agrofor. Syst.* **2007**, *71*, 1–17. [CrossRef]

76. Central Bureau of Statistics (CBS). *Nepal Living Standards Survey 2003–2004, NLSS Second*; Central Bureau of Statistics, Na-tional Planning Commission Secretariat, Government of Nepal: Kathmandu, Nepal, 2004. Available online: https://nada.cbs.gov.np/index.php/catalog/9 (accessed on 10 December 2020).

77. Pokharel, B.K.; Nurse, M. Forests and people's livelihood: Benefiting the poor from community forestry. *J. For. Livelihood* **2004**, *4*, 19–29.

78. Maid, M.; Tay, J.; Yahya, H.; Adnan, F.I.; Kodoh, J.; Chiang, L.K. The reliance of the forest community on forest for liveli-hood: A case of Kampung Wawasan Sook, Sabah, Malaysia. *Int. J. Agric. For. Plant.* **2017**, *5*, 110–118.

79. Iiyama, M.; Neufeldt, H.; Dobie, P.; Njenga, M.; Ndegwa, G.; Jamnadass, R. The potential of agroforestry in the provision of sustainable woodfuel in sub-Saharan Africa. *Curr. Opin. Environ. Sustain.* **2014**, *6*, 138–147. [CrossRef]

80. Murniati, D.; Garrity, P.; Gintings, A.N. The contribution of agroforestry systems to reducing farmers' dependence on the resources of adjacent national parks: A case study from Sumatra, Indonesia. *Agrofor. Syst.* **2001**, *52*, 171–183. [CrossRef]

81. Rahman, S.A.; Foli, S.; Pavel, M.A.; Mamun, M.A.; Sunderland, T. Forest, trees and agroforestry: Better livelihoods and ecosystem services from multifunctional landscapes. *Int. J. Dev. Sustain.* **2015**, *4*, 479–491.

82. Cedamon, E.D.; Nuberg, I.; Mulia, R.; Lusiana, B.; Subedi, Y.R.; Shrestha, K.K. Contribution of integrated forest-farm system on household food security in the mid-hills of Nepal: Assessment with EnLiFT model. *Aust. For.* **2019**, *82*, 32–44. [CrossRef]

83. Cedamon, E.; Nuberg, I.; Shrestha, K.K. How understanding of rural households' diversity can inform agroforestry and community forestry programs in Nepal. *Aust. For.* **2017**, *80*, 153–160. [CrossRef]

84. Government of Nepal (GoN). Presidential Chure Conservation Programme, Yearly Progress Report. (Nepali Version). President Chure Terai Madhesh Conservation Development Board, Kathmandu. 2020. Available online: https://chureboard.gov.np/wp-content/uploads/2020/10/Annual%20Progress%20Report_2076-2077_Chure%20final%20print.pdf (accessed on 5 December 2020).

85. Pandit, B.H. *Contribution of Community Forestry to Poverty Reduction: A Case Study from Inner Terai, Middle Hills and High Hills of Nepal*; Community Based Forest Management in the Himalayas-Phase III; Institute of Forestry: Pokhara, Nepal, 2012.

86. Montagnini, F.; Metzel, R. The Contribution of Agroforestry to Sustainable Development Goal 2: End Hunger, Achieve Food Security and Improved Nutrition, and Promote Sustainable Agriculture. In *Advances in Agroforestry*; Springer International Publishing: Cham, Switzerland, 2017; pp. 11–45.

87. Waldron, A.; Miller, D.C.; Redding, D.; Mooers, A.; Kuhn, T.S.; Nibbelink, N.; Roberts, J.T.; Tobias, J.A.; Gittleman, J.L. Reductions in global biodiversity loss predicted from conservation spending. *Nat. Cell Biol.* **2017**, *551*, 364–367. [CrossRef] [PubMed]

88. International Council for Research in Agroforestry (ICRAF). Agroforestry and ICRAF. 2016. Available online: https://www.worldagroforestry.org/about/agroforestry (accessed on 10 March 2021).

89. Katila, P.; de Jong, W.; Galloway, G.; Pokorny, B.; Pacheco, P. *Building on Synergies: Harnessing Community and Smallholder Forestry for Sustainable Development Goals*; IUFRO: Vienna, Austria, 2017.

90. De Jong, W.; Pokorny, B.; Katila, P.; Galloway, G.; Pacheco, P. Community Forestry and the Sustainable Development Goals: A Two Way Street. *Forests* **2018**, *9*, 331. [CrossRef]

91. Aryal, K.; Laudari, H.K.; Ojha, H.R. To what extent is Nepal's community forestry contributing to the sustainable de-velopment goals? An institutional interaction perspective. *Int. J. Sustain. Dev. World Ecol.* **2020**, *27*, 28–39. [CrossRef]

92. Kiyani, P.; Andoh, J.; Lee, Y.; Lee, D.K. Benefits and challenges of agroforestry adoption: A case of Musebeya sector, Nyamagabe District in southern province of Rwanda. *For. Sci. Technol.* **2017**, *13*, 174–180. [CrossRef]

93. Amatya, S.M.; Lamsal, P. Private Forests in Nepal: Status and Policy Analysis. *J. For. Livelihood* **2017**, *15*, 120–130. [CrossRef]

 *forests*

Opinion

Agroforestry: Opportunities and Challenges in Timor-Leste

Shyam Paudel [1], Himlal Baral [2,3], Adelino Rojario [4,5], Kishor Prasad Bhatta [6,7] and Yustina Artati [2,*

1   CTA, United Nation Environment Program (UNEP), Ministry of Natural Resources and Environment, Vientiane P.O. Box 7864, Laos; shyam.paudel@un.org
2   Center for International Forestry Research (CIFOR), Bogor 16115, Indonesia; h.baral@cgiar.org
3   School of Ecosystem and Forest Science, University of Melbourne, Parkville, VIC 3010, Australia
4   Faculty of Forestry and Environment, IPB University, Bogor 16680, Indonesia; roadelino56@gmail.com
5   Department of Reforestation Soil and Water Conservation, Directorate of Forestry Coffee and Industrial Plant, Ministry of Agriculture and Fisheries (GDFCIP), Dili P.O. Box No. 008, Timor-Leste
6   Faculty of Forest Science and Forest Ecology, Georg-August-Universität, Busgenweg 5, 37077 Gottingen, Germany; k.bhatta@stud.uni-goettingen.de
7   Research and Development Center (RDC), Durbarmarg, Kathmandu 44600, Nepal
*   Correspondence: y.artati@cgiar.org

Abstract: Agro forestry is a land management system that integrates trees, agriculture crops, and animal farming in order to provide a diverse range of ecosystem services. Timor-Leste, the newest country and one of the least developed counties, has faced multidimensional challenges on land use management, including deforestation, land degradation, and poverty. The agroforestry system is recognized as one of the viable options for balancing the socio-economic needs and ecological functions of the lands in Timor-Leste. The system has been practiced traditionally by farmers in the country; however, the lack of knowledge and experience, limited institutional capacity, and lack of funding have impeded the wider implantation of the agroforestry system in Timor-Leste. The Strategic Development Plan of Timor-Leste has recommended sustainable agriculture and natural resources management in the rural areas of the country to generate income and create employment for the youths. The paper presents the initiatives, challenges, and opportunities of agroforestry application in Timor-Leste to support sustainable forest management and livelihood improvement. Learning from existing initiatives, capacity building, market access, and financial incentives could promote the agroforestry system in the country.

Keywords: agroforestry; livelihood; food security; climate change; resilience; Timor-Leste

Citation: Paudel, S.; Baral, H.; Rojario, A.; Bhatta, K.P.; Artati, Y. Agroforestry: Opportunities and Challenges in Timor-Leste. *Forests* **2022**, *13*, 41. https://doi.org/ 10.3390/f13010041

Academic Editor: Bradley B. Walters

Received: 31 October 2021
Accepted: 25 November 2021
Published: 1 January 2022

Publisher's Note: MDPI stays neutral with regard to jurisdictional claims in published maps and institutional affiliations.

1. Background and Problem

The Democratic Republic of Timor-Leste is a small, new, developing nation that lies adjacent to Indonesia and Australia. Even though 80% of the population is involved in agriculture for their living [1], about 75% of the population in this state is under food insecurity of different intensities [2]. Low agricultural productivity, low value livelihood strategies, and high dependency on a single livelihood strategy were determined as the key factors causing food insecurity [2]. Furthermore, the country has been facing several development challenges including severe human and institutional capacity gaps, unstable economic growth, and heavy socio-economic pressure on the forest and natural resources due to the rapidly increasing population [3]. In addition, the state lies under a climatically vulnerable region prone to several disasters including flash floods, droughts, unpredictable rain, storm wind, river-bed collapse, paddy field destruction, landslides, and destructive winds [4,5]. Due to low economy of Timor-Leste, an integrated approach to ensure food security and to cope with climatic hazard is necessary.

Since mountainous landscape with more than 40% slope covers about 44% of the total area of the country, the amount of arable land is limited [6]. Only about 155,000 hectares of land, which forms 10.4% of the total land in Timor-Leste, is arable [7]. About 80% of households are engaged in crop production, mostly maize, legumes, and tubers, and 86% are engaged in livestock, rearing mainly chickens pigs, cattle, and goats [8]. The agriculture practice is primarily traditional, subsistence-based, and the productivity is low due to the variability of rainfall, lack of irrigation facilities, droughts, poor soil fertility, lack of organic fertilizers, and agriculture inputs [9]. The commercial farming is restricted by the small farm size and undeveloped marketing infrastructure [8]. On the other side, slash-and-burn agriculture—a traditional practice to compensate for reduced crop yields—is alleged to trigger deforestation and forest fire [10]. In addition, the unclear land tenure system due to resistance by elites to the full recognition of customary land management is claimed to delimit the commercial farming system in Timor-Leste [11].

Forest in Timor-Leste covers an area of 925,000 hectares i.e., about 61% of total land, but it undergoes an annual declination of 0.15% [12]. About 93% of households derive their energy from wood [13], whereas about 39% of the population collect food from the forest [14]. The population in Timor-Leste is increasing by 2.7% per year, and by 2030, it is estimated to reach 1.8 million [15]. The constant pressure on forest due to the increasing population has induced the natural hazards [16]. The vulnerability of communities in Timor-Leste to climate-induced disasters will further increase owing to maladaptive responses that cause environmental degradation [16,17]. These degradations resulting from unsustainable farming techniques such as slash-and-burn agriculture and the reduced practice of traditional terrace farming demand a sustainable farming strategy [18]. The pressure from environmental hazards and low food productivity has created a necessity to develop a strategy to manage land in such a way to maximize the profit. This opinion paper aims to understand the status, opportunities, and challenges of developing agroforestry as an integrated land use management system in Timor-Leste. The analysis draws from a literature review.

2. Agroforestry in Timor-Leste

There are various definitions of agroforestry from different experts; one [19] defines agroforestry as an integrated land use system in which social and ecological aspects are embedded in the system combining trees and agricultural crops and/or livestock, either simultaneously or successively. In another definition, agroforestry is a collective term for various land use systems and technologies designed for a single unit of land. It is applied by combining woody plants (trees, shrubs, palms, and bamboos) and agricultural crops or animals (livestock and/or fish), simultaneously or consecutively to promote ecological and economic interactions among the components [20]. Overall, it is a land management system that integrates trees, agriculture crops, and animal farming in order to provide a diverse range of ecosystem services by bridging the gaps between agriculture, forestry, and animal husbandry [21,22]. It involves economic and ecological interactions between two or more production systems such as tree–crop, tree–livestock, or tree–fish [22]. Different types of agroforestry practices such as agrosilviculture, silvopasture, agrosilvopasture, forest farming, etc. are adopted by farmers that provide different types of ecosystem services that include provisioning services: food fiber, freshwater, raw materials, fuelwood, non-timber forest products (NTFPs), medicinal resources, genetic resources, and ornamental resources; regulating ecosystem services such as erosion control, climate change moderation, nutrient retention, carbon storage and sequestration, and pest control; and habitat services including biodiversity enhancement and climate regulation [22–25]. Agroforestry can also increase profits by increasing plant diversity, improving agricultural productivity, preventing land degradation and soil erosion, and reducing the risk of crop failure and the need for fertilizers [26].

Timor-Lestes's Strategic Development Plan 2011 to 2030 identifies increasing agro-biodiversity with a focus on nutritious and high-yield crops including coffee, coconut, and other potential cash crops such as cocoa, cashews, hazelnuts, and spices [27]. The Ministry of Agriculture and Fisheries Strategic Plan has outlined the country's agricultural goals: (i) increase rural incomes and reduce poverty, (ii) promote environmental sustainability and conservation of natural resources, (iii) sustainably increase food production through improved crop varieties, forestry, livestock species, and fisheries, (iv) conserve, manage, and utilize natural resources [28].

The rehabilitation of degraded watersheds in Timor-Leste is prioritized by the National Biodiversity Strategy and Action Plan [29] whereas the Basic Environmental Law of Timor-Leste [30] aims to reduce the pressures on natural resources and promote landscape conservation for ecosystem services. Therefore, agroforestry has been identified as one of the key approaches to balance the agriculture product to sustain livelihoods and to rehabilitate degraded watersheds for enhanced ecosystem services [31]. Agroforestry has been providing a wide range of services and benefits to the local communities. It is traditionally used to control erosion and reduce flood in hilly regions. Agroforestry significantly contributes the supply of fodder for domesticated livestock and fuelwood for household energy. Local communities commonly prefer fruit trees such guava, avocado, etc. for domestic consumption as well as to generate cash income by selling in rural markets. However, there is no actual estimation of cash income generation from agroforestry fruit trees.

According to the UNDP [26], four different agroforestry models are common in Timor-Leste (Figure 1). (i) Alley cropping involves planting in the alley between rows of hedges/trees arranged according to contour lines. This model has been adopted mostly in hilly regions to reduce the occurrence of landslides and to facilitate the flow of water on a long-term basis. (ii) The trees-along border pattern involves planting trees/shrubs along the border (hedgerow). This is mostly practiced in steep hilly areas to conserve soil. (iii) Random mixers involves irregularly spacing trees while planting and simultaneously growing the annual crop in stratum underneath. This practice is mostly found in low land areas. (iv) Alternate rows involves planting trees in regular alternate rows and seasonal cultivation done in the space in between the rows. This pattern is practically suitable for flat and wide areas. This practice is not so common in Timor-Leste but can be seen in some places [26].

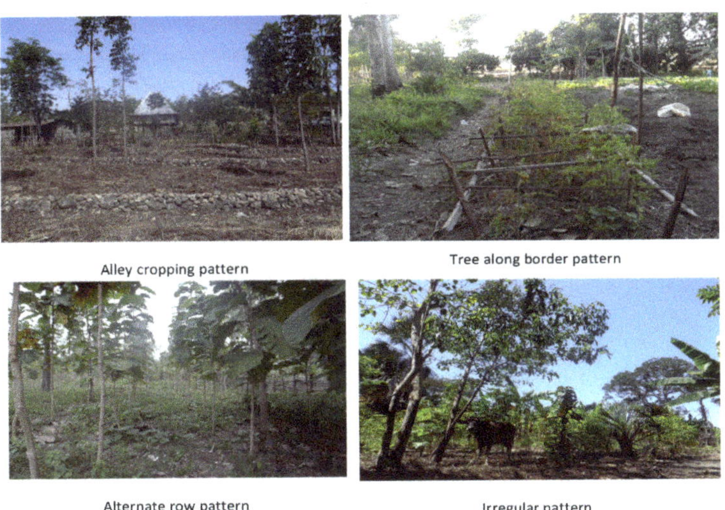

Figure 1. Various agroforestry models in Timor-Leste (Photos: Adelino/GDFCIP).

Fukuoka, a natural farming system, was introduced in Timor-Leste to plant trees in hilly agricultural marginal lands. About 200,000 Fukuoka seedballs were produced and distributed to rehabilitate larger vulnerable slopes previously damaged by slash-and-burn agriculture, erosion, and other forms of ecosystem degradation in hilly landscapes [32]. Fukuoka is a non-destructive type of reforestation method that can be used in hilly regions without tilling the ground. This has been widely practiced as it is easy, cheap, and non-destructive in erosion-prone landscapes [26]. However, the success rate was only about 50% mainly due to the lack of rain following the dispersal of the seed balls, browsing by animals in early stages, and poor seed quality.

According to Friday [33], fruit and fodder trees are among the major preferred species for agroforestry in Timor-Leste. In dryland, the preferred agroforestry species for forage include *Acacia leucophloea*, *Albizia lebbek*, *Hibiscus tiliaceus*, and *Prosopis* spp. Similarly, in areas with moderate rainfall, the preferred species are *Gliricidia sepium*, *Sesbania grandiflora*, *Samanea saman*, and *Leucaena leucocephala*. Other possible forage trees that exist in Timor-Leste that could be used for agroforestry include *Calliandra calothyrsus*, *Paraserianthes falcataria*, and *Albizia chinensis* [33].

3. Agroforestry: Initiatives, Challenges, and Opportunities

3.1. Current and Proposed Initiatives

Finding agroforestry as a viable option to restore landscape, sustain agriculture, and provide multiple livelihood benefits, several initiatives on agroforestry have been commenced in Timor-Leste funded by different donors. The United Nations Development Program's (UNDP) Global Environment Facility (GEF) funded projects on strengthening the community resilience along with the Dili to Anairo Road Development Corridor (DARDC) project, which was implemented from 2014 to 2019. The project aimed to build the resilience of watershed systems by improving natural resources management including forest, soil, and water in order to provide sustained ecosystem services to local communities. Agroforestry was considered as one of the key interventions to enhance ecosystem services [4]. The project introduced agroforestry systems in more than 200 ha of community land and introduced the Fukuoka system as a non-destructive and cost-effective way of plantation in hilly regions. Until 2018, 84.5 hectares of agroforestry had been completed. The UNDP worked closely with the National Directorate of Forestry within the Ministry of Agriculture and Fisheries (MAF) to reduce potential risks from disaster alongside building the resiliency of local people through reforestation and agroforestry [4]. The project used tara bandu (a local land law) to facilitate the planning and implementation of agroforestry at the community level [4,34]. The activity involved the collaboration of local MAF extension officers and NGOs during implementation to support the replication and sustainability of the interventions [4,34]. MAF has already been working with local and international NGOs to implement and promote reforestation interventions such as the planting of firewood trees (e.g., *Casuarina*), establishment of *Eucalyptus* nurseries, and restoration of mangroves [35]. Similarly, the European Union (EU) and Government of Germany through the Federal Ministry for Economic Cooperation and Development (BMZ) (in German: *Bundesministerium für wirtschaftliche Zusammenarbeit und Entwicklung*) have co-financed the project "Partnership for Sustainable Agroforestry" (PSAF), which has been implemented with the support of MAF in Timor-Leste from 2017 to 2022. The project aims to support more than 4000 households from four municipalities (Manatuto, Baucau, Viqueque, and Lautem) to adopt agroforestry and afforestation practice. The project has included a range of species such as Mahogany (*Swietenia* spp.), Teak (*Tectona grandis*), Sandalwood (*Santalum album*), ai-saria (*Cedrella toona*), Leucaena (*Leucaena* spp.), Rosewood tree (*Dysoxylum arborescens*), Orange (*Citrus sinensis*), Snake fruit (*Salacca edulis*), Mango (*Magnifera indica*), and Cashew nut (*Anarcadium occidentale*). The objective of the project is to provide an opportunity to marginalized people, in particular young men and women in rural areas, to benefit from better employment opportunities in the agroforestry systems of Timor-Leste. Specifically, this project aims to increase the productivity of agroforestry systems, strengthen the capacity of actors along with selected agroforestry value chains, improve the market access for selected agroforestry products (e.g., fruits and vegetables, raw and processed wood products), and improve the institutional and organizational framework for the promotion of agroforestry [36].

Furthermore, as a component of PSAF, ERA-agroforestry (Enhancing Rural Access Agroforestry Project) is being implemented with the support of the EU, Germany, and the International Labor Organization (ILO). The objective of this project is to implement a capacity building and labor-based program to rehabilitate rural roads so as to improve access to agroforestry areas, employment, and economic opportunities for local population [37].

The Agriculture Development Bank (ADB) [38] has proposed an initiative on Innovative Partnerships for coffee and agroforestry development. The project envisaged that improvements in coffee production and processing offers one of the clearest pathways for reducing poverty and growing Timor-Leste's non-oil economy. Coffee provides an important source of cash income for around 27.5% of Timorese households and is the largest non-oil export of the country [39]. However, the coffee sector is currently operating far below its long-term potential due to low and volatile production, inconsistent quality, and weak sector management. Strengthening sector management and providing targeted support to smallholder producers can generate sustained increases in household income. If implemented at scale, this could have a significant impact on the national poverty rate and growth of non-oil exports [39].

To help farmers defend their land ownership rights, adopt sustainable farming practices, and strengthen economic situations, an organization UNAER (Unian Agrikultor de Ermera) based in the Ermera municipality has been established in partnership with Union Aid Abroad-APHEDA [36]. With the support of this project, the farmers have grown a variety of alternative fruits including banana, papaya, lemon, mandarin, etc., in the non-coffee season and sold directly to buyers [40].

3.2. Key Challenges

Agriculture and natural resources are the main source of livelihood for rural communities [13]. About 75% of the total population face challenges of sustaining their livelihoods due to disaster risks, low food production, and lack of infrastructures [27]. Rural communities lack infrastructures such as water storage and irrigation facilities to increase their agriculture production and need to rely on rain-fed agriculture [41]. Unsustainable agricultural practice and the over-harvesting of natural resources are common in rural areas [42]. Deforestation, forest degradation, and the expansion of unsustainable subsistence farming are the major issues and the largest emitters of greenhouse gases in the country [38]. Agroforestry could be one of the viable options for balancing the socio-economic needs and ecological functions of the land [43]. However, a lack of knowledge and experience on improved agroforestry system impedes the wider implantation of an agroforestry system in the country. The incompetency of the government in providing technical inputs to farmers at the local level has also restricted the development of agroforestry in Timor-Leste [44].

A lack of cash flow and financial access is another major hurdle for agroforestry practice [45]. Based on the UNDP project experience, communities were motivated for agroforestry if provided with cash incentives for planting trees (25 cents per plant) besides providing free seedlings. They also expected cash enticements for protection and taking care of the plants [46,47]. Some donor-funded projects have been providing such incentives to farmers, but the approach has been under debate regarding its sustainability. However, according to the ADB's proposed project, assisting smallholder farmers to access formal financial services would enable them to build up savings and to access credit on more favorable terms [39]. This would complement and reinforce interventions to increase smallholder productivity.

Since 80% of the country is difficult to access due to a lack of good road networks, farmers have been facing challenges in selling their agricultural products [48]. The involvement of middlemen has delimited farmers from receiving fair prices for their agricultural products. This has demotivated farmers from investing in improved agricultural systems, including agroforestry practices [49]. Even though livestock forms one of the major sources of food and income for 80% of households in Timor-Leste, the current practice of allowing free grazing in the forest is a huge jeopardy for agroforestry [50,51]. Many agroforestry projects in past have faced this challenge and therefore failed to achieve their objective.

The major problems in developing agroforestry in Timor-Leste also include the lack of human resources and low level of education of formal institutions [52]. The agricultural extension services delivered by MAF staff and donor projects in Timor-Leste experienced difficulties because of the high expectations of landholders, lower education, lower trust in government, and poor skills among extension agents [53]. In addition, the lack of a proper agroforestry model as an alternative tool to improve community livelihood and improve the forest area and lack of legal forest management rights for communities are key challenges in developing the agroforestry in Timor-Leste [52].

Moreover, the unclear and complicated land tenure system of Timor-Leste has also discouraged private investors from investing in large-scale agroforestry [38,54]. There is still a great division between those who wish to exercise customary practices fully protected by law and those who wish to abolish customary rights and replace these by other forms of ownership that include state control and private property [11]. Reforming the law relating to land tenure is crucial for encouraging private investment for long terms.

3.3. Opportunities and Way Forward

The strategic development plan (SDP 2011–2030) of Timor-Leste has acknowledged the fact that the creation of local jobs through improved agriculture and natural resource management could be the best way to improve the livelihood of people living in rural areas [27]. Timor-Leste has an ample amount of land areas with suitable climate and soil to grow a wide range of valuable tropical trees in combination with agriculture. According to the ADB [38], around 32,500 hectares of suitable land that is not currently forested, protected or cultivated is available in Timor-Leste, which could be used for the large-scale implementation of agroforestry practices, which in turn would generate significant employment opportunities to 70% of its young population. There are opportunities to maximize the benefits by expanding the range of products from the agroforestry sector for the export market and natural resource restoration by developing community-based nurseries, training rural populations in agroforestry establishment.

The government of Timor-Leste is committed to protecting around 73% of land areas for the conservation of water resources, soil, and biodiversity by 2023, which includes 228,174.57 ha of dense forest cover, 278,999.19 ha of sparse forest cover, and 238,508.55 ha of non-forest areas [17]. Agroforestry could be one of the viable options for reversing land degradation. As Timor-Leste does not have a specific policy and strategy on agroforestry development., the country can learn a lot from its neighboring country Indonesia to develop and implement agroforestry strategy. Indonesia developed national strategy for agroforestry research (2013–2030) focusing on (i) smallholder production systems and markets, (ii) community-based forest management on state forest areas, (iii) the harmonization of agroforestry practices with global climate change, and (iv) enhancing agroforestry practices for environmental services [55]. Lessons from the adoption of agroforestry on state land in Indonesia, securing tenure right through agroforestry practice on state land, could be a reward of farmers to make the land more sustainable [56]. Timor-Leste can also learn from other regions such as planting trees and shrubs as nitrogen-fixing green fertilizers to increase average maize yields and to stabilize crop production in drought years and during other extreme weather events in southern Africa [57]. The Malawi Agroforestry Food Security Program is also a good example where improved green fertilizer technologies increased maize yields [58]. Timor-Leste also needs to prioritize diversification of the agriculture sector through a planned agroforestry system by planting high-value trees including sandalwood and teak, and non-timber products, such as bamboo, cocoa, coconuts, spices, coffee, perennial fruits, nuts, and animal fodder, integrated with livestock developments, need to be prioritized in MAF's strategy. Land tenure issues should be addressed to ensure maximum benefits for smallholders and to support the development of accessible agroforestry inputs by liaising with the private sector. A mixed agroforestry model for diversification would provide rural households to expand their sources of food and income and build their resiliency to climate change [59].

The current initiatives on agroforestry from different donor-funded projects have provided opportunities to learn skills and engage communities in the agroforestry system [36,40]. There is already a system-wide realization that agroforestry would be a viable option to build watershed resiliency and diversify income from agriculture and forestry. Furthermore, the government of Timor-Leste has intended to increase the number of national fisheries and aquaculture programs to sustainably improve livelihoods. With the support of the Food and Agriculture Organization (FAO), the government has expected to improve the livelihood of local people and build the capacity of fisherfolks and their supporting institutions, thus providing a huge opportunity for the development of silvo-aquaculture [60].

Agroforestry could also contribute to DRR and build community resiliency in Timor-Leste [61,62]. Timor-Leste is highly vulnerable to climate change and disaster risks due to its hilly geographical terrain [5]. Flood and erosions are common throughout the country during the rainy season [5]. Therefore, agroforestry could be a practical solution to control soil erosion and maintain land fertility in the hilly regions and thereby build the resilience of the local communities depending upon agriculture and natural resources for their livelihoods [62]. The fifth assessment report of Intergovernmental Panel on Climate Change (IPCC) has suggested an integrated strategy such as agroforestry to address climate change [63]. Agroforestry, the diversification of crops, and the use of A-frame terracing come under the most potential cost-effective climate change mitigation options in Timor-Leste [61]. The adoption of an agroforestry development strategy could facilitate the growth of agroforestry in Timor-Leste [52].

Following the requirement of resilience against natural calamities and the limited amount of arable land, agroforestry could be the best strategy to diversify the land use and prevent land degradation in Timor-Leste. An approach to develop agroforestry at a wider scale in Timor-Leste through technical support to maximize food production, financial investments, infrastructure development, policy intrusions, and land ownership rights is required [54,57,61].

4. Conclusions

Timor-Leste has been facing the problem of food insecurity alongside climatic catastrophes. With limited arable land and high population, agroforestry is recognized as a reliable strategy to maximize food production and cope with climatic vulnerabilities. Different kinds of agroforestry systems, including alley cropping, alternate row planting, random mixers, and trees along borders are adopted in Timor-Leste. The Fukuoka system is widely accepted to rehabilitate the degraded landscape. Fruit and fodder species were preferred for the agroforestry. Several projects such as DARDC, tara bandu, and PSAF have initiated the plantation programs for agroforestry, market development, and strengthening the capacity of institutions for the implementation of agroforestry projects. However, the lack of financial access, infrastructures, irrigation facilities, vulnerability to climatic hazards, and unclear land ownership rights have created a challenge for the wide acceptance of agroforestry system in Timor-Leste. On the other side, a support from donor agencies and the government have provided the local people with the skills and knowledge about agroforestry systems and its associated benefit in employing young people and building resilience against climate change, which has created an opportunity for the adoption of agroforestry at a larger scale. Capacity building, financial incentives, market accessibility, and infrastructure development are required to further foster agroforestry in Timor-Leste.

Author Contributions: Conceptualization, H.B., S.P. and A.R.; methodology, S.P. and A.R.; validation, S.P. and H.B.; investigation, S.P. and Y.A.; resources, S.P., A.R. and K.P.B.; data curation, S.P. and K.P.B.; writing—original draft preparation, S.P., K.P.B. and H.B.; writing—review and editing, H.B., S.P. and Y.A.; supervision, H.B.; funding acquisition, Y.A. and H.B. All authors contribute to the manuscript and approve final draft. All authors have read and agreed to the published version of the manuscript.

Funding: The publication of this manuscript was funded by the CGIAR research program on Forest, Tree, and Agroforestry (FTA).

Institutional Review Board Statement: Not applicable.

Informed Consent Statement: Not applicable.

Data Availability Statement: Not applicable.

Acknowledgments: The authors are thankful to a number of individuals at the General Directorate of Forestry, Coffee, and Industrial Plant of Ministry of Agriculture and Fisheries (GDFCIP), Timor Leste and UNDP Timor-Leste office. Many thanks are also extended to all colleagues who shared their precious time, thoughts and concerns on agroforestry and restoration in Timor Leste. Our gratitude is to the CGIAR Research Program on Forest, Tree, and Agroforestry (FTA) for funding support to publish the work.

Conflicts of Interest: The authors declare no conflict of interest.

References

1. General Directorate of Statistics (GDS). Timor-Leste Agriculture Census 2019, National Report on Final Census Results, General Directorate of Statistics, Ministry of Finance and Ministry of Agriculture and Fisheries. 2020. Available online: http://www.fao.org/fileadmin/templates/ess/ess_test_folder/World_Census_Agricuture/WCA_2020/WCA_2020_new_doc/TLS_REP_ENG_2019_01.pdf (accessed on 10 February 2021).
2. Ministry of Agriculture and Fisheries (MAF). The Integrated Food Security Phase Classification (IPC). 2019. Available online: http://www.ipcinfo.org/fileadmin/user_upload/ipcinfo/docs/3_IPC_Timor%20Leste_CFI_20182023_English.pdf (accessed on 12 February 2021).
3. United Nations Development Programme (UNDP). Country Program for Timor-Leste 2009–2013. 2008. Available online: http://www.undp.org/content/dam/rbap/docs/programme-documents/TL-CP-2009-2013.pdf (accessed on 14 February 2021).
4. United Nations Development Programme (UNDP). Strengthening Community Resilience to Climate-Induced Disasters in the Dili to Ainaro Road Development Corridor. 2019. Available online: https://www.tl.undp.org/content/timor_leste/en/home/all-projects/DARDC.html (accessed on 2 March 2021).
5. European Union (EU). Disaster Preparedness in Timor-Leste. Disaster Preparedness, Humanitarian Aid and Civil Protection Directorate General (DIPECHO), and Timor Leste Living Standards Survey. 2007. Available online: https://ec.europa.eu/echo/files/policies/dipecho/presentations/est_timor.pdf?fbclid=IwAR26DqR3-5WlKmBhmEs1jVVXfJAEJfDwn1Mk78PcU_Vw-U6PUrXf8o4uYv4 (accessed on 14 February 2021).
6. Barnett, J.; Dessai, S.; Jones, R.N. Vulnerability to climate variability and change in East Timor. *AMBIO* **2007**, *36*, 372–378. [CrossRef]
7. World Bank. World Development Indicators. Arable Land (hectares) Timor Leste (DataBank). 2018. Available online: https://data.worldbank.org/indicator/AG.LND.ARBL.HA?locations=TL (accessed on 13 February 2021).
8. Ministry of Finance, Timor-Leste. Ministry of Finance (MOF)Launch of the Main Results of the 2015 Census of Population and Housing. 2016. Available online: https://www.statistics.gov.tl/wp-content/uploads/2015/10/1-Preliminary-Results-4-Printing-Company-19102015.pdf (accessed on 15 September 2021).
9. Cruz, C.J. Improving food security through agricultural development in Timor-Leste: Experiences under 13 years of democratic government. In *Australian Center for International Agricultural Research (ACIAR) Proceedings Series*; Australian Centre for International Agricultural Research (ACIAR): Canberra, Australia, 2016; pp. 30–37.
10. Food and Agricultural Organization (FAO). FAO in Timor-Leste. 2019. Available online: http://www.fao.org/timor-leste/news/detail-events/fr/c/1197518/ (accessed on 17 February 2021).
11. Batterbury, S.; Palmer, L.; Reuter, T.; de Carvalho, D.D.A.; Kehi, B.; Cullen, A. Land access and livelihoods in post-conflict Timor-Leste: No magic bullets. *Int. J. Commons* **2015**, *9*, 619–647. [CrossRef]
12. Food and Agricultural Organization (FAO). Global Forest Resource Assessment. Report Timor-Leste, Food and Agriculture Organization of the United Nations, Rome. 2020. Available online: http://www.fao.org/3/cb0077en/cb0077en.pdf (accessed on 21 February 2021).
13. Gonzaga Fraga, L.; Teixeira, C.F.; Ferreira, E.C. The potential of renewable energy in Timor-Leste: An assessment for biomass. *Energies* **2019**, *12*, 1441. [CrossRef]
14. Hosgelen, M.; Saikia, U. Forest reliance as a livelihood strategy in Timor-Leste. *Underst. Timor-Leste* **2014**, *II*, 66–73.

15. National Statistics Directorate (NDS). Population and Housing Census. Government of Timor-Leste, Ministry of Finance, General Directorate of Statistics; and United Nations Population Fund. 2010. Available online: https://www.statistics.gov.tl/wp-content/uploads/2013/12/English_20Census_20Preliminary_20Results_202010.pdf?fbclid=IwAR3ZunGqy8AJj4RgRbe6USXecddjRW13I4HX15ThavpXFQGL0Au1XdlPxKY (accessed on 16 February 2021).
16. Molyneux, N.; Da Cruz, G.R.; Williams, R.L.; Andersen, R.; Turner, N.C. Climate change and population growth in Timor Leste: Implications for food security. *AMBIO* **2012**, *41*, 823–840. [CrossRef]
17. Ministry of Agriculture and Fisheries (MAF). Final Country Report of the Land Degradation neutrality Target Setting Programme in Timor-Leste. 2018. Available online: https://knowledge.unccd.int/sites/default/files/ldn_targets/2019-01/Timor-Leste%20LDN%20TSP%20Country%20Report.pdf (accessed on 21 February 2021).
18. Asian Development Bank (ADB). Country Partnership Strategy for Timor Leste 2016–2020. 2016. Available online: https://www.adb.org/sites/default/files/institutional-document/183715/cps-tim-2016-2020.pdf (accessed on 26 February 2021).
19. Nair, P.R. *An Introduction to Agroforestry*; Springer: Dordrecht, The Netherlands, 1983; Available online: http://apps.worldagroforestry.org/Units/Library/Books/PDFs/32_An_introduction_to_agroforestry.pdf?n=161 (accessed on 15 September 2021).
20. Lundgren, B.O.; Raintree, J.B. Sustained Agroforestry. In *Agricultural Research for Development: Potentials and Challenges in Asia*; Nestel, B., Ed.; ISNAR: The Hague, The Netherlands, 1982; pp. 37–49.
21. Oelbermann, M.; Voroney, R.P.; Gordon, A.M. Carbon sequestration in tropical and temperate agroforestry systems: A review with examples from Costa Rica and southern Canada. *Agric. Ecosyst. Environ.* **2004**, *104*, 359–377. [CrossRef]
22. Shin, S.; Soe, K.T.; Lee, H.; Kim, T.H.; Lee, S.; Park, M.S. A Systematic Map of Agroforestry Research Focusing on Ecosystem Services in the Asia-Pacific Region. *Forests* **2020**, *11*, 368. [CrossRef]
23. de Oliveira, R.E.; Carvalhaes, M.A. Agroforestry as a tool for restoration in atlantic forest: Can we find multi-purpose species? *Oecol. Aust.* **2016**, *20*, 425–435. [CrossRef]
24. Jose, S. Agroforestry for ecosystem services and environmental benefits: An overview. *Agrofor. Syst.* **2009**, *76*, 1–10. [CrossRef]
25. Chang, C.H.; Karanth, K.K.; Robbins, P. Birds and beans: Comparing avian richness and endemism in arabica and robusta agroforests in India's Western Ghats. *Sci. Rep.* **2018**, *8*, 3143. [CrossRef]
26. United Nations Development Programme (UNDP). Community Agroforestry Guide. UNDP-DARDC Project. 2018. Available online: https://www.tl.undp.org/content/timor_leste/en/home/library/resilience/community-agroforestry-guide.html (accessed on 12 January 2021).
27. Government of Timor-Leste (GoTL). Strategic Development Plan 2011 to 2030, Government of Timor-Leste, Dili, Timor-Leste. 2011. Available online: https://www.adb.org/sites/default/files/linked-documents/cobp-tim-2014-2016-sd-02.pdf (accessed on 19 September 2021).
28. Ministry of Agriculture and Fisheries (MAF). Ministry of Agriculture and Fisheries Strategic Plan 2014 to 2020, Ministry of Agriculture and Fisheries, Government of Timor-Leste, Dili, Timor-Leste. 2012. Available online: http://extwprlegs1.fao.org/docs/pdf/tim149148.pdf (accessed on 20 September 2021).
29. Ministry of Economy and Development (MED). The National Biodiversity Strategy and Action Plane of Timor-Leste (2011–2020). 2011. Available online: https://www.tl.undp.org/content/timor_leste/en/home/library/environment_energy/the-national-biodiversity-strategy-and-action-plan-of-timor-lest.html (accessed on 10 February 2021).
30. Democratic Republic of Timor-Leste (RDTL). Environment Basic Law. Decree Law No 26/2012 of 4 July 2012. Unofficial Translation from Protuguese of the Decree Law as published in the Official Gazette 4 July 2012, Series 1 No 24. 2012. Available online: https://www.laohamutuk.org/Agri/EnvLaw/2012/DL26EnvBasicLaw4Jul2012en.pdf (accessed on 17 February 2021).
31. Garrity, D.P. Agroforestry and the achievement of the Millennium Development Goals. *Agrofor. Syst.* **2004**, *61*, 5–17.
32. United Nations Development Programme (UNDP). Local Famers Use Innovative Fukuoka Method to Cultivate Land. 2018. Available online: https://www.tl.undp.org/content/timor_leste/en/home/stories/local-farmers-Fukuoka-method-cultivate-land.html (accessed on 27 February 2021).
33. Friday, J.B. Agroforestry opportunities for Timor-Leste. Forage banks and forage gardens. In *PowerPoint Presentation, Timor-Leste Agricultural Rehabilitation, Economic Growth, and Sustainable Natural Resources Management Project*; University of Hawaii: Honolulu, HI, USA, 2005.
34. The Asia Foundation. Tara Bandu: Its Role and Use in Community Conflict Prevention in Timor-Leste. Belun. 2013. Available online: https://asiafoundation.org/resources/pdfs/TaraBanduPolicyBriefENG.pdf (accessed on 28 January 2021).
35. World Vision International. Planting Trees to Re-Green Timor-Leste. 2019. Available online: https://www.wvi.org/timor-leste/article/planting-trees-re-green-timor-leste (accessed on 28 January 2021).
36. Deutsche Gesellschaft für Internationale Zusammenarbeit (GIZ). Ai ba Futuru-Partnership for Sustainable Agroforestry. 2020. Available online: https://www.giz.de/en/downloads/giz2020_en_PSAF_Factsheet.pdf (accessed on 10 March 2021).
37. International Labour Organization (ILO). Enhancing Rural Access Agroforestry Project (ERA Agro-forestry). 2019. Available online: https://www.ilo.org/jakarta/whatwedo/projects/WCMS_553152/lang--en/index.htm (accessed on 17 February 2021).
38. Asian Development Bank (ADB). *Asian Development Outlook (ADO) 2019: Strengthening Disaster Resilience*; ADB: Manila, Philippines, 2019.
39. Asian Development Bank (ADB). Timor-Leste: Innovative Partnership for Coffee and Agroforestry Development. Concept paper. 2018. Available online: https://www.adb.org/sites/default/files/project-documents/51396/51396-001-cp-en.pdf (accessed on 15 March 2021).

40. Union Aid Board APHEDA. Sustainable Farming and Organizing farmers in Timor Leste. 2018. Available online: https://www.apheda.org.au/sustainable-farming-timor-leste/ (accessed on 12 January 2021).

41. Noltze, M.; Schwarze, S.; Qaim, M. Impacts of natural resource management technologies on agricultural yield and household income: The system of rice intensification in Timor Leste. *Eco. Econ.* **2013**, *85*, 59–68. [CrossRef]

42. Moore, A.; Dormody, T.; VanLeeuwen, D.; Harder, A. Agricultural sustainability of small-scale farms in Lacluta, Timor Leste. *Int. J. Agric. Sust.* **2014**, *12*, 130–145. [CrossRef]

43. Gangadharappa, N.R.; Shivamurthy, M.; Ganesamoorthi, S. Agroforestry–A viable alternative for social, economic and ecological sustainability. In Proceedings of the XII World Agroforestry Congress, Quebec, QC, Canada, 21–28 September 2003.

44. National Action Programme. Timor-Leste National Action Programme to Combat Land Degradation. 2008. Available online: http://www.fao.org/fileadmin/templates/cplpunccd/Biblioteca/bib_TL_/Timor-Leste_NAP_Revised_Draft.pdf (accessed on 3 March 2021).

45. Zinngrebe, Y.; Borasino, E.; Chiputwa, B.; Dobie, P.; Garcia, E.; Gassner, A.; Kihumuro, P.; Komarudin, H.; Liswanti, N.; Makui, P.; et al. Agroforestry governance for operationalising the landscape approach: Connecting conservation and farming actors. *Sustain. Sci.* **2020**, *15*, 1417–1434. [CrossRef]

46. Bond, J.; Millar, J.; Ramos, J. Livelihood benefits and challenges of community reforestation in Timor Leste: Implications for smallholder carbon forestry schemes. *For. Trees Livelihoods* **2020**, *29*, 187–204. [CrossRef]

47. Enters, T.; Durst, P.B.; Brown, C. What Does it Take? The Role of Incentives in Forest Plantation Development in the Asia-Pacific Region. 2004. Available online: http://www.fao.org/forestry/5247-021bfef4098d1413fbbcd9a64c8103fbd.pdf (accessed on 24 March 2021).

48. Rola-Rubzen, M.F.; Janes, J.A.; Correia, V.P.; Dias, F. Challenges and constraints in production and marketing horticultural products in Timor Leste. In *III International Symposium on Improving the Performance of Supply Chains in the Transitional Economies*; International Society for Horticultural Science: Leuven, Belgium, 2010; Volume 895, pp. 245–253.

49. Lundahl, M.; Sjöholm, F. Improving the Lot of the Farmer: Development Challenges in Timor-Leste during the Second Decade of Independence. *Asian Econ. Pap.* **2013**, *12*, 71–96. [CrossRef]

50. Bettencourt, E.M.V.; Tilman, M.; Narciso, V.; Carvalho, M.L.D.S.; Henriques, P.D.D.S. The livestock roles in the wellbeing of rural communities of Timor-Leste. *Rev. Econ. Sociol. Rural* **2015**, *53*, 63–80. [CrossRef]

51. Coimbra, L.; Deus, P.D.; Supriyadi, M. Developing Strategies for Improving Bali Cattle Productivity in Timor-Leste. In Proceedings of the 15th AAAP Animal Science Congress, Pathum Thani, Thailand, 26–30 November 2012.

52. da Silva, F.; Dassir, M.; Mujetahid, M.; Nadirah, S. Collaboration Based Agroforestry Development Strategy in Laubonu Village, Atsabe Subdistrict, District Ermera, Timor Leste. *Adv. Environ. Biol.* **2019**, *13*, 10–16.

53. Reid, R. Developing farmer and community capacity in Agroforestry: Is the Australian Master TreeGrower program transferable to other countries? *Agrofor. Syst.* **2017**, *91*, 847–865. [CrossRef]

54. Thu, P.M. Access to land and livelihoods in post-conflict Timor-Leste. *Aust. Geogr.* **2012**, *43*, 197–214. [CrossRef]

55. Rohadi, D.; Herawati, T.; Firdaus, N.; Maryani, R.; Permadi, P. National Strategy for Agroforestry Research in Indonesia 2013–2030. Center for Research and Development on Forest Productivity Improvement, Forestry Research and Development Agency, Bogor, Indonesia. 2013. Available online: http://simlit.puspijak.org/files/buku/Book_NSAF_English.pdf (accessed on 18 September 2021).

56. Suyanto, S.; Permana, R.P.; Khususiyah, N.; Joshi, L. Land tenure, agroforestry adoption, and reduction of fire hazard in a forest zone: A case study from Lampung, Sumatra, Indonesia. *Agrofor. Syst.* **2005**, *65*, 1–11. [CrossRef]

57. Sileshi, G.W.; Akinnifesi, F.K.; Ajayi, O.C.; Muys, B. Integration of legume trees in maize-based cropping systems improves rain use efficiency and yield stability under rain-fed agriculture. *Agric. Water Manag.* **2011**, *98*, 1364–1372. [CrossRef]

58. Centre for Independent Evaluations (CIE). Evaluation of ICRAF's Agroforestry Food Security Programme (AFSP) 2007–2011. In *Final Report Submitted to IRISH AID*; Center for Independent Evaluations: Lilongwe, Malawi, 2011.

59. Fanzo, J.; Boavida, J.; Bonis-Profumo, G.; McLaren, R.; Davis, C. Timor Leste strategic Review: Progress and Success in Achieving the Sustainable Development Goal 2. Centre of Studies for Peace and Development (CEPAD) Timor-Leste and John Hopskin University. 2017. Available online: https://www.interpeace.org/wp-content/uploads/2017/06/TL-SR-May-15-V2-21_-May-24-FINAL.pdf (accessed on 20 September 2021).

60. Food and Agriculture Organization (FAO). Evaluation of FAO's Contribution to the Democratic Republic of Timor Leste 2015–2018. Food and Agriculture Organizations of the United Nations, Rome, Italy. 2019. Available online: http://www.fao.org/3/ca5653en/ca5653en.pdf (accessed on 12 February 2021).

61. Chandra, A.; Dargusch, P.; McNamara, K.E. How might adaptation to climate change by smallholder farming communities contribute to climate change mitigation outcomes? A case study from Timor-Leste, Southeast Asia. *Sustain. Sci.* **2016**, *11*, 477–492. [CrossRef]

62. Ismail, C.J.; Takama, T.; Budiman, I.; Knight, M. Comparative study on agriculture and forestry climate change adaptation projects in Mongolia, the Philippines, and timor leste. In *Climate Change-Resilient Agriculture and Agroforestry*; Castro, P., Azul, A., Leal Filho, W., Azeiteiro, U., Eds.; Springer: Cham, Switzerland, 2019; pp. 413–430.
63. Intergovernmental Panel on Climate Change (IPCC). Summary for policy makers. In *Climate Change 2014: Impacts, Adaptation, and Vulnerability*; In IPCC Working Group II contribution to AR5; IPCC: Geneva, Switzerland, 2014. Available online: http://ipcc-wg2.gov/AR5/images/uploads/IPCC_WG2AR5_SPM_Approved.pdf (accessed on 3 March 2021).

MDPI

St. Alban-Anlage 66

4052 Basel

Switzerland

Tel. +41 61 683 77 34

Fax +41 61 302 89 18

www.mdpi.com

Forests Editorial Office

E-mail: forests@mdpi.com

www.mdpi.com/journal/forests

9 783036 556802